Modern Power Engineering (MPMB 2020)

IOP Conference Series: Materials Science and Engineering
Volume 963

Moscow, Russia
8 September 2020

ISBN: 978-1-7138-3201-0
ISSN: 1757-8981

Printed from e-media with permission by:

Curran Associates, Inc.
57 Morehouse Lane
Red Hook, NY 12571

CURRAN ASSOCIATES INC. proceedings .com

Some format issues inherent in the e-media version may also appear in this print version.

Institute of Physics
Dirac House, Temple Back
Bristol BS1 6BE UK

Phone: 44 1 17 929 7481
Fax: 44 1 17 920 0979

techtracking@iop.org

Additional copies of this publication are available from:

Curran Associates, Inc.
57 Morehouse Lane
Red Hook, NY 12571 USA
Phone: 845-758-0400
Fax: 845-758-2633
Email: curran@proceedings.com
Web: www.proceedings.com

TABLE OF CONTENTS

MPMB-2020

The conference "MPMB-2020" is held annually at the Bauman Moscow State Technical University. The Department of Fluid Mechanics, Hydraulic Machines and Hydraulic and Pneumatic Automation is the conference organizer. The best conference reports are published as articles in the journal "Hydraulics", indexed in the Russian Science Citation Index. Every year, representatives of various scientific, educational and industrial organizations of Russia and other countries come to us with reports and speeches. The main topics of the reports are hydromechanics, hydraulic machines, and hydraulic drives. A part of the reports is devoted to industrial ecology issues. We are pleased to see at our conference scientists from any countries and from any organizations devoting their work to industrial hydraulics, ecology and solving fundamental issues in these areas.

Sincerely, A. Protopopov

Stanislav Semenov
Bauman Moscow state technical University
a1e10@rambler.ru
semenov_se@bmstu.ru

IOP Conf. Series: Materials Science and Engineering **963** (2020) 011002 doi:10.1088/1757-899X/963/1/011002

Peer review declaration

All papers published in this volume of *IOP Conference Series: Materials Science and Engineering* have been peer reviewed through processes administered by the Editors. Reviews were conducted by expert referees to the professional and scientific standards expected of a proceedings journal published by IOP Publishing.

- **Type of peer review: Open**
- **Conference submission management system: Manual mailing from the editor**
- **Number of submissions received: 42**
- **Number of submissions sent for review: 42**
- **Number of submissions accepted: 35**
- **Acceptance Rate (Number of Submissions Accepted / Number of Submissions Received X 100): 83%**
- **Average number of reviews per paper: 1**
- **Total number of reviewers involved: 4**
- **Contact person for queries:**
  S Semenov, e-mail: a1e10@rambler.ru and A Protopopov, proforg6@yandex.ru

Surface modification of low carbon steel to improve corrosion resistance

S Pakhomova

Bauman Moscow State Technical University, 5 Second Baumanskaya Street, Moscow, 105005, Russian Federation

E-mail: mgtu2013@yandex.ru

Abstract. Issues of improving the operational properties of low-carbon steels by coating are considered. It is shown that aluminum and chromium coatings increase the corrosion resistance of precision alloys without deterioration of their magnetic properties. It has been established that chromium coatings are more promising.

Introduction

Unalloyed electrical steels have the same chemical composition as technically pure iron - they contain the same amount of carbon and impurities [1, 2]. Electrical steels are used as low-frequency soft-magnetic materials with high saturation magnetic induction for the manufacture of a wide range of magnetic conductors: low-voltage relays, switches, and the like [3, 4]. Electrical steels are supplied with guaranteed magnetic characteristics: the value of the coercive force *Hc* and magnetic saturation induction *B*. The most used steel grades are 20895, 20864 and 20848, which have *Hc* respectively 95, 64 and 48 *A/m*, and B — 2.05 *T* (GOST 3836–83). A significant disadvantage of this class of steels is their low corrosion resistance in atmospheric corrosion conditions. To increase the corrosion resistance, chemical heat treatment of steels is necessary, forming single-component or multi-component coatings that are resistant to the environment without deterioration of the soft-magnetic properties [5-8]. There are different types of chemical-thermal treatment, such as carburization and nitriding. However, it is known that after carburization, parts have low corrosion resistance [9, 10], and after nitriding — low operating temperatures.

Low electrical resistivity (less than 0.1 $\mu Om{\cdot}m$) of these steels leads to an increase in heat losses when the frequency of remagnetization increases, while the loss of hysteresis depends on the magnitude of the coercive force [11]. Therefore, different types and modes of coating formation should not be accompanied by an increase in the coercive force of electrical steels, which leads to an increase in hysteresis losses.

It is known that one of the ways to increase the corrosion resistance of steels of this class is the diffusion saturation of their surface with aluminum or chromium [12, 13]. Previous studies have shown that one of the most modern and effective is the circulation method of alitizing and chrome-plating [14, 15]. Carrying out the surface deformation hardening after chemical-thermal treatment can not increase the corrosion resistance, since it will lead to an increase in stress in the surface layer [16, 17].

Improvement of the circulation method required further studies of the influence of technological factors of the process, both on the corrosion resistance of coatings and on the magnetic properties of electrical steels [18].

In this work, the **purpose** of applying the diffusion coatings was to provide corrosion resistance in a humid industrial air atmosphere while maintaining the magnetic properties of electrical steels of grades 20864 and 20895 (Table 1).

Table 1. **Chemical composition of both steels 20864 and 20895**

Content of elements, %						
No	C	Si	Mn	S	P	Cu
1	<0,035	<0,3	<0,3	<0,03	<0,02	<0,3

They were used to create single-component alitized or chrome-plated coatings using the circulation method. The initial microstructure of both steels where shown on Fig. 1.

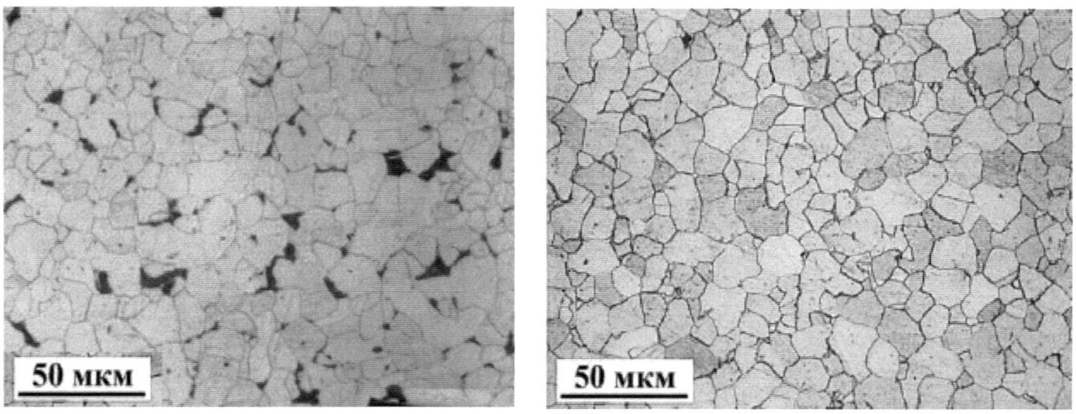

a *b*

Fig. 1. The initial microstructure of steels 20864 (*a*) and 20895 (*b*)

Research methods

The influence of technological factors, such as process temperature, duration of the process and the composition of the saturating medium on the coating thickness Δh, the coercive force *Hc* and the corrosion resistance of steels, has been studied.

To preserve the soft-magnetic properties of the coating structure over the entire thickness should be single-phase. For an alitized coating, this structure is preserved if the concentration of aluminum in the steel is not more than 7 % (*wt.*). it is obvious that the concentration of aluminum in the alitized layer is higher. Therefore, it was necessary to evaluate the effect of annealing (before or after alitizing) on the possibility of improving the soft-magnetic properties.

For chrome-plating coatings, there are no such restrictions, since the solubility of chromium in steel is unlimited. In this case, the chromium content for high corrosion resistance in humid air should be at least 12.5...13.0 % (*wt.*).

The alitizing process was carried out by the circulation method at a temperature of 950 °C, the exposure during the diffusion saturation process was two or four hours; cooling was carried out at a certain controlled speed of 40 or 100 °*C/h*. For comparison, the process of alitizing combined with annealing in different sequences.

The thickness of the alitized coating was determined by metallographic method on cross-sections after etching in a 5 % alcohol solution of nitric acid.

Chrome-plating was performed by the same circulation method. The composition of the sending gas medium was characterized by the amount of chromium metal and ammonium chloride NH4Cl. The influence of the process temperature, the duration of exposure at diffusion saturation, and the composition of the sending gas medium on the thickness of the chrome coating was studied.

The thickness of the chrome-plating coating was also measured metallographically on strips etched first with a 5 % solution of nitric acid to reveal the structure of the base, and then - in a special reagent (a mixture of picric and hydrochloric acids) to reveal the structure of the coating itself. In some cases, the thickness of the coating was estimated by the value of microhardness.

The influence of heat treatment on the structure and properties of single-component coatings was studied by x-ray and microstructure studies.

Experimental results and their discussion

Studies have shown that alitized layer has a sufficient thickness and uniform at different sites of the coating (Fig. 2*a*). The coating has good corrosion resistance. X-ray studies have revealed the presence of several phases in the structure that contribute to a decrease in soft-magnetic properties and an increase in the coercive force (Table 2).

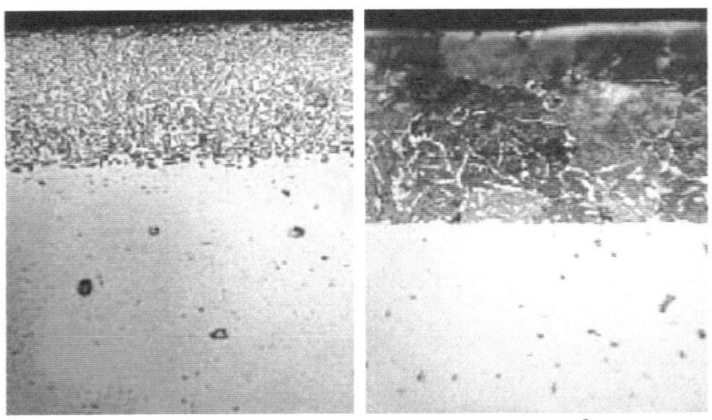

a *b*

Fig. 2. The microstructure of steel 20864 after the alitizing 950°C, 4 h (a) and chrome-plating 950°C, 4 h (b)

Table 2. **Technological modes of alitizing of electrical steel 20864**

Chemical heat treatment			**Cooling speed, °C/*h***	**Layer thickness Δ*h*, μ*m***
Type of processing	***t*, °C**	**τ, *h***		
		2	100	150
Alitizing		2	40	130
	950	4	40	210
Alitizing + annealing		2 + 2	40	210
Annealing + alitizing		2 + 2	40	200

Increasing the exposure time from 2 *h* to 4 *h* increases the thickness from 130 μ*m* to 200 μ*m*. The effect of the cooling rate on the coating thickness is negligible. The effect of annealing in the same installation without supplying a saturating mixture was studied. Annealing for 4 *h*, only due to the remains of saturating mixtures in the chamber created a coating with a thickness of 75 μ*m*. Annealing before and after of alitizing effect on the properties as well as increasing the exposure time.

At the obtained thicknesses, we should expect significant concentrations of aluminum in the layer, and, consequently, the presence of Fe_3Al, FeAl or $FeAl_2$ type aluminides. The presence of such phases increases the coercive force, while reducing the magnetic properties. For corrosion resistance, such phases are not dangerous, since they are cathodes that accelerate the passivation of the aluminum coating. However, too large thicknesses are still undesirable, because the outer surface of the layer becomes loose and therefore less resistant to corrosion [19].

IOP Conf. Series: Materials Science and Engineering **963** (2020) 012001 doi:10.1088/1757-899X/963/1/012001

Studies of chrome coatings have shown that the microhardness of the coating is higher than that of the base. This is the result of the formation of a solid solution-chromic ferrite, since the dissolution of chromium in ferrite creates stresses in the crystal lattice.

The obtained values of layer thicknesses during chrome-plating are significantly less than during alitizing, although the chromium saturation process was performed at higher temperatures and longer exposures (Fig. 2*b*). At the same time, even at a thickness of 90 *μm*, the outer surface of the coating was loose.

The effect of chrome plating temperature on the coating thickness is more effective than the duration of exposure (Fig. 3). The intensity of chromium saturation increases with an increase in the amount of ammonium chloride in the saturating gas medium.

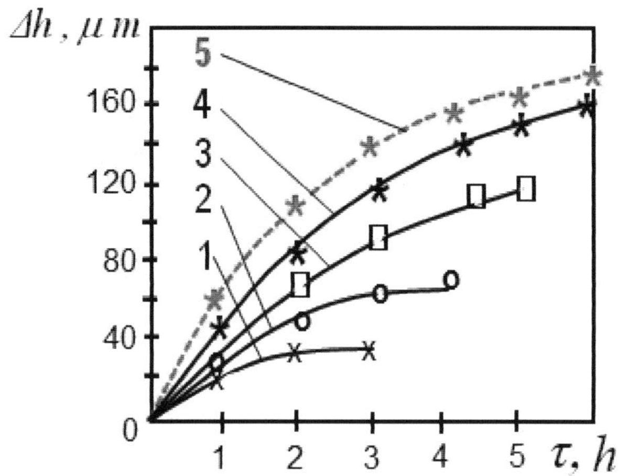

Fig. 3. Depends of thickness Δ*h* from electrical steel 20864 chrome-plating technological modes: 1 — 850°C, 2 — 950°C, 3 — 1050°C, 4 and 5 — 1150°C; composition of the medium: 1 — 4 — (Cr 100 *g* + NH4Cl 25 *g*), 5 — (Cr 100 *g* + NH4Cl 50 *g*).

The measured coating thicknesses are in good agreement with the basic laws of diffusion and kinetics of chemical reactions: exponential temperature dependence and power-time dependence [20, 21].

Based on the diffusion representations [3, 12], it becomes understandable that at the same temperatures and time parameters of diffusion saturation, the thickness of the layers in alitizing is significantly greater than in chrome-plating. It is known that the diffusion mobility of aluminum in steel is much greater than that of refractory chromium [22]. This confirms the reliability of the results obtained.

X-ray study of the effect of heat treatment on the structure and properties of single-component coatings showed that alitized coatings in all treatment modes have multiphase structures consisting of a solid solution of aluminum in a low-temperature modification of iron and aluminides. The presence of such intermediate phases is favorable for corrosion resistance and undesirable for the soft-magnetic properties of steel.

Cold-rolled steel 20864 with traces of riveting was used for alitizing, which adversely affected the magnetic properties, which were determined on ring samples by the ballistic method (GOST 11036–75). The use of vacuum pre-annealing significantly improved the magnetic characteristics by eliminating the hardening. Since the altering was carried out practically at annealing temperatures, this made it possible to assume that the slope would be eliminated during the altering process. In order to find out whether pre-annealing can be canceled, alitizing was performed before and after annealing (see Table 2 and Table 3).

Table 3. **Effect of heat treatment on the layer thickness and properties of steel 20864**

Annealing			**Type of coating**	**Layer thickness Δh, μm**	***Hc, A/m***
***t*, °C**	***τ*, *h***	**Composition of the medium**			
950	2	–	Al	130	110
1150	3	NH4CL	Cr	90	63
1000	5	NH4CL	Si	45	82
1040	8	Si/Al=1	Al +Si	7	65
1100	8	Si/Al=1/2	Al +Si	20	72

The results obtained showed that heat pre-treatment does not significantly change either the layer thickness or properties, and an increase in the annealing temperature affects the properties as well as an increase in the exposure time during the alitizing process.

The study of the coating structure after the chrome-plating showed that it is mainly single-phase due to the unlimited solubility of chromium in the low-temperature modification of iron. X-ray structural analysis showed the traces of chromium carbide and σ–phase on the outer surface of the chrome-layer. This structure is favorable for obtaining high corrosion resistance and preserving the magnetically soft properties of steel. At the same time, the thicker the dense part of the coating and more chromium in it, the higher the corrosion resistance and magnetic properties.

Compared with aluminum, the diffusive mobility of chromium is less, so to obtain larger thicknesses, the chrome plating process requires higher temperatures and exposures. Such modes create a large-crystal structure, which significantly improves the magnetic properties.

Table 4. **Effect of chrome-plating modes in a chloride environment on the layer thickness and magnetic properties of steel 20895**

Chrome-plating			**Layer thickness Δh, μm**	***Hc, A/m***
***t*, °C**	***τ*, *h***	**NH_4CL, *g* on 100 *g* Cr**		
850	3	12,5	15	90
950	1	25	25	76
	3	25	30	86
	6	25	38	77
	9	25	35	87
1000	1	25	25	87
	3	25	50	78
	6	25	75	78
	9	25	120	72
1040	3	25	40	73
	3,5	10	–	60
	3,5	25	65	69
	2,5	20	65	72
	2,5	50	65	58
1100	1	50	100	84
	3	50	138	69
	6	50	180	68
1150	3	25	88	69
	3	50	165	62

The study of chrome-plating modes on the coating thickness and magnetic characteristics of steel 20895 was conducted. The first series of tests included circulation chrome-plating in the medium of solid chromium and ammonium chloride (Table 4). The composition of the mixture was studied by changing the ratio of chromium and ammonium chloride. It was found that the thickness and properties of coatings are less affected by changes in the ratio of the mixture than the exposure time and, especially, the temperature of chrome-plating.

The possibility of chrome-plating in an iodide medium containing chromium and crystalline iodine was studied. Table 5 shows the amount of iodine in grams per 100 g of chromium.

Table 5. **Effect of chrome-plating modes in an iodide medium on the layer thickness and magnetic characteristics of steel 20895**

Chrome-plating			**Layer thickness Δh, µm**	***Hc*, *A/m***
***t*, °C**	**τ, *h***	***I*, *g* on 100 *g* Cr**		
1100	3	17	–	55
1100	3	20	–	58
1150	3	20	76	63

The process of chrome-plating in an iodide medium has shown the better magnetic properties, but chrome-plating in chlorides is a more environmentally friendly process.

Research has shown that circulation chrome-plating of electrical steels in an active chloride medium gives a uniform diffusion layer, which significantly improves both corrosion resistance and soft-magnetic properties. Thus, electrical steel 20895 in the initial state had *Hc* 140 *A/m*, and after chrome-plating in a mixture of 25% NH_4CL per 100 *g* (for 3 *h* at $t = 1150$°C) — 63 *A/m*.

Conclusions

1. Alitizing and chrome-plating of electrical steels is accompanied by an increase in their corrosion resistance.

2. Chrome-plating coatings with a thickness of 30...60 µ*m* and chromium content of 50...58 % have the highest corrosion resistance.

3. Chrome-plating of electrical steels improves their soft-magnetic properties due to the refining action of chromium.

4. Alitizing increases the coercive force and reduces the magnetic properties as a result of the formation of Fe_3Al and FeAl type aluminides.

References

[1] Qin J, Yang P, Mao W, Ye F. Effect of texture and grain size on the magnetic flux density and core loss of cold-rolled high silicon steel sheets. J Magn Magn Mater, 2015, 393:537–43.

[2] Yajiang L., Juang V., Puchkov Yu.A., Gerasimov S.A. Structure and composition of the transition zone of welded joints of Fe_3Al intermetallic and low-carbon steel // Metal Science and Heat Treatment. 2008, vol. 50, No 3–4, pp 200–203. DOI: 10.1007/s11041-008-9026-3.

[3] Evangelista L.R., Lenzi E.K. Fractional diffusion equations and anomalous diffusion. Publisher: Cambridge University Press, 2018, 358 p. DOI: 10.1017/9781316534649

[4] Suslov A.G. Inzheneriya poverkhnosti detaley [Engineering of surface parts]. M.: Mashinostroyeniye, 2008, 320 p.

[5] Kablov E.N., Muboyadzhyan S.A. Erosion-resistant coatings for gas turbine engine compressor blades // Russian metallurgy (Metally). 2017, vol. 2017, No 6 , pp 494–504. DOI: 10.1134/S0036029517060118

[6] Muboyadzhyan S.A., Gorlov D.S., Egorova L.P., Bulavintseva E.E. Corrosion-resistant antifretting coating for the protection of blade locking pieces in gte compressors and fans // Russian metallurgy (Metally). 2014, vol. 2014, No 9, pp 725–732. DOI: 10.1134/S0036029514090134

[7] Bakulo A.V., Yakushin B.F., Puchkov Y.A. Structure and corrosion resistance of welded joints of alloy 1151 in marine atmosphere // Metal Science and Heat Treatment. 2 017, vol. 59, No 3–4, pp 218–222. DOI: 10.1007/s11041-017-0132-y.

[8] Polyanskiy V.M., Puchkov Y.A., Orlov M.R., Napriyenko S.A., Lavrov A.V. Influence of tensile stresses on the corrosion resistance of titanium alloy VT22 in aqueous solution of NaCl // Inorganic Materials: Applied Research. 2017, vol., No 1, pp 94–99. DOI: 10.1134/S2075113317010294.

[9] Pakhomova S., Fakhurtdinov R., Tsinkolenko O., Zolotov B. The influence of carburization technology on performance properties of high-loaded gear wheels // IOP Conf. Series: Materials Science and Engineering.2020, **747** (2020), pp 012126. DOI:10.1088/1757-899X/747/1/012126.

[10] Pakhomova S.A., Unchikova M.V., Fakhurtdinov R.S. Gear wheels surface engineering by deformation hardening and carburization // Materials Science Forum. 2016, T. 870. pp 383–391. DOI: 10.4028/www.scientific.net/MSF.870.383.

[11] Dems M., Komeza K., Szulakowski J., Kubiak W. Modeling of core loss for non-oriented electrical steel // The International Journal for Computation and Mathematics in Electrical and Electronic Engineering. 2019. T. 38, No 6, pp 1874–1884. DOI: 10.1108/COMPEL-09-2018-0339.

[12] Wei-Jen Cheng, Chaur-Jeng Wang. Study of microstructure and phase evolution of hot-dipped aluminide mild steel during high-temperature diffusion using electron backscatter diffraction // Appl. Surf. Sci. 2011, No 257, pp 4663–4668. DOI: 10.1016/j.apsusc.2010.12.118.

[13] Wei-Jen Cheng, Chaur-Jeng Wang, Growth of intermetallic layer in the aluminide mild steel during hot-dipping // Surf. Coat. Technol. 2009, No 204, pp 824–828. DOI: 10.1016/j.surfcoat.2009.09.061.

[14] Arzamasov, B.N., Simonov, V.N. Circulation method for depositing diffusion coatings // Metal Science and Heat Treatment. 2011. vol. 52, No 9–10, pp 403–407. DOI: 10.1007/s11041-010-9291-9.

[15] Simonov, V.N., Unchikova, M.V., Shkretov, Yu.P. One-stage process of chromoaluminizing of gas turbine blades by the method of circulation // Metal Science and Heat Treatment. 2009, vol. 52, No 9–10, pp 501–504. DOI: 10.1007/s11041-010-9201-1.

[16] Pakhomova S. A. , Manayev O. I. Effect of Heat Shotblast Treatment Exerted on the Contact Fatigue of Carburised Heat-Resistant Steel C0.12Cr2NiWV // Inorganic Materials: Applied Research. July. 2018. T 9, No 4, pp 732–735. DOI: 10.1134/S2075113318040251.

[17] Pakhomova, S.A., Ryzhov, N.M., and Vasil'ev, V.R., Changes in the structure of marten site of iron-nickel alloys under the action of thermal shot blast treatment // Met. Sci. Heat Treat. 2001, vol. 43, No. 11–12, pp. 438–439. DOI: 10.1023/A:1014855712535.

[18] Simonov, V.N., Abraimov, N.V., Shkretov, Yu.P., Lukina, V.V., Terekhin, A.M. Chromo-aluminizing of cooled blades of gas turbines by circulation method // Metal Science and Heat Treatment. 2007, vol. 49, No 7–8, pp 362–365. DOI: 10.1007/s11041-007-0066-x.

[19] Muboyadzhyan S.A., Galoyan A.G. Diffusion aluminide coatings for protecting the surface of the internal space of single-crystal turbine blades made of rhenium- and rhenium-ruthenium-containing high-temperature alloys: part 2 // Russian metallurgy (Metally). 2013 vol. 2013, No 3, pp 198–205. DOI: 10.1134/S0036029513030075.

[20] Budinovskii S.A., Muboyadzhyan S.A., Gayamov A.M., Stepanova S.V. Ion-plasma heat-resistant coatings with composite barrier layer for protecting alloy ZHS36VI from oxidation // Metal Science and Heat Treatment. 2011, vol. 53, No 1–2, pp 32–38. DOI: 10.1007/s11041-011-9336-8.

[21] Brodova I.G., Shirinkina I.G., Yablonskikh T.I., Astaf'ev V.V., Zaikov Y.P., Kovrov V.A., Shtefanyuk Y.M., Pingin V.V., Vinogradov D.A., Golubev M.V. Structure and phase composition of protective coatings on steel produced by methods of liquid-phase calorizing // The Physics of Metals and Metallography. 2015, vol. 116, No 9, pp 879–887. DOI: 10.7868/S0015323015090041.

[22] Świercz R., Oniszczuk-Świercz D. Experimental investigation of surface layer properties of high thermal conductivity tool steel after electrical discharge machining // Metals. 2017, vol. 7. No 12, pp 550. DOI: 10.3390/met7120550.

Improved handling of 4x4 vehicles with differential transmission in curved driving on support surfaces with low coupling properties

M Zhileykin

Bauman Moscow State Technical University, 5 Second Baumanskaya Street, Moscow, 105005, Russian Federation

E-mail: jileykin_m@mail.ru

Abstract. Studying the trends in the development of modern automotive industry, you can see that manufacturers are constantly increasing the level of control over the parameters of the movement of wheeled vehicles, achieving the maximum level of stability and controllability of cars. The purpose of this work is to improve the handling of vehicles with a 4x4 wheel formula with a differential transmission in curved motion on support surfaces with low coupling properties. An algorithm is proposed to improve the handling of vehicles with a 4x4 wheel formula with a differential transmission in curved motion on support surfaces with low coupling properties due to braking of individual wheels. The efficiency and efficiency of the proposed algorithm is proved by the methods of simulation of the movement of two-axle cars with a 4x4 wheel formula with a differential transmission.

Introduction

Studying the trends in the development of modern automotive industry, we can see that manufacturers are constantly increasing the level of control over the parameters of the movement of wheeled vehicles, achieving the maximum level of stability and controllability of cars.

Currently, the creation of active safety systems for cars that provide increased stability and manageability is actively engaged abroad. Research is conducted in two directions. First, dynamic stabilization systems (DSS) are created, the principle of operation of which is based on changing the torques supplied to the driving wheels, including by braking individual wheels. The use of various all-wheel drive systems is becoming more relevant and appropriate. The distribution of traction on all wheels allows you to use the entire weight of the car as a hitch, which has a positive effect on the dynamic qualities of the car, cross-country, and allows you to implement control algorithms that improve the handling and course stability of the car. The development of these methods is devoted, for example, to the work [1–6].

Secondly, increasing stability and controllability is provided by the introduction of automatic corrective changes in the angle of rotation of the controlled wheels (steering) [7–10]. Combined management based on these approaches is possible [11].

The purpose of this work is to improve the handling of 4x4 vehicles with differential transmission in curved motion on support surfaces with low coupling properties. To achieve this goal, you need to solve the following tasks:

IOP Conf. Series: Materials Science and Engineering **963** (2020) 012002 doi:10.1088/1757-899X/963/1/012002

- development of a mathematical model of curvilinear motion of 4x4 vehicles with differential transmission on support surfaces with low coupling properties;
- development of an algorithm for the operation of a 4x4 vehicle motion stabilization system with a differential transmission;
- checking the performance of the 4x4 vehicle traffic stabilization system with differential transmission using simulation methods.

Mathematical model of the differential transmission of a 4x4 car with a differential transmission
Consider the diagram (fig. 1) differential transmission of a four-wheel drive two-axle car.

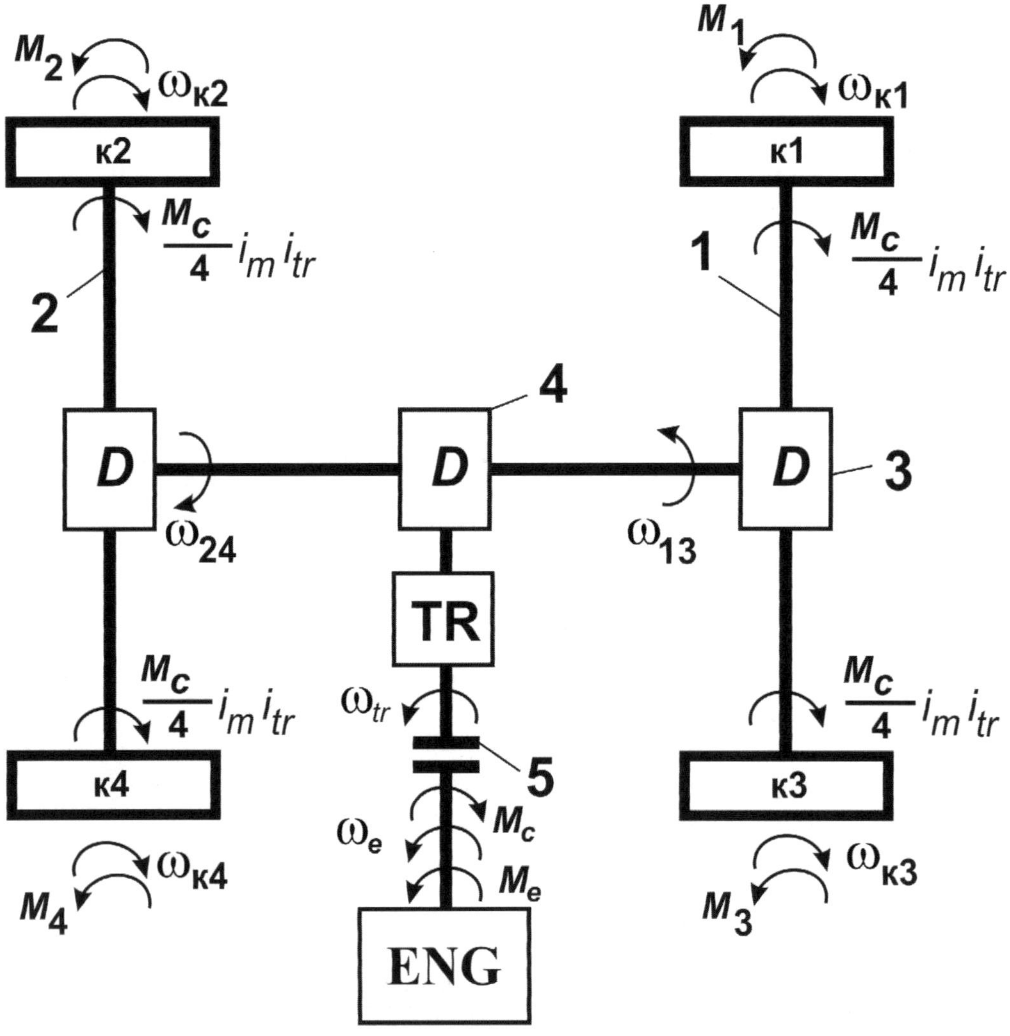

Fig. 1. Differential transmission scheme for a four-wheel drive two-axle car:1, 2 — front and rear axles; 3, 4 (D) — symmetrical inter-wheel differential; 5 — clutch; ω_{13},ω_{24} — angular rotation speeds of the front and rear transmission shafts, respectively;ω_{tr}– angular rotation speed of the primary transmission shaft; M_c — torque by the clutch; M_i, i=1,…,4 — resistance moments on the i-thwheel; M_e — torque by the engine; i_{tr} — the gear ratio of the transmission; i_m — the gear ratio of the main transmission; $\omega_{\kappa i}$, i=1,…,4 — angular speed of rotation of the i-th wheel; ω_e – angular rotation speed of the engine crankshaft; ENG — internal combustion engine; TR — transmission

The operation of the transmission can be described by the following system of equations:

$$\begin{cases} J_\kappa \dot{\omega}_{\kappa 1} = \dfrac{M_c}{4} i_{tr} i_m - M_1 \\ J_\kappa \dot{\omega}_{\kappa 2} = \dfrac{M_c}{4} i_{tr} i_m - M_2 \\ J_\kappa \dot{\omega}_{\kappa 3} = \dfrac{M_c}{4} i_{tr} i_m - M_3 \\ J_\kappa \dot{\omega}_{\kappa 4} = \dfrac{M_c}{4} i_{tr} i_m - M_4 \\ \dot{\omega}_{tr} = i_{tr} \dfrac{\dot{\omega}_{13} + \dot{\omega}_{24}}{2} \\ \dot{\omega}_{13} = i_m \dfrac{\dot{\omega}_{\kappa 1} + \dot{\omega}_{\kappa 3}}{2} \\ \dot{\omega}_{24} = i_m \dfrac{\dot{\omega}_{\kappa 2} + \dot{\omega}_{\kappa 4}}{2} \\ J_e \dot{\omega}_e = h_{dr} M_e - M_c \end{cases}$$

where J_κ — moment of inertia of the wheel; J_e — the moment of inertia of moving engine parts brought to the crankshaft; $h_{dr} = [0 \ldots 1]$ — the degree of pressing a pedal of submission of fuel.

Mathematical model of a single-disc friction clutch

In mathematical models of friction clutches, the dependence of the dry friction force F_{mp} on the relative movement speed V_n of the contacting surfaces is given in the form shown in fig. 2.

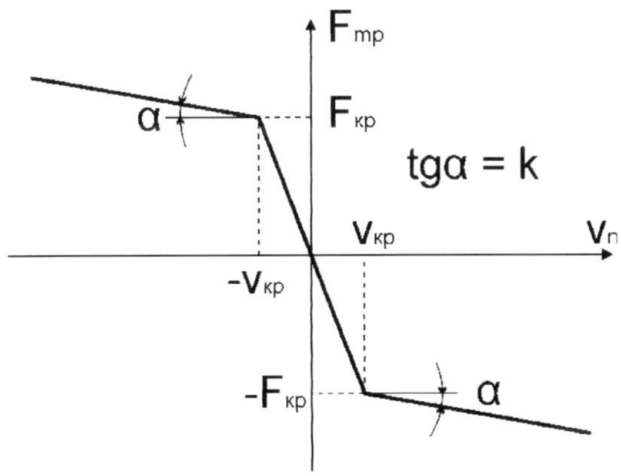

Fig. 2. Characteristics of dry friction

The torque M_c transmitted by the clutch is determined by the formula

$$M_c = F_{mp} r_{ef}$$

$$r_{ef} = \frac{\pi\left(D_{nkl}^3 - d_{nkl}^3\right)}{12 S_{nkl}}$$

$$S_{nkl} = \frac{0{,}94\pi\left(D_{nkl}^2 - d_{nkl}^2\right)}{4}$$

where r_{ef} — effective friction radius; D_{nkl}, d_{nkl} — accordingly, the external and internal diameters of the clutch friction plates; S_{nkl} — area of clutch friction plates.

The strength of F_{mp} is approximated using the following algorithm. If $|\omega_e - \omega_{tr}| \geq \Delta\omega_n$ (where $\Delta\omega_n$ — threshold value), that

$$F_{mp} = F_c\left[1+(k_b-1)e^{-c_w|\omega_{dv}-\omega_{kp}|}\right]sign(\omega_{dv}-\omega_{kp}) \quad (1)$$

$$F_c = F_{pr} + f_{cfr}\frac{h_{sc}N_{\max}}{\pi r_{эф}^2} \quad (2)$$

$$F_{pr} = \mu h_{sc} N_{\max} \quad (3)$$

$$N_{\max} = \frac{1{,}1 M_{e\max}}{r_{ef}\mu} \quad (4)$$

where F_c — Coulomb friction; k_b — the rate of change of force breakaway; c_w — conversion factor; F_{pr} — rest friction force; f_{cfr} — the coefficient of Coulomb friction forces; μ — the coefficient of static friction; N_{max} — maximum compression force of the clutch friction discs; $M_{e\max}$ — maximum torque developed by the engine; $h_{sc} = [0 \ldots 1]$ — the position of the pedal clutch control: 0 — clutch off; 1 — clutch on.

If $|\omega_e - \omega_{tr}| < \Delta\omega_n$, that

$$F_{mp} = K_x(\omega_e - \omega_{tr})$$

where K_x— coefficient of proportionality.

$$K_x = \frac{F_c\left[1+(k_b-1)e^{-c_w|\omega_e-\omega_{tr}|}\right]}{\Delta\omega_n}$$

The numerical values of the parameters in formulas (1)–(4) used in the simulation are shown in the table 1.

Table 1. Numerical values of the parameters adopted in the simulation

Parameter	Designation	Dimension	Value
The coefficient of Coulomb friction forces	f_{cfr}	N/Pa	0,00033
The rate of change of force breakaway	k_b	–	1,1
Conversion factor	c_w	s	1,24
Threshold value	$\Delta\omega_n$	1/s	0,5
The coefficient of static friction	μ	–	0,4

Features of the mathematical model of motion are considered in [12–17].

Algorithm of operation of the 4x4 vehicle stabilization system with differential transmission due to individual braking of individual wheels

The operation of the system is based on the recognition of the rear axle skidding (fig. 3a) or demolition of the front axle (fig. 3b) of the car.

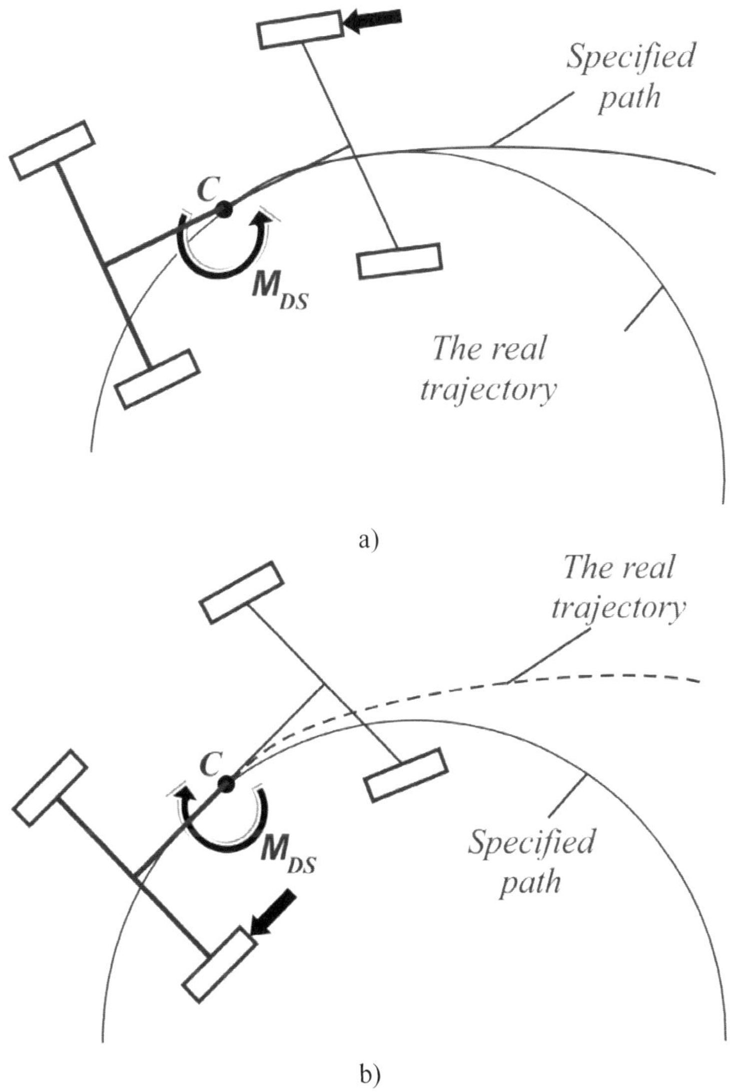

Fig. 3. Braking of the outer front wheel in case of skidding of the rear axle (a) and the inner rear wheel in case of demolition of the front axle (b): C — center of mass of the vehicle; M_{DS} — dynamic stabilization moment

To prevent the development of the front axle demolition, it is necessary to first recognize the occurrence and development of this process. To do this, use the results of the work [18], where the parameter is used as a diagnostic sign of the onset of front or rear axle drift $\delta_V = \left\| \boldsymbol{V}_{C1} \right| - \left| \boldsymbol{V}_{C2} \right\|$, representing the difference in the estimation of linear speeds of the center of mass of the car, first using the linear speed of the center of the front axle (vector $\boldsymbol{V}_{C1}$), and then — the linear speed of the center of the rear axle (vector $\boldsymbol{V}_{C2}$).

In the event of a skid the rear axle is braked the outside front wheel (fig. 3A). If the front axle is demolished, the inner rear wheel is braked (fig. 3b) [19, 20].

The formula for determining the braking torque on each wheel is as

$$M_{mi} = h_{ESPi} T_{\max}, i = 1, \ldots, N,$$

where $T_{\max}$ — maximum braking torque developed by the wheel brake mechanism; N — number of car wheels; $h_{ESPi} = [0...1]$— the degree of reduction of the effective braking torque on the *i*-th wheel due to the operation of the algorithm to counteract the demolition of the front axle during braking.

Algorithm for determining the value $h_{ESPi}, i = 1,...,N$ taking into account the sign rule adopted in the simulation, it should be as follows.

If $\Theta_1 > 0°$ (turn left) & $\delta_V > 0$ (demolition of the front axle), that $h_{ESP1} = h_{ESP3} = h_{ESP4} = 0$; $h_{ESP2} = C_1|\delta_V| + C_2\dot{\delta}_V$.

If $\Theta_1 > 0°$ (turn left) & $\delta_V < 0$ (the drift of a back axis), that $h_{ESP1} = h_{ESP2} = h_{ESP4} = 0$; $h_{ESP3} = C_1|\delta_V| + C_2\dot{\delta}_V$.

If $\Theta_1 < 0°$ (turn right) & $\delta_V > 0$ (demolition of the front axle), that $h_{ESP1} = h_{ESP2} = h_{ESP3} = 0$; $h_{ESP4} = C_1|\delta_V| + C_2\dot{\delta}_V$.

If $\Theta_1 < 0°$ (turn right) & $\delta_V < 0$ (the drift of a back axis), that $h_{ESP4} = h_{ESP2} = h_{ESP3} = 0$; $h_{ESP1} = C_1|\delta_V| + C_2\dot{\delta}_V$.

In the above formulas C_1 and C_2 — regulator gain factors that can be adjusted individually for each vehicle.

Performance criteria for the 4x4 vehicle stabilization system with differential transmission due to individual braking of individual wheels

As criteria for the effectiveness of the stabilization system of a 4x4 vehicle with a differential transmission due to individual braking of individual wheels, you can take the following.

1) Type of vehicle trajectory (no rear axle skid or front axle drift).

2) The case when $\delta_V = 0$, corresponds to the vehicle's neutral turnability (ideal value). Therefore, the diagnostic parameter δ_V it can be used to formulate a criterion for the efficiency *KE* operation of motion stabilization algorithms:

$$KE = \sqrt{\frac{1}{n_r - 1}\sum_{i=1}^{n_r}\delta_{Vi}^2},$$

where n_r — the number of points in the implementation process δ_{Vi}.

Study of the effectiveness of the proposed algorithm for the system of stabilization of a 4x4 car with a differential transmission due to individual braking of individual wheels

The health check algorithm of the system stabilization system of a vehicle 4x4 with differential transmission due to the individual braking of each wheel were conducted on the example of vehicle 2400 kg.

Was carried out a theoretical study of the motion of the vehicle using simulation mathematical modeling.

We study the movement on the support base "ice with snow" (with the coefficient of interaction of the engine with the support base when fully skidding $\mu_{s\max}$ = 0,35). Note that the term "support base" refers only to a solid non-deformable support surface. The front wheels of the car are controllable. The steering angle changes from zero to the set value within 3 seconds and then remains unchanged.

The mode of entering a turn and moving in a turn with a fixed radius is simulated, while the speed *V*=25 km/h, which the driver tries to keep constant during the movement.

4x4 car with no controls braking of the wheels (the base for comparison)

The trajectory of the 4x4 vehicle on the ice when making a turn is shown in fig.4.

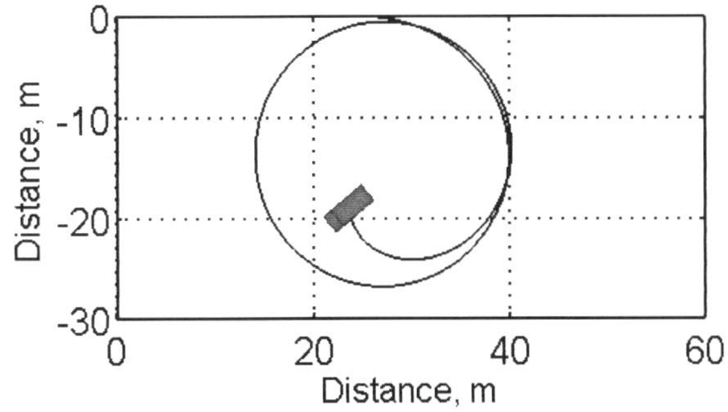

Fig. 4. The trajectory of the 4x4 vehicle in a turn on the ice in a turn

Analysis of vehicle trajectories (fig. 4) shows that when performing a maneuver on the support surface "ice with snow" the car loses control (develops a skid of the rear axle).

The value of the *KE* efficiency criterion for motion stabilization algorithms is 1,00 m/s.

4x4 vehicle with wheel braking control

The path of movement on the ice of a 4x4 vehicle with the control of braking the wheels when making a turn is shown in the fig. 5.

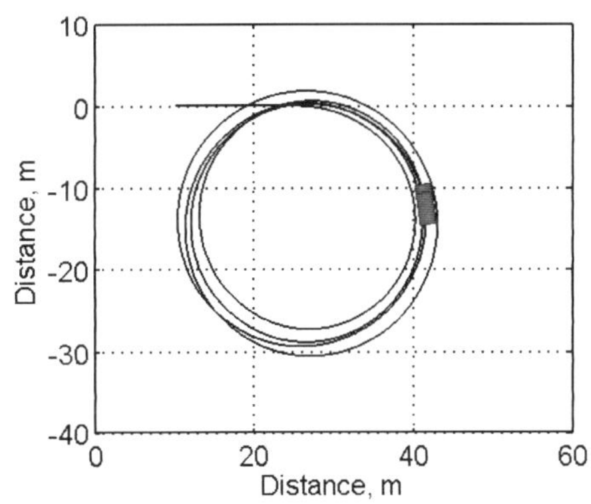

Fig. 5. The trajectory of the car 4x4 with control braking wheels in a turn on the ice

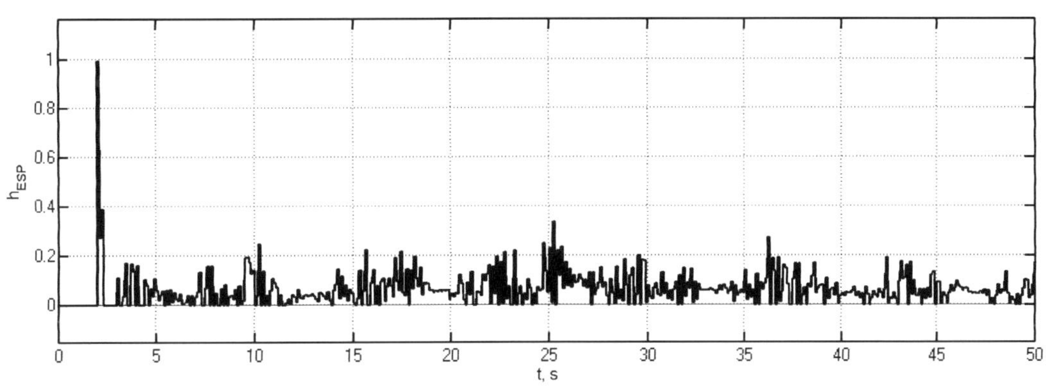

Fig. 6. The time variation of the parameters h_{ESP} for the front left wheel of a 4x4 vehicle with wheel braking control when cornering

Analysis of the movement paths of a 4x4 vehicle with wheel braking control (fig. 5) shows that when performing a maneuver on the "ice with snow" support surface, the car remains controllable.

Onfig.6 shows the change in time of the control parameter h_{ESP} for the front left wheel of a 4x4 vehicle with wheel braking control when driving in a turn.

As seen in fig. 6, braking of the front left wheel occurs all the time of movement in the turn to prevent the development of skidding of the rear axle, the level of braking torque does not exceed 10...20% of its maximum value.

The value of the *KE* efficiency criterion for motion stabilization algorithms is 0,39 m/s.

Conclusions

1. An algorithm is proposed to improve the handling of 4x4 vehicles with differential transmission in curved motion on support surfaces with low coupling properties due to braking of individual wheels.
2. The efficiency and efficiency of the proposed algorithm is proved by the methods of simulation of the movement of two-axle 4x4 vehicles with differential transmission.

References

[1] Diez, D.R., Velenis, E., Tavernini, D., Smith, E.N., Siampis, E., Soltani, A. Front/Rear Axle Torque Vectoring Control for Electric Vehicles (2019) Journal of Dynamic Systems, Measurement and Control, Transactions of the ASME, 141.

[2] Wu, Z., Xie, B., Chi, R., Du, Y., Mao, E. Active modulation of torque distribution for dual-motor front- and rear-axle drive type electric vehicle based on slip ratio (2018) Nongye Gongcheng Xuebao/Transactions of the Chinese Society of Agricultural Engineering, 34 (15), pp. 66–76.

[3] Zhao, B., Xu, N., Chen, H., Guo, K., Huang, Y. Stability control of electric vehicles with in-wheel motors by considering tire slip energy (2019) Mechanical Systems and Signal Processing, 118, pp. 340–359.

[4] Xu, W., Chen, H., Zhao, H., Ren, B. Torque optimization control for electric vehicles with four in-wheel motors equipped with regenerative braking system (2019) Mechatronics, 57, pp. 95–108.

[5] Song, Y., Shu, H., Chen, X., Luo, S. Direct-yaw-moment control of four-wheel-drive electrical vehicle based on lateral tyre-road forces and sideslip angle observer (2019) IET Intelligent Transport Systems, 13 (2), pp. 356–366.

[6] Zhao, X., Guo, G. Braking torque distribution for hybrid electric vehicles based on nonlinear disturbance observer (2019) Proceedings of the Institution of Mechanical Engineers, Part D: Journal of Automobile Engineering, .

[7] Zhang, J., Zheng, H., Zhao, M. Analysis of Vehicle Steering Stability of Nonlinear Four Wheel Steering Based on Sliding Mode Control (2018) SAE Technical Papers, 2018-August (August).

[8] Ma, X., Wong, P.K., Zhao, J., Xie, Z. Cornering stability control for vehicles with active front steering system using T-S fuzzy based sliding mode control strategy (2019) Mechanical Systems and Signal Processing, 125, pp. 347–364.

[9] Hu, C., Qin, Y., Cao, H., Song, X., Jiang, K., Rath, J.J., Wei, C. Lane keeping of autonomous vehicles based on differential steering with adaptive multivariable super-twisting control (2019) Mechanical Systems and Signal Processing, 125, pp. 330–346.

[10] Chu, Z., Wu, C., Sepehri, N. Automated steering controller design for vehicle lane keeping combining linear active disturbance rejection control and quantitative feedback theory (2018) Proceedings of the Institution of Mechanical Engineers. Part I: Journal of Systems and Control Engineering, 232 (7), pp. 937–948.

[11] Mokhiamar O., Abe M. Active wheel steering and yaw moment control combination to maximize stability as well as vehicle responsiveness during quick lane change for active vehicle handling safety. // J Automobile Eng. 2002, V.216(2). — P.115–124.

[12] Belousov, B., Ksenevich, T.I., Vantsevich, V., Komissarov, D. 8×8 platform for studing terrain mobility and traction performance of unmanned articulated ground vehicles with steered wheels (2013) SAE Technical Papers, 9.
[13] Gorelov, V.A., Komissarov, A.I., Miroshnichenko, A.V. 8×8 wheeled vehicle modeling in a multibody dynamics simulation software (2015) Procedia Engineering, 129, pp. 300–307.
[14] Keller, A.V., Gorelov, V.A., Vdovin, D.S., Taranenko, P.A., Anchukov, V.V. Mathematical model of all-terrain truck (2015) Proceedings of the ECCOMAS Thematic Conference on Multibody Dynamics 2015, Multibody Dynamics 2015, pp. 1285–1296.
[15] Gorelov, V.A., Komissarov, A.I. Mathematical model of the straight-line rolling tire — Rigid terrain irregularities interaction (2016) Procedia Engineering, 150, pp. 1322–1328.
[16] Vol'skaya, N.S., Zhileykin, M.M., Zakharov, A.Y. Mathematical model of rolling an elastic wheel over deformable support base (2018) IOP Conference Series: Materials Science and Engineering, 315 (1), article № 012028.
[17] Wong, J.Y. Theory of Ground Vehicles / J.Y. Wong. — New York: Wiley IEEE, 2001. — 560 p.
[18] Ankinovich G.G., Verzhbitskiy A.N., Zhileykin M.M. A Method of Determining Motion Parameters of Two-Axle Wheeled Vehicles to Ensure the Operation of the Dynamic Stability System. (2017) BMSTU Journal of Mechanical Engineering, 4 (685), p.p. 11–20.
[19] H. Fennel, R. Gutwein, A. Kohl, M. Latamik, G. RollDas modulare Regler- und Regelkonzept beim ESP von ITT Automotive 7. Aachener Kolloquium Fahrzeug- und Motortechnik, 5.–7. Oktober, Aachen (1998), pp. 409–431
[20] K. LangwiederMit ESP schwere Unfalle vermeiden oder mildern ESP-Workshop, November 10, Boxberg, Germany (1999).

IOP Conf. Series: Materials Science and Engineering **963** (2020) 012003 doi:10.1088/1757-899X/963/1/012003

Research of dependence centrifugal pump efficiency of impeller slit seal design parameters

A Protopopov[1,2], N Belov[1], M Kalugina[1], and A Mukhlaeva[1]

[1]Bauman Moscow State Technical University, 5 Second Baumanskaya Street, Moscow, 105005, Russian Federation

[2]Email: proforg6@yandex.ru

Abstract. Efficiency and resource are the two main characteristics of a centrifugal pump. In present study, the dependence of these characteristics from impeller sealing wear ring configuration was considered. Range of models with various configurations has been analyzed using numerical flow simulation. Compromise curves nave been obtained as a result. This information allows select an optimal ratio of dimensions of wear ring hydrodynamic slit seal to achieve a balance between efficiency and reliability of centrifugal pump.

Introduction

Double suction centrifugal pumps (type D pumps) are one of the most common types of centrifugal pumps. Studies of these pumps characteristics using CDF simulation methods are presented in [1]–[6]. Studies results of centrifugal pumps of similar subjects are described in [7]–[11].

The employ of the LP-tau search as a method for analyzing a number of models with various parameters and determining their optimal values is given in [12]–[15].

Sealing wear ring configuration of pump's impeller has significant impact on general efficiency and reliability of the centrifugal pump. Studies of parameters of contained fluid in slit gaps using hydrodynamic modeling are described in [16]–[20].

In type D centrifugal pumps, one of the most actual issues is the selection of optimal design parameters of impeller gap seal from the point of view of volumetric efficiency and possible hydrodynamic reaction in it. Force of hydrodynamic reaction occurs due to drive shaft deflection and as a consequence an uneven distribution of pressure over the surface of the rotor with eccentricity. Volumetric efficiency is determined predominantly by the amount of leakage.

First of all, the geometrical parameters of the gap seal are a radial clearance δ and length L which significantly influence the value of overflows and the lifting force. The criteria under study also depend on the total radial force that arises during pump operation and leads to shaft deflection.

Method

Two methods were used to calculate the leakage and lifting force in a gap seal:

1. CFD simulation employed the STAR-CCM+ software package to calculate values of leakage (volumetric efficiency) and lift force;

2. The LP-Tau search to calculate all values of varied parameters and determine the objective function.

IOP Conf. Series: Materials Science and Engineering **963** (2020) 012003 doi:10.1088/1757-899X/963/1/012003

The LP-Tau search generates points in a quasi-random manner in the interval for two parameters. For considering pump the radial clearance δ can vary from 0,3 to 0,6 mm and the length of the gap seal L can vary from 10 to 20 mm.

It is not correct to vary shaft deflection value by the LP-Tau search method, since the pump does not always work in an optimal mode (especially when using the frequency regulation method). Accordingly, the radial force also changes during operation which leads to determine only the radial clearance and the length of the gap seal by the LP-Tau search method. In this case, pump shaft deflection varies for each calculated point obtained (for fixed values of δ and L) as a percentage of the gap: 0%, 30%, 60%. Thus, it becomes possible to analyze the dependence of the studied criteria on pump shaft's deflection for each specific model.

In this case, generated operating points field can be visualized as a rectangle (Fig. 1).

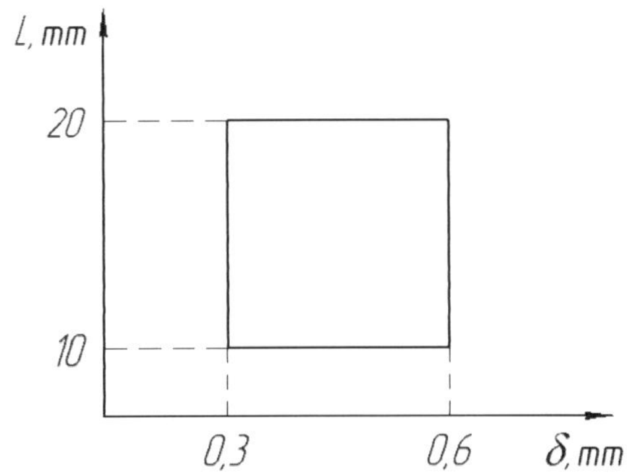

Fig. 1. Points generation area

Obtained using the LP-Tau search control points are summarized in table 1.

Table 1. Control points for the LP-Tau search

Point №	δ, mm	L, mm
1	0,3	12
2	0,45	15
3	0,525	12,5
4	0,375	17,5
5	0,4875	11,25
6	0,3375	16,25
7	0,4125	13,75
8	0,5625	18,75
9	0,58125	10,625
10	0,43125	15,625
11	0,35625	13,125
12	0,50625	18,125
13	0,39375	11,875
14	0,54375	16,875
15	0,46875	14,375
16	0,31875	19,375
17	0,459375	10,3125

Point №	δ, mm	L, mm
18	0,309375	15,3125
19	0,384375	12,8125
20	0,534375	17,8125
21	0,346875	11,5625
22	0,496875	16,5625
23	0,571875	14,0625
24	0,421875	19,0625
25	0,440625	10,9375
26	0,590625	15,9375
27	0,515625	13,4375
28	0,365625	18,4375
29	0,553125	12,1875
30	0,403125	17,1875
31	0,328125	14,6875
32	0,478125	19,6875
33	0,5390625	10,15625

In accordance with the method of considering varying pump drive shaft deflection for each control point, three models were computed (a total number of models is 99).

To obtain the results, the pressure difference at impeller gap inlet and impeller gap outlet is of primary importance. Pressure difference value can be computed from impeller gap flow part CFD simulation method employing STAR-CCM+. Also, this method allows evaluate the quality of design.

A three-dimensional model appearance of the computational domain is shown in Fig. 2. The computational domain is includes two elements: a gap between pump casing and the impeller and cylindrical slit seal.

Fig. 2. 3-D model of computational domain

IOP Conf. Series: Materials Science and Engineering **963** (2020) 012003 doi:10.1088/1757-899X/963/1/012003

Results

The following describes computation results of first model (this is control point №1, pump shaft deflection is 0%). A leakage graph through the slit seal is depicted in Fig. 3.

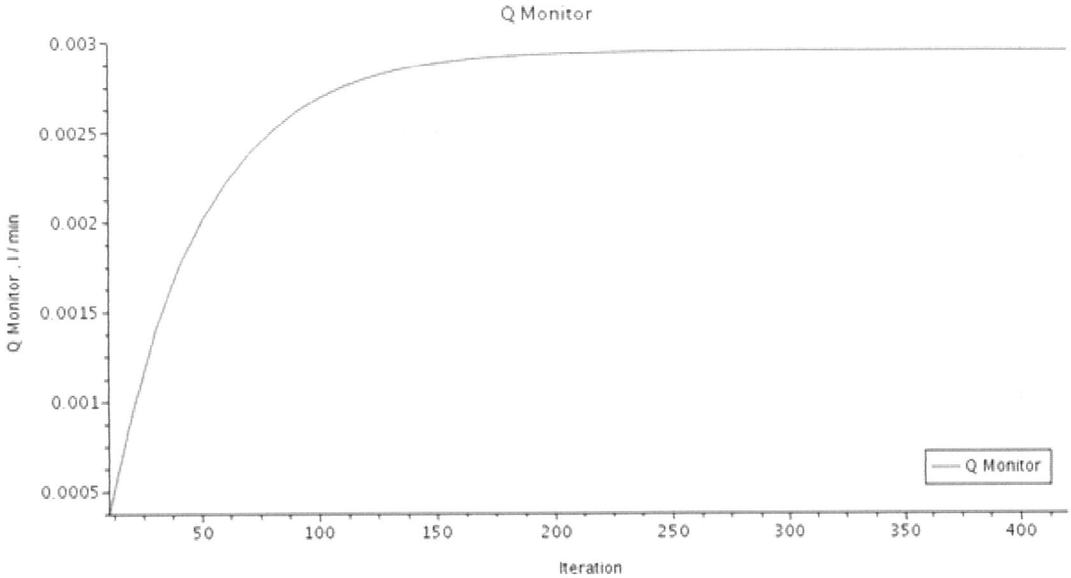

Fig. 3. Flow rate through slit seal

The radial force is 0,0107 N according to the calculation results.

The remaining models are calculated in a similar method in accordance with the data in Table 1. The calculation results are summarized in Table 2. It should be noted that the radial force in all models with pump shaft deflection absence assumes fairly small values of the order of 0,01 N. Therefore, it can be neglected.

Table 2. Calculation results

Pump shaft deflection, %	$Q_{ут}\cdot10^{-3}$, m^3/s	$\eta_{об}$, %	F_R, N
Model № 1, 2, 3: δ = 0,3 mm, L = 12 mm			
0	2,951	90,1	0,0
30	3,071	90,0	33,5
60	3,325	89,3	70,6
Model № 4, 5, 6: δ = 0,45 mm, L = 15 mm			
0	4,584	85,7	0,0
30	4,772	85,2	37,4
60	5,167	84,2	78,1
Model № 7, 8, 9: δ = 0,525 mm, L = 12,5 mm			
0	5,697	83,0	0,0
30	5,927	82,4	24,5
60	6,419	81,2	51,2
Model № 10, 11, 12: δ = 0,375 mm, L = 17,5 mm			
0	3,581	88,5	0,0
30	3,726	88,1	53,7
60	4,036	87,2	112,1

Pump shaft deflection, %	$Q_{ут}\cdot10^{-3}$, m^3/s	$\eta_{об}$, %	F_R, N
Model № 13, 14, 15: δ = 0,4875 mm, L = 11,25 mm			
0	5,251	84,0	0,0
30	5,464	83,4	21,7
60	5,918	82,3	45,3
Model № 16, 17, 18: δ = 0,3375 mm, L = 16,25 mm			
0	3,213	89,6	0,0
30	3,334	89,2	50,7
60	3,606	88,4	105,9
Model № 19, 20, 21: δ = 0,4125 mm, L = 13,75 mm			
0	4,201	86,7	0,0
30	4,371	86,3	34,2
60	3,733	85,3	71,5
Model № 22, 23, 24: δ = 0,5625 mm, L = 18,75 mm			
0	5.741	82,7	0,0
30	5,973	82,2	46,8
60	6,469	81,1	97,7
Model № 25, 26, 27: δ = 0.58125 mm, L = 10,625 mm			
0	6,448	81,0	0,0
30	6,709	80,4	17,2
60	7,266	79,1	35,9
Model № 28, 29, 30: δ = 0,43125 mm, L = 15,625 mm			
0	4,329	86,4	0,0
30	4,504	85,9	41,1
60	4,879	84,9	85,8
Model № 31, 32, 33: δ = 0,35625 mm, L = 13,125 mm			
0	3,562	88,5	0,0
30	3,706	88,1	34,8
60	4,013	87,3	72,8
Model № 34, 35, 36: δ = 0.50625 mm, L = 18,125 mm			
0	5,099	84,4	0,0
30	5,305	83,8	47,4
60	5,746	82,7	98,8
Model № 37, 38, 39: δ = 0,39375 mm, L = 11,875 mm			
0	4,075	87,1	0,0
30	4,240	86,6	27,6
60	4,592	85,7	57,7
Model № 40, 41, 42: δ = 0,54375 mm, L = 16,875 mm			
0	5,614	83,0	0,0
30	5,841	82,5	40,1
60	6,326	81,3	83,8

Pump shaft deflection, %	$Q_{ут} \cdot 10^{-3}$, m^3/s	$\eta_{об}$, %	F_R, N
Model № 43, 44, 45: δ = 0,46875 mm, L = 14,375 mm			
0	4,844	85,0	0,0
30	5,039	84,5	33,9
60	5,458	83,4	70,8
Model № 46, 47, 48: δ = 0,37875 mm, L = 19,375 mm			
0	3,547	88,6	0,0
30	3,692	88,2	62,6
60	3,997	87,3	130,9
Model № 49, 50, 51: δ = 0,459375 мм, L = 10,3125 мм			
0	4,964	84,7	0,0
30	5,165	84,2	19,5
60	5,594	83,1	40,6
Model № 52, 53, 54: δ = 0,309375 mm, L = 15,3125 mm			
0	2,915	90,4	0,0
30	3,033	90,1	48,5
60	3,285	89,3	101,4
Model № 55, 56, 57: δ = 0,384375 mm, L = 12,8125 mm			
0	3,913	87,5	0,0
30	4,071	87,1	31,9
60	4,409	86,2	66,6
Model № 58, 59, 60: δ = 0,534375 mm, L = 17,8125 mm			
0	5,452	83,5	0,0
30	5,672	82,9	44,4
60	6,143	81,7	92,8
Model № 61, 62, 63: δ = 0,346875 mm, L = 11,5625 mm			
0	3,529	88,6	0,0,
30	3,671	88,2	28,8
60	3,976	87,4	60,1
Model № 64, 65, 66: δ = 0,496875 mm, L = 16,5625 mm			
0	5,066	84,4	0,0
30	5,271	83,9	41,3
60	5,709	82,8	86,2
Model № 67, 68, 69: δ = 0,571875 mm, L = 14,0625 mm			
0	6,116	81,8	0,0
30	6,363	81,2	28,4
60	6,892	80,1	59,4
Model № 70, 71, 72: δ = 0,421875 mm, L = 19,0625 mm			
0	4,059	87,1	0,0
30	4,223	86,7	57,4
60	4,574	85,7	119,9

Pump shaft deflection, %	$Q_{ут}\cdot 10^{-3}$, m^3/s	$\eta_{об}$, %	F_R, N
Model № 73, 74, 75: δ = 0,440625 mm, L = 10,9375 mm			
0	4,697	85,4	0,0
30	4,887	84,9	22,9
60	5,292	83,9	46,4
Model № 76, 77, 78: δ = 0,590625 mm, L = 15,9375 mm			
0	6,236	81,5	0,0
30	6,488	80,9	34,4
60	7.026	79,6	71,9
Model № 79, 80, 81: δ = 0,515625 mm, L = 13,4375 mm			
0	5,465	83,4	0,0
30	5,686	82,9	28,3
60	6,158	81,7	59,1
Model № 82, 83, 84: δ = 0,365625 mm, L = 18,4375 mm			
0	3,434	88,9	0,0
30	3,573	88,5	59,1
60	3,871	87,7	123,5
Model № 85, 86, 87: δ = 0,553125 mm, L = 12,1875 mm			
0	6,102	82,1	0,0
30	6,243	81,5	22,7
60	6,761	80,3	47,4
Model № 88, 89, 90: δ = 0,403125 mm, L = 17,1875 mm			
0	3,923	87,5	0,0
30	4,082	87,1	50,1
60	4,421	86,2	104,4
Model № 91, 92, 93: δ = 0,328125 mm, L = 14,6875 mm			
0	3,159	89,7	0,0
30	3,286	89,3	43,9
60	3,559	88,5	91,8
Model № 94, 95, 96: δ = 0,478125 mm, L = 19,6875 mm			
0	4,691	85,4	0,0
30	4,884	84,9	56,1
60	5,285	83,9	117,3
Model № 97, 98, 99: δ = 0,5390625 mm, L = 10,15625 mm			
0	5,956	82,2	0,0
30	6,617	81,6	16,8
60	6,712	80,4	35,1

For the possibility of analyzing obtained results, values of volumetric efficiency and hydrodynamic radial force are plotted on the η-P coordinate plane and compromise curves are constructed (Fig. 4, Fig. 5).

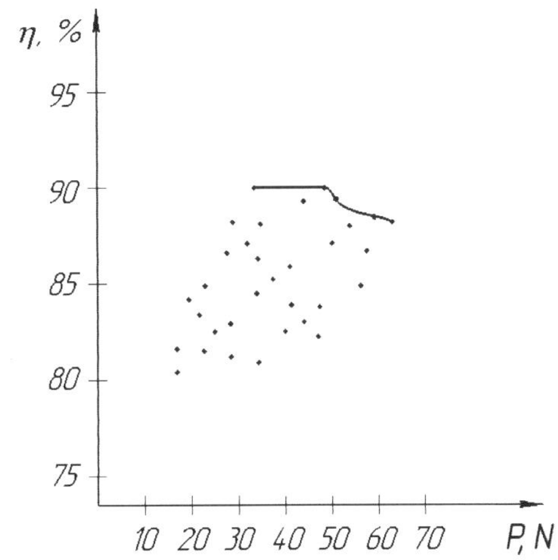

Fig. 4. Compromise curve for shaft deflection 30%

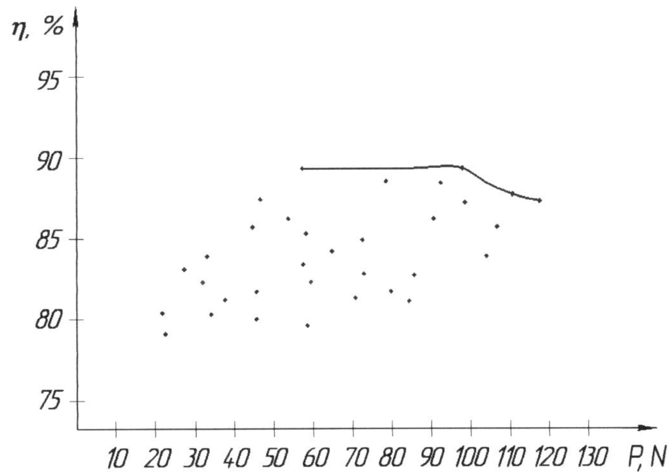

Fig. 5. Compromise curve for shaft deflection 60%

Conclusion

If there is a significant load on pump shaft, efficiency of the pump is reduced, since if a shaft deflection occurs, leakage through the gap seal increases. The compromise curves (Fig. 4, Fig. 5) show that the greatest efficiency is achieved by increasing the length of the gap seal and reducing the radial clearance. From the point of view of reliability, the radial clearance value does not make sense to choose excessively small, since in this case wear of gap seal ring occurs more intensively over time. Thus, for considering pump parameters values of the gap seal are recommended L = 14…20 mm, δ = 0,4…0,5 mm, provided that there is no pump shaft load that can establish a shaft deflection more 60%. Also, the obtained data allow select optimal range of frequency regulation application.

Reference

[1] Y. Zhang, S. Hu, Y. Zhang, and L. Chen, ''Optimization and analysis of centrifugal pump considering fluid-structure interaction,'' *Sci. World J.*, vol. 2014, Aug. 2014, Art. no. 131802. DOI: 10.1155/2014/131802

[2] S. Hatano, D. Kang, S. Kagawa, M. Nohmi and K. Yokota, Study of cavitation instabilities in double-suction centrifugal pump, International Journal of Fluid Machinery and Systems, 7 (2014) 94–100. DOI: 10.5293/IJFMS.2014.7.3.094

[3] Yao, Z.;Wang, F.; Zhang, Z.; Xiao, R.; He, C. Numerical and experimental investigation on the radial force characteristic of a large double suction centrifugal pump in a real pumping station. In Proceedings of the ASME-JSME-KSME 2015 Joint Fluids Engineering Conference, Korea, Seoul, 26–31 July 2015 DOI: 10.1115/AJKFluids2015-33468
[4] A Škerlavaj et al 2017 *J. Phys.: Conf. Ser.***796** 012007
[5] Gonzalez, J., Oro, J. M. F. ,Arglles-Diaz, K. M. , and Santolaria, C., 2009, "Flow Analysis for a Double Suction Centrifugal Machine in Pump and Turbine Operation Modes," Int. J. Numer. Methods Fluids, 61, pp. 220–236. DOI: 10.1002/fld.1951
[6] Tao R, Xiao R, Zhu D Multi-objective optimization of double suction centrifugal pump. ProcIMechE, Part C: J Mechanical Engineering Science 2018; 232: 1108–1117. DOI: 10.1177/0954406217699020
[7] Gülich J.F. Centrifugal Pumps. Springer-Verlag Berlin Heidelberg, 2010, 964 p.
[8] Bellarya S.A.I., Husain A., Samadc A., KanaiaR.A.. Performance Optimization of Centrifugal Pump for Crude Oil Delivery. The Journal of Engineering Research (TJER), Vol. 15, No. 1 (2018) 88–101 DOI: 10.1155/2018/3987594
[9] A Petrov et al 2019 *IOP Conf. Ser.: Mater. Sci. Eng.* **492** 012036
[10] V Cheremushkin and V Lomakin 2019 *IOP Conf. Ser.: Mater. Sci. Eng.***492** 012039
[11] T Valiev and APetrov 2019 *IOP Conf. Ser.: Mater. Sci. Eng.***492** 012038
[12] Lomakin V O, Chaburko P S and Kuleshova M S 2017 Multi-criteria Optimization of the Flow of a Centrifugal Pump on Energy and Vibroacoustic Characteristics *Procedia Engineering***176** 476–482
[13] Lomakin V.O., Kuleshova M.S., Chaburko P.S., Baulin M.N. Complex wet end part optimization of hermetic pump with LP-TAU method. Nasosy. Turbiny. Sistemy [Pumps. Turbines. Systems], 2016, no. 1, pp. 55–61.
[14] V Tkachyk et al 2019 *IOP Conf. Ser.: Mater. Sci. Eng.***589** 01201
[15] P Chaburko and Z Kossova 2019 *IOP Conf. Ser.: Mater. Sci. Eng.* 492 012011
[16] Nakashima, C.Y. and Caetano, E.F. "Calculation Of Pressure Drop In Narrow Rotating Annular Clearances," "*EngenhariaTérmica (Thermal Engineering)*" 2008, pp. 27–34.
[17] Lomakin V O, Kuleshova M S and Kraeva E A 2015 Fluid Flow in the Throttle Channel in the Presence of Cavitation Procedia Engineering 106 27–35
[18] Kun Wang et al 2017 *J. Phys.: Conf. Ser.***916** 012025 DOI :10.1088/1742-6596/916/1/012025
[19] Babayigit, O.; Kocaaslan, O.; Aksoy, M.H.; Guleren, K.M.; Kocaaslan, O. Experimental and CFD investigation of a multistage centrifugal pump including leakages and balance holes. Desalin. Water Treat. 2017, 67, 28–40. DOI: 10.5004/dwt.2017.20153
[20] Prausova, H., Bublík, O., Vimmr, J., Luxa, M., Hála, J., Clearance gap flow: Simulations by discontinuous Galerkin method and experiments, EPJ Web of Conferences 92 (2015) 02073. DOI:0.1051/epjconf/20159202073.

Improving the material properties of vacuum devices electrodes by technology rational improvement

Yu Kyrganova[1,3], Yu Nikolukin[2] and K Panina[1]

[1]Bauman Moscow State Technical University, 5 Second Baumanskaya Street, Moscow, 105005, Russian Federation
[2]OOO «Kurs» Leading Specialist

[3]Email: kurganova_ya@mail.ru

Abstract. The material of the electrodes of high-current vacuum and plasma electronics is traditionally obtained by powder metallurgy methods (mixing components, pressing blanks, hydrogen sintering in a given mode). However, the material of the cathodes produced according to the adopted technology, has a number of significant limitations. They are caused by the impossibility of a uniform distribution and grinding of the initial powder components by the adopted grinding in globe mills, as well as by abundant gas emission during the sintering time. The nonuniformity of the distribution of the composition in the structure, low density, high porosity, low mechanical strength of the emitting material with a metal matrix give a decrease in the operating properties of the cathodes, which leads to a decrease in the lifetime of electronic tubes. This problem is proposed to be solved by directional changes in the physical and mechanical properties of the electrode material by selecting the gravimetric consist of the base material components and improving the existing production process. The article presents the analysis of research results, changes in the structure of the material by optical and scanning microscopy methods, and also describes a complex of acquired physical and mechanical properties of experimental samples obtained by modernized technology. Thus, the developed innovative techniques allowed to increase the hardness by 36%; density by 19%, and to reduce porosity by 25% in relation to traditional cathodes. Such changes in the physical and mechanical properties make it possible to predict an increase in the service life of produced and developed vacuum tubes operating in current flow regimes up to 100 kA and in highly stressed electromagnetic fields up to 500 kV.

Introduction

High-vacuum tubes are used as key elements in plasma installations, in medical equipment, in oil production and in other fields of the national economy. They serve for transformation of high frequency and high power electromagnetic signals [1]. The operation of such devices is based on the phenomenon of electron emission from the electrodes surface[2, 3]. The cathode serves as the electrode, the main purpose of which is to produce electrons (Fig. 1).

The lifetime of high-vacuum tubes depends on the life of the cathodes. The main operating characteristics of the devices determine the cathodes emissivity. Init's turn, the stability and high level of cathodes emissivity largely depend on the structure and physical and mechanical properties of the materials from which they are made of. Thus, the task of finding ways to increase the emissivity and the lifetime of the electrodes of high-current vacuum and plasma electronics is relevant.

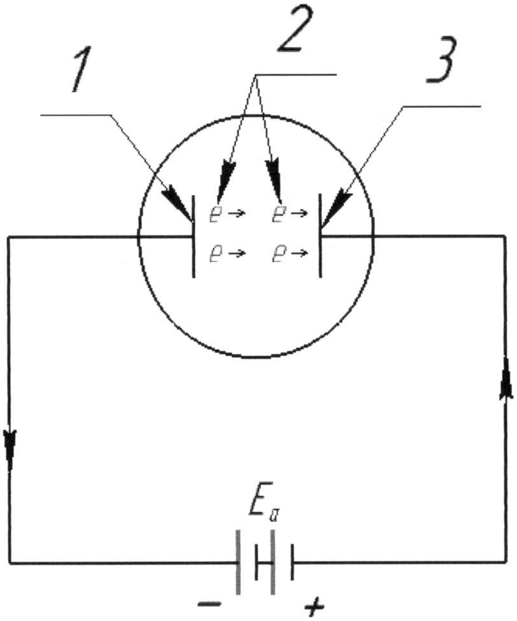

Fig.1. Scheme of an electrons flow formation:
1— cathode; *2*— electron flow; *3*— anode

The oxide low-temperature cathodes are most widely used according to the monitoring data on the use of high-current high-vacuum tubes cathodes. Their level of effective efficiency is 3...5 W/cm^2, and the efficiency of use— 100...200 mA/cm^2[3].

The cathode material consists of a refractory base and an emission component representing a mixture of oxides of elements of II group. The most common are alkaline earth metals barium, calcium, strontium[4–6].

There are various methods for producing oxide cathodes. One of the common methods is to obtain dispenser cathodes by impregnating a metal sponge of a refractory metal with an emission material [4, 5]. The limitations of such cathodes are low mechanical strength and low stability, strong response of the emissivity to the gravimetricconsist of the original powders, to the shape and number of pores of the metal sponge [7–9], the complexity of the technical process [10]and others.

High-emission electrodes obtained by powder metallurgy methods are more durable and stable duringoperation in high voltage fields of vacuum tubes. Micron and submicron powders of the materials which are electrodecomponents are mixed, compacted and sintered in inert gasesenvironment. The authors of [11, 12]receive electrodes for plasmatrons using a similar technology.

Traditionally, the cathode material composition is a high-melting base of tungsten, a nickel (preferably) metal bond, and an emission material — strontium carbonate. An abundant gassing occurs in the material due to the evaporation of gases on the surface of the initial powders and the primary change in thecarbonates structure during the componentssintering as the temperature rises. The carbonate synthesis to oxide occurs together with the sintering process, since the oxides of earth metals are unstable in the air —about 1 Pa·m^{3}of gas from 1 cm^2 of coating is released during the carbonates decomposing.

The technology does not allow to obtain a uniform structure of electrode materials, which leads to its final high porosity, and consequently, low density, which in its turn leads to a decrease in the mechanical resistance. Some cavities are formed in which the component of the emission material freely "lies". The weak mechanical connection of the electrode base with the component of the emission material contributes to its "tearing" out of the electrode body on exposure to strong current flows — up to 100 kA and high voltages of electromagnetic fields — up to 500 kV.

These negative technology aspects reduce theelectrode performance characteristics and the service life of high-current vacuum and plasma electronics devices in general.

In this paper, the possibility of changing the fractional composition of the components initial mixture and the sintering mediumreplacement are considered.

1. The methodology

A well-proven alloy system W-SrO-Ni, which has the tungsten is the base, and the mass fraction of its components corresponds to the following values, %: Ni — 2...3.6; SrO— 3...4.5, is chosen as the object of research.

According to traditional sintering technology, micron powders of components are used. This project proposes a transition to nanoparticles of nickel and strontium oxide powders in the range from 10 to 100 nm. The authors of a number of papers [13–16] showed the positive effect of the nanoscale components introduction on the physical and mechanical properties of various materials. Fractional composition of the developed technology is presented inTable1.

Table1. The powders fractional composition

Powder	Particle size
Tungsten	0,4...0,6μm
Nickel	10...100nm
Strontium carbonate	10...100 nm

The results of the study of the powders composition by x-ray microanalysis are presented in Table. 2. X-ray microanalysis was performed by a Jeol JSM-6490 electron microscope with an INCA 350 energy-dispersive microanalyzer.

Table2. The powders chemical composition, %

Powder	Ni	Sr	O	W
Tungsten	–	–	13,27	86,73
Nickel	100.00	–	–	–
Strontium carbonate	–	51,91	51,91	–

The obtained results show that a significant amount of oxygen is observed in the composition of the tungsten powder. Any other impurities were not detected.

Photomicrographs of the initial loose powders are shown in Fig. 2–4. The photographs were taken in the laboratory of thin physical methods for studying the structure of materials at the Moscow Bauman State Technical University on a scanning electron microscope REM VEGA II LMH.

IOP Conf. Series: Materials Science and Engineering **963** (2020) 012004 doi:10.1088/1757-899X/963/1/012004

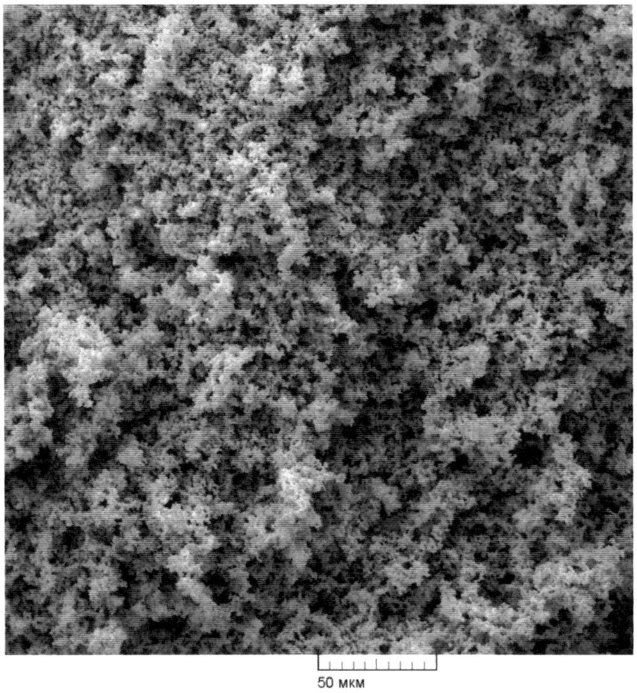

Fig.2. The tungsten powder (scaled-up 1000)

Fig. 3. The nickel powder (scaled-up 1000)

Fig. 4. The strontium carbonate powder (scaled-up 1000)

According to the traditional technology of producing electrode blanks, the initial powders are mixed with a plasticizer, which is polyvinyl alcohol, compacted and sintered in a hydrogen medium. The sintering mode is shown in fig. five

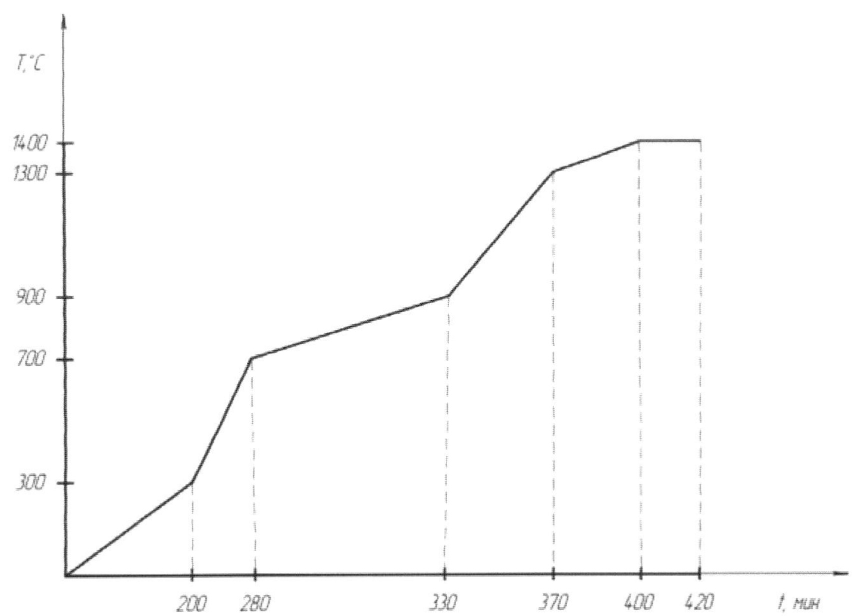

Fig. 5. The sintering mode according to traditional technology

The following consecutive processes take place during sintering:
atT< 100°C—evaporation of moisture from the powders surface;
atT = 100…300°C— plasticizer burnout;
atT = 700…900°C — strontium carbonate structure change with active gas emission;
atT = 1300 °C— strontium carbonate decompounding into strontium oxide; the chemical reaction of strontium carbonate decompoundinginto strontium oxide is described by the equation

$$SrCO_3 = SrO + CO_2\uparrow ,$$

this produces CO_2 gas that promotes pore formation;
atT = 1400 °C— sintering with a liquid nickel phase formation.

The analysis shows that all processes are accompanied by abundant gas emission, which causes the final high porosity of the electrode, and hence its low density and low mechanical strength.

The stages of the technological process of traditional and improved technology are presented in Table 3.

Table3. Stages of obtaining electrodes

Technologic aloperation	I	II	III	IV	V	VI	VII
Traditiontec hnology	Preparati onofpow ders, weighing	Mixing + adding plasticizer	Dehydration	Separation	Molding	Placing samples on a pallet	Hydrogen sintering
Improved technology	Preparati on of powders, weighing	***Mechanicala ctivation***	Dehydration	Separation	Molding	Placing samples on a pallet	***Vacuumsi ntering***

Differential characteristics of the proposed technology are the following:

- transition to nanoparticles of the initial binder powders and emission material;
- mechanicalactivationadding;
- sinteringinvacuum.

Mechanical activation is carried out in a planetary mill (2SL mechanical activator), where the processes of mixing powders and plasticizers, additional grinding of powders by grinding media and the mixture mechanical activation due to high kinetic energy of grinding bodies [17], resulting from significant overloads, are combined.

Mechanical activation affects the microstructural properties of powders.The significant particle size reduction [18–22] and active mixing powders occur under the influence of this technological stage. That provides a more uniform and crushed structure during sintering.

The proposed transition from sintering the mixture of the initial powder components in a hydrogen medium to sintering one in vacuum allows to increase their density and, consequently, their mechanical strength. In addition, the possiblility to eliminate the effect of a "shell" formation around the emission material component appears.

Obtaining a uniform electrode structure as a result of using binder powders and emission material in the range of 10 ... 100 nm, on condition that the powder mixture is mechanically pre-activated, improves the performance characteristics of the electrode material, i.e., increases the emission ability of the electrodes and the operation life of high-current vacuum and plasma electronics devices.

2. Results

Table. 4 shows the characteristics of an electrode material for a comparative analysis of the cathodes samples properties obtained by traditional and improved technology.

Table4. Samples properties

Production technology	Density, г/см^3	Pore volume, %	Hardness, HV
Traditional	12,84	28	310
Improved	15,34	7	422

The measurement of density and pore volume was carried out by hydrostatic weighing and performed according to State Standard GOST 2409–2014.

Hardness measurement by the Vickers method was carried out on the EMCO-TEST hardness machine according to State Standard GOST 2409–2014.

An illustrative comparative analysis of these properties is presented at Fig. 6.

The studies of the surface of electrodes obtained by scanning electron microscopy are shown at fig. 7, 8.

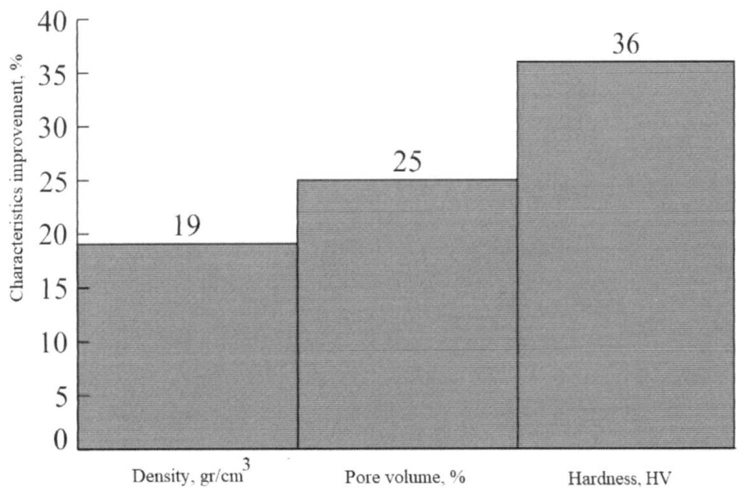

Fig.6. The relative improvement in performance characteristics as a result of the technology improvement

IOP Conf. Series: Materials Science and Engineering **963** (2020) 012004 doi:10.1088/1757-899X/963/1/012004

ab

cd

Fig. 7. The surface of the electrodes produced by the traditional (a, b) and improved (c, d) technology

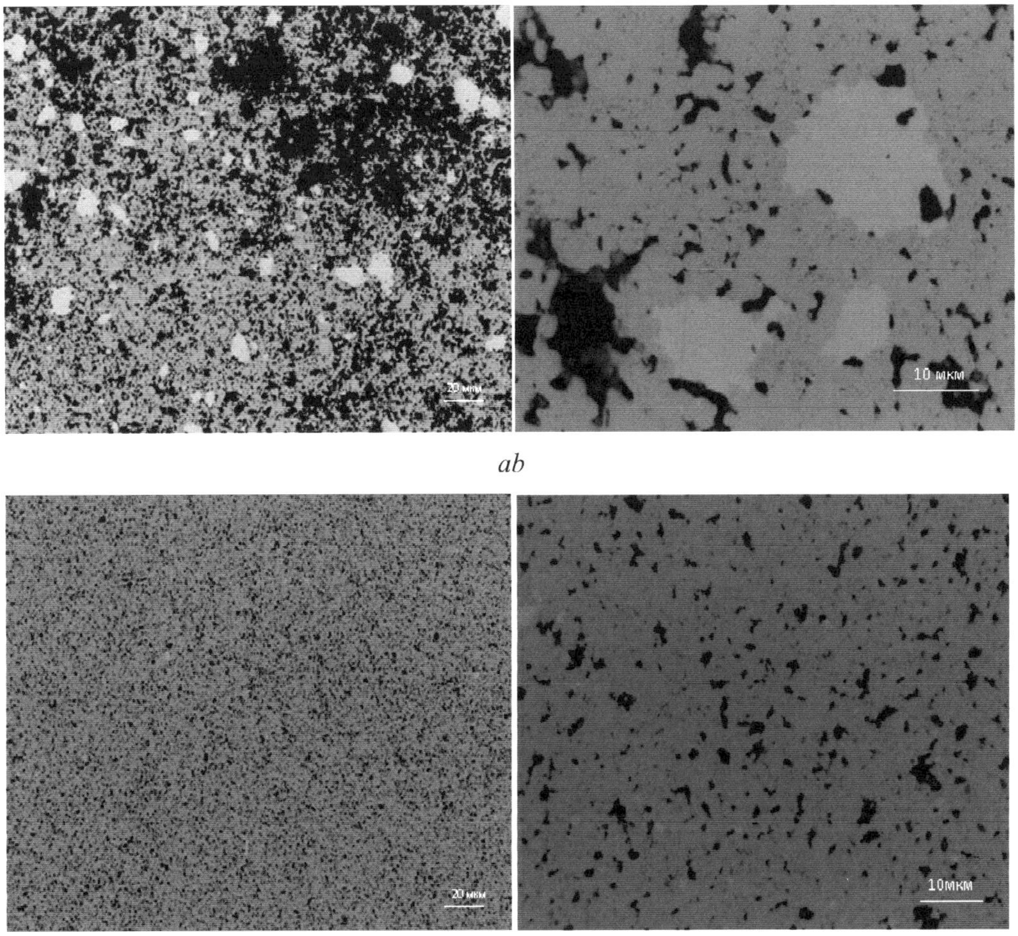

ab

cd

Fig. 8. The microstructure of electrodes produced according to traditional (a, b) and improved (c, d) technology

3. Discussion

The photographs of the surface in the microstructure of the samples microsections, obtained by the improved technology, show a uniform structure, a reduced number and size of pores in the structure alternatively to the traditional implementation — with an uneven distribution of pores and large bright nickel inclusions. In addition, it is possible to determinevisually the grinding of grain about 10 times.

4. Conclusion

There is an increase in physical and mechanical properties and the structure uniformity is ensured as the result of innovative techniques in the developed technology for an electrode materialproduction, namely: the use of nickel and strontium carbonate powders in the range of 10...100 nm, the introduction of theoperation of micro- and nanopowders mixture mechanical activation, the transition from sintering the mixture of the initial powder components in a hydrogen medium to sintering one in vacuum. The findings demonstrate the effectiveness of the advanced technology in the increasing of the operational life of electrode and device as a whole.

IOP Conf. Series: Materials Science and Engineering **963** (2020) 012004 doi:10.1088/1757-899X/963/1/012004

References

[1] Ashkinazi L.A. Materials of electronic emitters. Moscow, Moscow State Institute of Electronics and Mathematics (Technical University), 2007, 150 p.

[2] Bondarenko G.G., Kristya V.I., Savichkin D.O. Modeling of the effect of field electron emission from the cathode with a thin insulating film on its emission efficiency in gas discharge plasma.Vacuum, 2018,vol. 149,p. 114–117.

[3] Popov V.S., Nikolaev S.A. General electrical engineering with the basics of electronics. Moscow: Energy, 1972, 507 p.

[4] Mishra K.C., Garner R., Schmidt P.C. Model of work function of tungsten cathodes with barium oxide coating. J. Appl. Phys., 2004,vol. 95, no. 6, pp. 3069–3074.

[5] Melnikova I.P., Lyasnikov V.N., Lyasnikova A.V. Physics of Wave Processes and Radio Engineering Systems, 2012, vol. 15, No. 2, p. 84–90.

[6] Brodie I. A new model for the mechanism of operation of scandate and refractory oxide cathodes .IEEETrans. Electron Devices., 2011, vol. 58, no. 4,pp. 1247–1254.

[7] Getman O.I., Skorokhod V.V., Krylova N.A. Stabilization of the microstructure of tungsten frames of impregnated metal-porous cathodes. Electrical contacts and electrodes. Ser. Composite, layered and gradient materials and coatings, 2014, p. 102–111.

[8] Smirnov V.A., Akimov P.I., AleksandrovV.Yu. et al. Study of metal-porous cathodes with a slotted pore structure. Microwave Electronics and Microelectronics, 2016, vol. 2, No. 1, p. 115–119.

[9] Kapustin V.I., Lee I.P., Shumanov A.P. and others. The mechanism of formation and properties of crystallites of barium oxide in a metalloporous cathode. Promising Materials, 2016, vol. 7, p. 5–15.

[10] JiangD., SikongH., ChenfengZ. etal. Preparation of Impregnated Barium Scandate Cathode and Its Applhcatiqn. 1982. P. 206–207.

[11] Ledentsova N.E., Li I.P., Petrov V.S. and other promising technologies of oxide-nickel cathodes of microwave devices of the centimeter wavelength range. Fine Chemical Technologies, 2016, vol. 11, No. 3, p. 74–81.

[12] KuchinaI.Y., PolushinN.I., ZaharovaE.S. etal. Experimental Support of Magnetron Nickel Oxide Cathode Fabrication Process.Izv. Vyss. UchebnykhZaved. Mater. Elektron. Tekhniki = Mater. Electron. Eng, 2015, vol. 18, no. 4,pp. 285–290.

[13] Sevost'yanov M.A., Nasakina E.O., Baikin A.S. et al. Biocompatibility of new materials based on nano-structured nitinol with titanium and tantalum composite surface layers: experimental analysis in vitro and in vivo. J. Mater. Sci. Mater. Med. Springer US, 2018,vol. 29, no. 3.

[14] Yagodnikov D.A., Ignatov A. V., Gusachenko E.I. Ignition and combustion of pyrotechnic compositions based on microsized and ultra-nanosized aluminum particles in a moist medium in a two-zone gas generator. Combust. Explos. Shock Waves, 2017, vol. 53, no. 1, pp. 15–23.

[15] Kalashnikov I.E., BolotovaL. K., Bykov P. A. et al. Tribological properties of the babbit B83-based composite materials fabricated by powder metallurgy. Russ. Metall, 2016, vol. 2016, no. 7, pp. 669–674.

[16] Silkin A.A., Linnik A.A.,PankratovetA.S. etal. Formation of the structure of the weld metal upon the introduction of nanoparticles into the weld pool. Russ. Metall, 2016, vol. 2016, no. 13, pp. 1253–1256.

[17] Bulgarevich S.B., Boiko M.V., Tarasova E.N. et al. Kinetics of mechanoactivation of tribochemical processes. J. Frict. Wear, 2012,vol. 33, no. 5, pp. 345–353.

[18] Gaffet E., Bernard F., NiepceJ-C.et al. Some recent developments in mechanical activation and mechanosynthesis. J. Mater. Chem, 1999, vol. 9, no. 1, pp. 305–314.

[19] Živojinović J., Pavlović V., Kosanović D. et al. The influence of mechanical activation on structural evolution of nanocrystallineSrTiOpowders. J.AlloysCompd,2017,vol.695, pp.1–10.

[20] Mostaan H., Mehrizi M.Z., Rafie M. et al. Contribution of mechanical activation and annealing in the formation of nanopowders of Al(Cu)/TiC-Al2O3hybrid nanocomposite. Ceram. Int, 2017, vol. 43, no. 2, pp. 2680–2685.
[21] Padhan A.M., SathishM., SaravananP.et al. Mechanical activation on aluminothermic reduction and magnetic propertiesofNiO powders. J. Phys. D. Appl. Phys. IOP Publishing, 2017, vol. 50, no. 21.
[22] Vega L.E.R., LeivaD.R., Leal NetoR.M.et al. Mechanical activation of TiFe for hydrogen storage by cold rolling under inert atmosphere. Int. J. Hydrogen Energy, 2018, vol. 43, no. 5, pp. 2913–2918.

Optimization of the flow part of the pump for abrasive-containing liquids by hydrodynamic modeling methods

O Martynyuk[1] **and A Petrov**[1,2]

[1]Bauman Moscow State Technical University, 5 Second Baumanskaya Street, Moscow, 105005, Russian Federation

[2]E-mail: alexeypetrov@bmstu.ru

Abstact. In this article the severe operating conditions of pumps for abrasive media, as well as the complexity of obtaining an analytical solution to the wear problem are described. Optimization criteria for such pumps are selected. A mathematical model for calculating a ground pump is presented. The STAR CCM+ software package modeled 15 models with different geometries and identified the main geometry parameters that affect the rate of erosion. A weight function was created that characterizes the success of model optimization based on the selected criteria. According to the accepted parameters, 20 models were modeled and the best geometric dimensions of the flow part were selected. The results are analyzed and conclusions are drawn.

Introduction

Pumps for abrasive media have a very wide range of applications. They are used for pumping waste and polluted water, in quarries, mines, mining and processing plants, and dredgers. The pumped liquid contains a lot of solid inclusions, which leads to erosion of the pumps. The service life of pumps is estimated in hundreds of hours and rarely exceeds 4000–5000 hours, so they try to achieve primarily a reduction in abrasive erosion, but also to obtain the maximum possible efficiency in their design. The length of the life cycle in such conditions is mainly affected by the rate of erosion [1]. Therefore, it was decided to optimize the pump for abrasive-containing hydraulic mixtures for these two parameters using hydrodynamic modeling [2]. The main difficulty of this method is that the existing wear models are purely empirical and require verification. Although a large amount of information has been collected on the operation of pumps for abrasive-containing liquids, the problem of flow wear and simulation of two - and three-phase flows has not yet been solved analytically and numerically [3].

Methods

Hydrodynamic modeling of the existing model ground centrifugal pump was performed in the STAR CCM+ software package [4,5]. This research method allows solving various problems of continuum mechanics [6–10].

Mathematical model:

- Continuity equation (mass conservation equation):

$$\frac{\partial \rho}{\partial t} + div(\rho \cdot \vec{u}) = 0,$$

where ρ — density;

t — time;
$\vec{u}$ — velocity vector

- Reynolds equations (equations of conservation of the amount of motion averaged over time):

$$\frac{\partial}{\partial t}(\rho \cdot u_j) + \frac{\partial}{\partial x_j}(\rho \cdot u_i \cdot u_j) = -\frac{\partial p}{\partial x_i} + \frac{\partial}{\partial x_i}\left[T_{ij} - \rho \cdot \overline{u'_i u'_j}\right],$$

where $T_{ij} = 2\mu s_{ij} - \frac{2}{3}\frac{\partial u_j}{\partial x_j}$ — the tensor of viscous stresses;

μ — dynamic coefficient of liquid viscosity, Pa·s;

$s_{ij} = \frac{1}{2}\left(\frac{\partial u_i}{\partial x_j} + \frac{\partial u_j}{\partial x_i}\right)$ — strain rate tensor;

$\rho\overline{u'_i u'_j}$ — Reynolds stresses;

u'—pulsation component of the speed;

p — the averaged value of the pressure.

- The transport equation of turbulent kinetic energy:

$$\frac{\partial}{\partial t}(\rho k) + \frac{\partial}{\partial x_j}(\rho u_j k) = P_k - \rho\beta^* k\omega + \frac{\partial}{\partial x_j}\left[(\mu + \sigma_k \mu_t)\frac{\partial k}{\partial x_j}\right],$$

where k — kinetic energy of turbulence;

P_k — energy generation of turbulence by the velocity shear;

σ_k, β^* — the coefficients of the circuit;

μ_t — coefficient of turbulent viscosity.

- Transfer equation for the relative rate of dissipation of this energy:

$$\frac{\partial}{\partial t}(\rho\omega) + \frac{\partial}{\partial x_j}(\rho u_j \omega) = a_k^\omega p_k - \rho\beta\omega^2 + \frac{\partial}{\partial x_j}\left[(\mu + \sigma_\omega \mu_t)\frac{\partial \omega}{\partial x_j}\right] + 2\cdot(1 - F_i)\sigma_{\omega 2}\cdot\frac{1}{\omega}\frac{\partial k}{\partial x_i}\frac{\partial \omega}{\partial x_i},$$

Where:

ω — velocity of kinetic energy dissipation of turbulence;

a, β, σ_ω — the coefficients of the circuit;

F_i— weighting function.

The constructed grid of 362,330 cells is shown in figure 1.

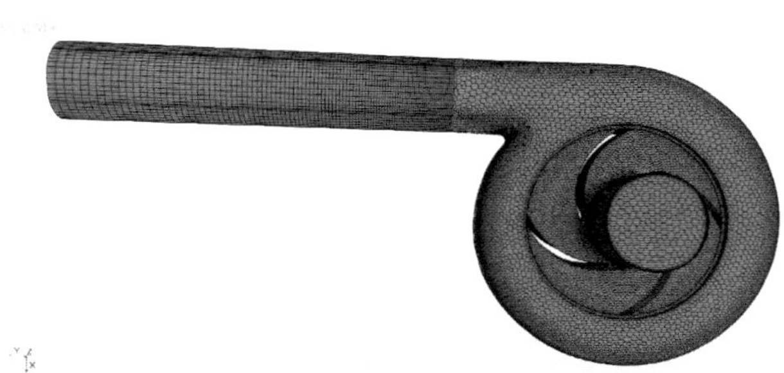

Fig. 1. Source pump grid

Further, a simulation of the movement of the second phase of the flow (solid) was added to the calculation model, for which the trajectories of the particles were determined based on the Lagrange method.

The selected particle material is sand with a density of 2650 kg / m3. Particles with a diameter of 2.75·10-4 m.

The model of the particle drag force is determined by the drag coefficient, which is the Schiller-Nauman correlation.

Restitution coefficients (tangential and normal recovery coefficients) are used to predict the angle at which a particle bounces off a solid wall. For steel, they are defined using FORDER's formulas:

$$e_T = 1 - 0,78\alpha_p + 0,84\alpha_p^2 - 0,21\alpha_p^3 + 0,028\alpha_p^4 - 0,022\alpha_p^2$$

$$e_N = 0,988 - 0,78\alpha_p + 0,19\alpha_p^2 - 0,24\alpha_p^3 + 0,027\alpha_p^4$$

The disadvantage of this calculation is that the equations for these coefficients are derived only for carbon steel (based on experimental dependencies).

A method of modeling erosion of an Eye.

An injector generating particles with a resolution of 50x50 is installed at the entrance to the supply.

To assess the quality of the simulated flow parts of the pump, a weight function was created that characterizes how low the erosion rate is and how high the efficiency is in the designed flow part [11, 12].

$$f = 1 - 0,7 \cdot v + 0,3 \cdot \eta,$$

where f — weighting function;

$$v = \frac{v_m}{v_1}$$

v — dimensionless coefficient;

v_m — the rate of erosion of the model in question;

v_1 — rate of erosion of the original model;

η — efficiency of the model under consideration.

The coefficients of the target function in this case are taken for reasons of prevailing low abrasive erosion over the requirement to obtain maximum efficiency, and can be changed to solve other similar problems.

Results

To determine the parameters that most affect the rate of erosion, models with variable geometry were previously modeled before optimization, and table 1 was constructed based on the simulation results [13].

Table 1. Influence of geometric parameters of the pump on its main characteristics

№	Type of change	Erosion rate, g / hour	The efficiency	Force, H	The pressure of the fluid, м	Weighting function
1	Source model	0,813	0,77	3900	68	0,531
2	Increased inlet to the pump	0,816	0,74	3880	65	0,519
3	Ring outlet	0,767	0,74	13125	67	0,562
4	Round section of the tap	0,939	0,78	4380	68	0,426

Окончание таблицы

IOP Conf. Series: Materials Science and Engineering **963** (2020) 012005 doi:10.1088/1757-899X/963/1/012005

№	Type of change	Erosion rate, g / hour	The efficiency	Force, H	The pressure of the fluid, м	Weighting function
5	Increased tongue	0,897	0,75	3000	68	0,453
6	The increased width of the outlet at the same bandwidth	0,901	0,77	2300	68	0,455
7	Increased bandwidth	0,8976	0,74	2600	67	0,449
8	The increased angle of the blade	0,903	0,71	3080	61	0,436
9	The entrance to the blade is sharper	0,759	0,73	4000	65	0,565
10	Larger angle β1	0,734	0,73	3750	65	0,587
11	Larger angle β2	0,794	0,53	2300	63	0,475
12	The thickness of the blades at the outlet	0,803	0,72	3300	64	0,525
13	The thickness of the blades at the inlet	0,892	0,73	3700	65	0,451
14	Another form of the blade	0,697	0,74	4200	65	0,622
15	The combined outlet	0,913	0,73	8582	67	0,433

It turned out, the erosion rates are mostly affected by the use of ring devices instead of helically-ring, the shape of the blade, the angle β1, the login form on the blade, the fillet radius and the width of removal and angle of the blades. Graphs of the dependence of the efficiency and rate of erosion on some parameters are shown in figures 2–5.

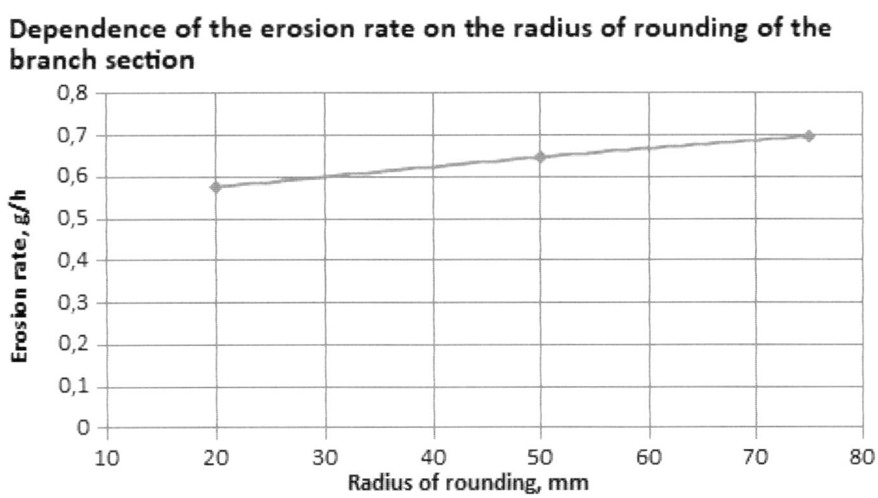

Fig.2 Dependence of the erosion rate

IOP Conf. Series: Materials Science and Engineering **963** (2020) 012005 doi:10.1088/1757-899X/963/1/012005

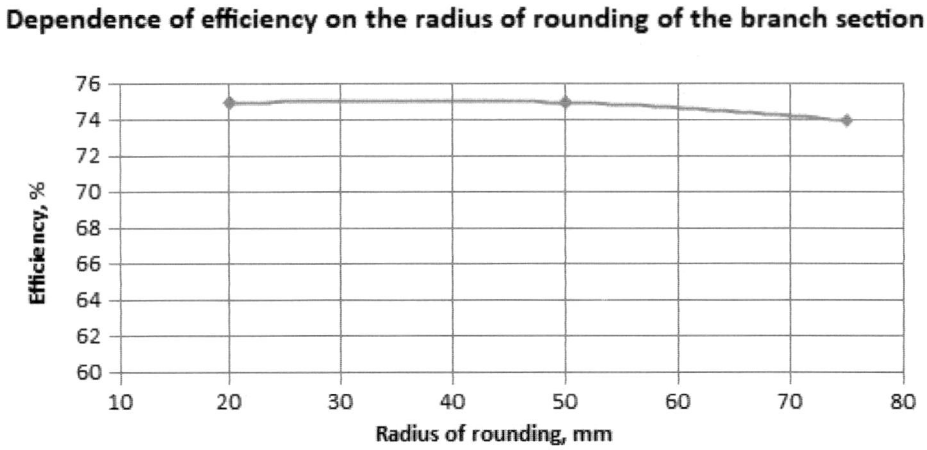

Fig.3 Dependence of efficiency on radius

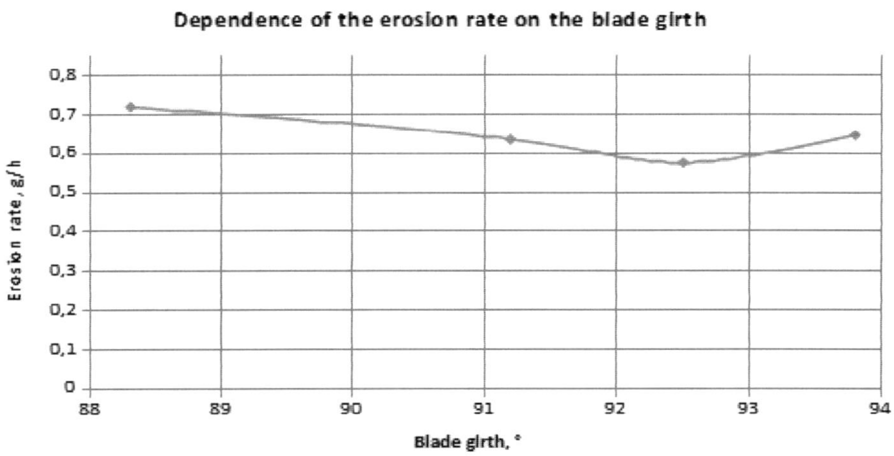

Fig.4 Dependence of the erosion rate

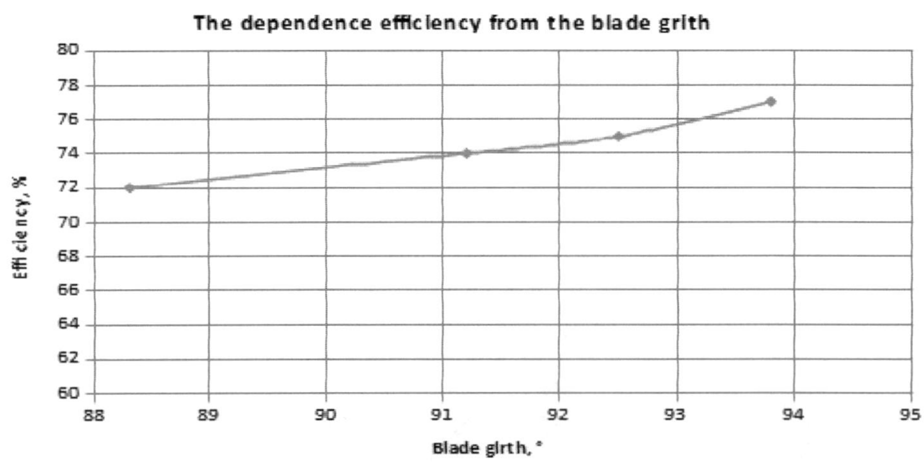

Fig.5 Dependence of efficiency from the blade grith

IOP Conf. Series: Materials Science and Engineering **963** (2020) 012005 doi:10.1088/1757-899X/963/1/012005

However, as an object of optimization, the annular tap is not suitable because of the large radial force on the wheel rotor, which in this case acted as a restriction for the pump under study. Therefore, the wheel with the best shape of the blade and the shape of the blade at the inlet is used as a model for optimization [14]. In further calculations, the following parameters were used as variable parameters: the blade coverage, the angle β1 at the inlet, the width of the branch and the radius of rounding of its section. The blade coverage varied from 88 to 94°. When changing it, it was necessary to take into account the value of the minimum conditional passage, which is 80 mm. The angle β1 at the inlet varied from 36° to 40°, the width of the branch at the same throughput - from 342 to 390, the radius of the rounding section of the branch-from 75 mm to 20 mm [15]. The results are shown in table 2.

Table 2. The results obtained.

№of model	Erosion rate, g / hour	The efficiency	Force, H	The pressure of the fluid, м	Weighting function
1	0,647	0,75	3982	67,5	0,668
2	0,72	0,72	4843	63	0,596
3	0,627	0,70	2880	64	0,670
4	0,577	0,75	3971	68	0,728
5	0,576	0,72	2904	64	0,720
6	0,697	0,71	3916	63	0,613
7	0,656	0,75	4840	69	0,660
8	0,631	0,77	3804	68	0,688
9	0,561	0,72	3900	64	0,733
10	0,578	0,70	3788	62	0,712
11	0,715	0,72	1241	64	0,600
12	0,555	0,75	1700	65	0,747
13	0,58	0,74	3874	67	0,723
14	0,605	0,74	4750	65	0,701
15	0,573	0,73	4391	65	0,726
16	0,636	0,74	3726	66	0,674
17	0,647	0,77	4164	68	0,674
18	0,572	0,73	3604	65	0,727

The highest weight coefficient was found in the model number 12. It is shown in figure 6, and the results in 7.8 and 9.

Its parameters are:
Radius of rounding of the branch section 20 mm;
Tap width 380 mm;
Blade coverage angle 92.5;

IOP Conf. Series: Materials Science and Engineering **963** (2020) 012005 doi:10.1088/1757-899X/963/1/012005

The angle β1 of 38°.
Model characteristics:
Efficiency 75%;
Erosion rate 0.555 g / hour;
Head 65 m;
Radial force 1700 H.

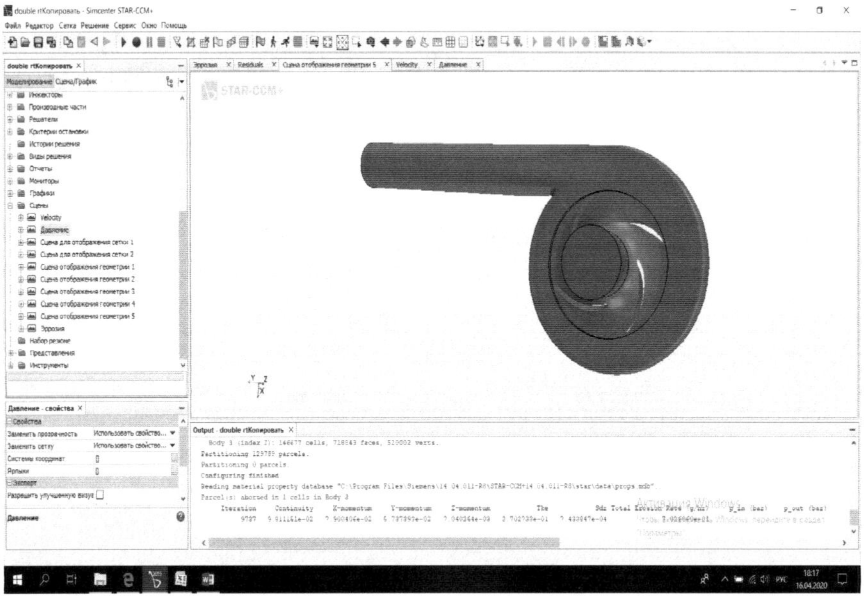

Fig. 6. Geometry of the model № 12

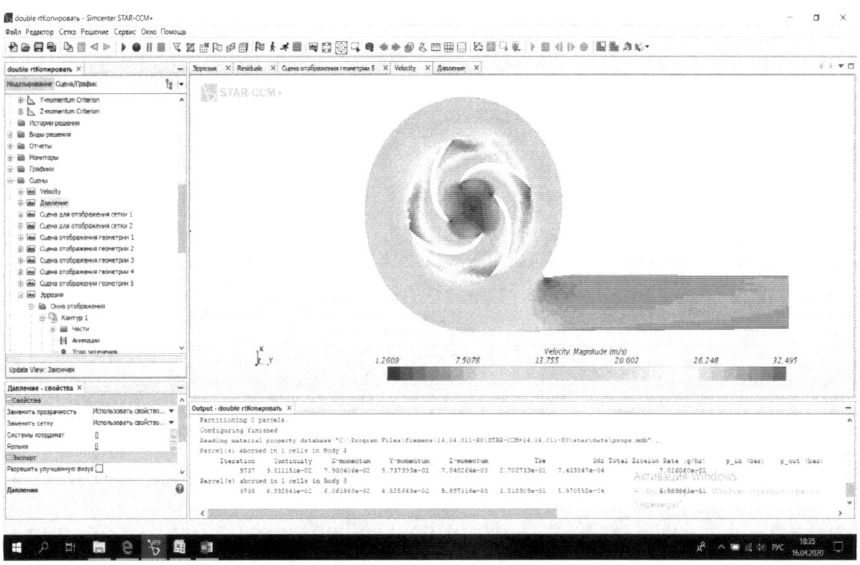

Fig. 7. Speed distribution in the flow section № 12

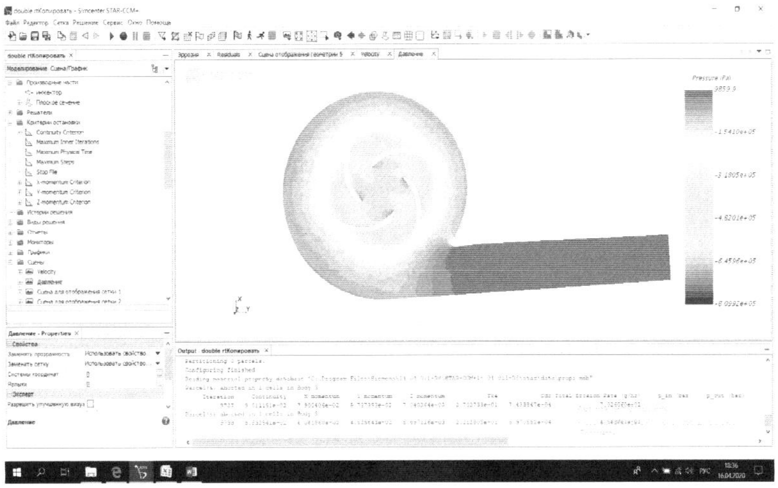

Fig. 8. Pressure distribution in the flow part № 12

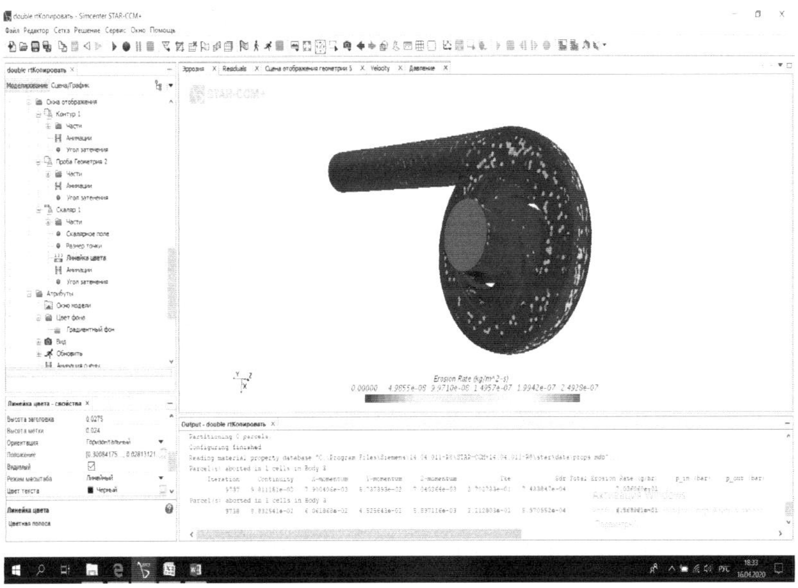

Fig.9. Rate of erosion in model №12

Discussions

Preliminary results of the study showed that for the lowest rate of erosion, it is necessary to design wider blades in the meridional section. Reduce the surface areas perpendicular to the flow, namely: use a slightly elongated inlet to the blade and a square section of the outlet with the minimum allowable radii of rounding the corners. In this case, it was possible to achieve a significant reduction in the calculated erosion rate at an acceptable efficiency value. However, the results obtained require experimental verification.

References

[1] Vane pumps for abrasive hydraulic mixtures Zhivotovsky L. S. Smolovskaya L. A. m: Mashinostroenie, 1978. — 223 p., Il.

[2] Lomakin V. O. Petrov A. I. Verification of calculation results in the package of hydrodynamic modeling of the ster CCM + flow part of the centrifugal pump Ah 50-32-200 / /

Izvestiyahigher educational institutions. Mechanical engineering 2012. —Spets.the release of the Spec.issue" Works of students and young scientists of Bauman Moscow state technical University". — P. 6–9

[3] Lomakin V. O. Petrov A. I. Kuleshova N. S. Investigation of two-phase flow in an axecenter wheel by hydrodynamic modeling methods / / Science and education: scientific publication of Bauman Moscow state technical University 2014. — № 9

[4] Lomakin V. O. Chaburko P. S. Numerical simulation of liquid flow in a jet pump / / mechanical engineering: network electronic scientific journal 2014. — No. 3. — P. 55–58

[5] n. Isaev et al. 2019 VGD Conf. Serial.: Mother. Sci. Eng. 589 012009

[6] eMorozova et al. 2019 IOP Conf. Serial.: Mother. Sci. Eng. 589 012008

[7] A Protopopov et al 2019 IOP Conf. Serial.: Mother. Sci. Eng. 589 012003

[8] K Abramov 2019 IOP Conf. Serial.: Mother. Sci. Eng. 589 01201

[9] Petrov A. I. Fraz A. S. Gavryushina O. S. Hydrodynamic modeling of flow in the flow part of a centrifugal pump running on a high-viscosity liquid / / Hydraulics 2018.- #5

[10] Lomakin V. O. Petrov A. I. Stepanyuk A. I. Optimization of geometric parameters of the oil main pump discharge of the NM type / / Science and education: scientific publication of Bauman Moscow state technical University 2012-no. 3

[11] Lomakin V. O. Artemov A.V. Petrov A. I. Determining the influence of the main geometric parameters of the pump discharge NM 10000-210 on its characteristics / / Science and education: scientific publication of Bauman Moscow state technical University 2012. — № 8

[12] m. Saprykina and V. Lomakin 2019 IOP Conf. Serial.: Mother. Sci. Eng. 589 012017

[13] Lomakin V. O. Cheremushkin V. A. Influence of the shape of the impeller blades on the pressure of the centrifugal pump // Engineering Bulletin 2016. - no. 1

[14] inCheremushkin and Polyakov 2019 VGD Conf. Serial.: Mother. Sci. Eng. 589 012001

[15] withBoyarshinov et al. 2019 VGD Conf. Serial.: Mother. Sci. Eng. 589 012014

Technological bases for increasing the durability of aviation parts by chemical heat treatment

S Pakhomova[1] and S Karpukhin[1,2]

[1]Bauman Moscow State Technical University, 5 Second Baumanskaya Street, Moscow, 105005, Russian Federation

[2]E-mail: mgtu2013@yandex.ru

Abstract. The paper presents studies of the structure and properties of low-carbon alloy steels after carburizing. A comparison of four types of carburizing (batch carburization, conveyor carburization, fluidized bed carburization and ionic carburization) is made. The influence of technological modes of chemical-thermal treatment on the phase composition and amount of carbide phase in a carburized layer is considered. It is shown that the type of carburizing affects the operational properties of parts of aircraft mechanisms. The best results were shown by the details after ion vacuum carburizing. The increase in wear resistance and contact fatigue is associated with the formation of a high-quality surface layer having a higher hardness and a higher proportion of the carbide phase. Such properties cannot be obtained after conventional types of carburizing in a shaft or conveyor furnace. In this case, ionic carburizing and carburizing in a fluidized bed make it possible to halve the technological regime of carburizing in comparison with two other methods.

Introduction

For parts of aviation equipment operating under variable loads and intensive wear, chrome-nickel carburizedsteels of the C0.12CrNi3, C0.12Cr2Ni2WV, C0.2CrNi3, C0.2Cr2Ni4 and C0.2Cr3Ni2MoW types are used [1–3]. Despite the hardening processes used, the durability of these parts is insufficient. There are any methods of producing high-load parts, such as composite materials [4, 5] or ceramics [6]. Recently, methods of surface hardening by electro-mechanical processing [7–9] or plastic deformation [10] have been widely used.

The working surfaces of carburized teeth of high-loaded gears experience high contact stresses and sliding speeds. This is caused by a complex mechanism of high-speed wear. As a result, the matrix is first destroyed in the carburized layer, and then solid and brittle particles of carbides are chipped off.

To increase the resistance to contact destruction of gears, a strong surface layer is required, in which there is a developed carbide phase. The diffusion layer formed during gas carburization in the mine furnace does not provide such increased requirements. Higher layer properties can be obtained after carburization in a fluidized bed or by vacuum ion carburization. As is known, ion saturation processes provide the ability to control the phase and chemical composition of the layer. An additional important advantage is the reduction of the duration of chemical and heat treatment [11]. In [12, 13] it is shown that the equipment used for vacuum ion carburization allows you to quickly change the temperature of the process and its carburizing ability, and therefore adjust the effective thickness of the carburized layer.

Therefore the authors work on optimizing the steel grade and their surface hardening modes to increase the durability of parts by increasing both contact fatigue strength and wear resistance.

The purpose of this work is to improve the technology of surface hardening of high-loaded gears made of high-nickel carburized steels.

Materials and equipment of the experiment

Authors are working to optimize the steel grade and its hardening modes to increase the durability of parts by increasing both contact fatigue strength and wear resistance. The study used [Table 1] steel C0.12CrNi3 (analogue in the USA — 3415 and Germany — 12NiCr) and C0.2CrNi2MoW (analogue in the USA — 4320 and Germany — 17CrNiMo).

Table 1. Chemical composition of steels C0.12CrNi3 (1) and C0.2CrNi2MoW (2)

Contentofelements, %								
No	C	Si	Mn	Cr	Ni	W	Mo	Cu
1	0.09–0.16	0.17–037	0.3–0,6	0.6–0.9	2.75–3.15	–	0.4–0.6	<0,3
2	0.15–0.22	0.17–037	0.4–0.7	0.4–0.6	1.6–2.0	1.0–1.4	0.5–0.8	<0,3

Two steels were subjected to carburization and subsequent heat treatment according to various technological modes. The quality of carburization is determined by the technological level of the process and the chemical composition of the steel. The chemical composition of steel for inherently coarse-grained steels, such as those of the studied, requires several high-temperature heats during the hardening process because the grain size in the steel affects not only the strength characteristics but also the technological properties of the steel. With the enlargement of the grain and an increase in different grain sizes, warping and the tendency to form cracks increases. Heterogeneous grain is accompanied by a decrease in impact strength after all types of heat treatment.

Therefore, to ensure the quality of parts, initial control of the structure of the steel coming for the production of parts is necessary. The microstructure of investigated steels is characterized by the presence of pronounced banding and non-metallic (carbide) inclusions. To eliminate such a structure, pre-heat treatment is necessary, which can be used as diffusion annealing or normalization [14]. The use of diffusion annealing eliminates streakiness and heterogeneity, but leads to grain growth, which affects the properties after the final heat treatment, in addition, it is accompanied by obtaining a ferrite structure on the surface and, consequently, large allowances for machining, which is unacceptable for these steels. Therefore, for recrystallization it is better to use the normalization. After that the field of the size of the austenite grain is 10–12 *points* with an average conventional diameter of 9.9–4.9 μm. The final chemical heat treatment should ensure a rational carbon content in the surface layer and microhardness of the layer and of the core.

It should be borne in mind that the greater the difference in the carbon content of the surface and the core, the greater in the surface layer the compressive stresses, which inhibitthe development of fatigue cracks. However, if the difference is too large, the stresses can be reduced due to the deformation of the core and this contributes to the appearance of cracks. Therefore it is necessary to observe the optimal amount of carbon in the surface layer. For steel C0.12CrNi3 carbon concentration should not exceed 0.75–0.80%, and for steel C0.2Cr3Ni2MoW — 0.90–1.0%. Both increasing and decreasing the concentration of carbon is reduced flexural strength, although the tensile strength and yield strength are growing.

The core strength of carburized parts must be within 1200–1500 *MPa*, while the bending strength is within 2100–2500 *MPa*. The core hardness of 34–42 *HRC* is considered acceptable. The core structure in parts that work for bending and torsion in the presence of shock loads should represent fine-needle martensite 2–3 *points* with the size of the largest length of the martensite needle 2–4 μm. The structure allows 10–15% excess ferrite, which slightly reduces the hardness of the core [15, 16].

Experimental results and discussion

Based on these considerations, authors analyzed the impact of various technological methods of carburization on the quality of carburized layers and the durability of parts. Carburization was carried out in four ways: 1) batch carburization (*BC*) in shaft mine furnace «MIMP-SSh-3»; 2) conveyor carburization (*CC*) in conveyor furnace «RG-805G» with a controlled supply of saturating gas; 3) fluidized bed carburization (*FC*) in chamber furnace «Holcroft»; 4) ionic carburization (*IC*) in «Seco Warwick» firms equipment.

Microstructure of carburized layer (Fig. 1), microhardness, depth of the carburized layer, hardness and core structure were studied. Carrying out the *CC* process of two steels eliminates the saturation of the surface with carbon, reduces the density of carbides, and increases the effective thickness of the layer (Fig. 1). Their distribution and diffusion redistribution of carbon increase the doping of the solid solution and the uniformity of the structure of the carbide zone: the difference in the size of the carbide particles decreases, and the density of their distribution is equalized.

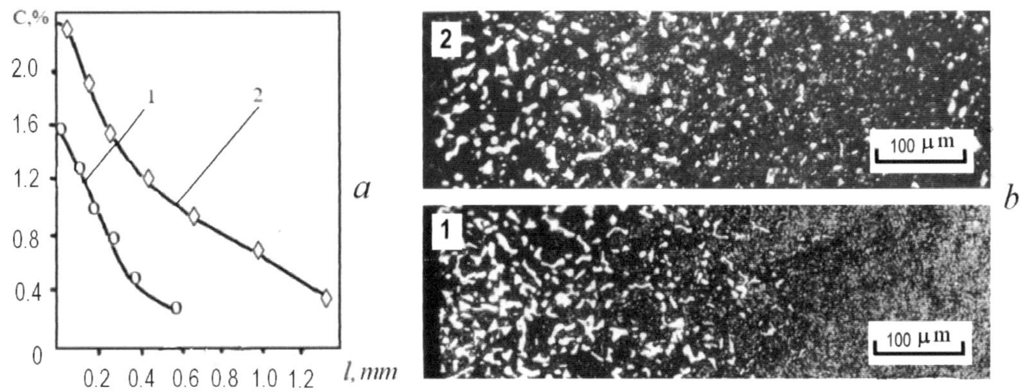

a)b)

Fig. 1. Carbon distribution *C* by thickness (*a*) and microstructure of carburized layer (*b*) after the *CC* of steel: 1 —C0.12CrNi3; 2 —C0.2CrNi2MoW

As can be seen from the above data, after *CC* of steel C0.12CrNi3 at a depth of 0.2 *mm*, there is some decarburization of the surface with a decrease in its hardness $H_{0,2}$to 6000-6300 *MPa*, the carburization depth is 1.10–1.20 *mm*. The microstructure of the carburized layer is fine-needle martensite and 10-15% residual austenite in steel C0.12CrNi3 and 8-10% - in steel C0.2CrNi2MoW.

Along with this, a study of the hardness and wear resistance of parts was conducted (Fig. 2).

The Fig. 2 shows the distribution of hardness after four ways carburization and heat treatment carried out after each method. As can be seen from the above data, the *BC* in shaft mine furnaces does not provide a satisfactory quality of the carburized layer. There is decarburization on the surface, in addition, when the depth of carburization is 1.8–2.0 *mm*, there is a broken carbide mesh in the surface layer. *IC* with followed heat treatment showed the best quality of the carburized layer (see Fig. 2). There is no decarburization on the surface, but the high hardness at a depth of up to 0.25 *mm* is due to the presence of scattered carbides in the surface layer. This hardness and structure is optimal for aviation equipment parts. Carburization *CC* and *FC* have satisfactory results. To reduce the time of carburization and increase the productivity of carburization equipment, we have studied carburization in the fluidized bed. When the part was carburized in a fluidized bed, a high hardness is obtained at a depth of 0.4 *mm*. the Structure of the carburized layer has a layer with a carbide mesh. The structure of the core — large needle-like martensite.

Studies have shown that different carbon concentrations in the carburized layer of steel C0.2CrNi2MoW caused different effects on surface quality characteristics: surface hardness, wear resistance andand contact fatigue strength (Tab. 2). The differences are due to the different proportion of the formed carbide phase [17, 18]. The volume fraction of the latter affects the degree of localization of microplastic deformation and the level of development of local stress relaxation processes in the martensitic matrix [19]. It is determined that the *IC* method has higher indicators (see Tab. 2) [20].

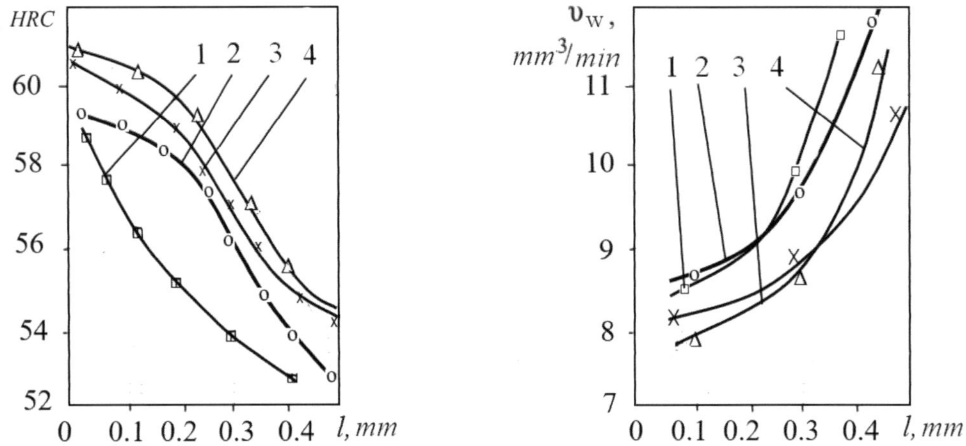

a *b*

Fig. 2. Hardness *HRC (a)* and wear rate υ_w (*b*) change over the thickness of the diffusion layer *l* of steel C0,12CrNi3 after chemical heat treatment: 1 — *BC*, 2 — *CC*, *3* — *FC* or 4 — *IC*.

Table 2. The properties of steel C0.2CrNi2MoW after chemical heat treatment (*BC*, *CC*, *FC* or *IC*)

No	Chemical heat treatment	Volumetric share carbides in the layer 0-20 μ*m*, %	Surface hardness, *HRC*	Wear rate v_w, *mm³/min*	Contact fatigue strength, $N_{50}\cdot10^{-6}$ *cycles*
1	*BC*, *t*=950°C, τ =5 *h*	12	59-61	25	4
2	*CC*, *t*=950°C, τ=5 *h*	15	60-62	16	12
3	*FC*, *t*=950°C; τ=2,5 *h*	20	62-63	18	24
4	*IC*, *t*=950°C; τ=2,5 *h*	25	63-64	11	36

Conclusions

Thus, the research shows that the optimal process for conducting chemical heat treatment is the process of ionic carburization treatment in chamber furnaces with a controlled carbon potential.

The carbon distribution should correspond to the values shown in Fig. 2 (curve 4) and the microstructure of the surface and core correspond to the structures: in the carburized layer fine-needle martensite 2–3 *points*, residual austenite 10–15%, volumetric share carbides in the layer 0–20 μ*m* in the amount of 20–25%.

References

[1] Kablov E.N., Bakradze M.M., Gromov V.I., Voznesenskaya N.M., Yakusheva N.A. New high-strength structural and corrosion-resistant steels for aerospace engineering developed by «VIAM» (review) // Aviation materials and technologies [*AviatsionnyyeMaterialy I Tekhnologii*]. 2020, No 1 (58), pp 3–11.

[2] Suslov A.G. Engineering of surface parts [*Inzheneriyapoverkhnostidetaley*]. M.: Mashinostroyeniye, 2008, 320 p.

[3] Zinchenko V. M. The Engineering of the Surface of Gear Wheels by the Methods of Chemical Heat Treatment [*Inzheneriyapoverkhnostizubchatykhkolesmetodamikhimiko-termicheskoiobrabotki*]. Moscow: MGTU im N. E. Baumana Press, 2001, 303 p.

[4] Silkin A.A., Linnik A.A., Pankratov A.S., Kurganova Y.A., Kobernik N.V., Mikheev R.S. Formation of the structure of the weld metal upon the introduction of nanoparticles into the weld pool // Russian metallurgy [*Metally*]. 2016. T. 2016. № 13. C. 1253–1256. DOI: 10.1134/S0036029516130206

[5] Berezovskii V.V., Shavnev A.A., Solyaev Y.O., Lur'e S.A., Babaitsev A.V., Kurg–nova Y.A. Mechanical properties of a metallic composite material based on an aluminum alloy reinforced by dispersed silicon carbide particles // Russian metallurgy [*Metally*]. 2015. T. 2015. № 10. C. 790–794. DOI: 10.1134/S0036029515100055

[6] Pakhomova S.A. and Povalyayev A.I. Silicon nitride-based ceramic composite materials for corrosion-resistant rolling bearings // IOP Conf. Series: Materials Science and Engineering. 683 (2019) 012040. DOI:10.1088/1757-899X/683/1/012040

[7] Fedorova L.V., Fedorov S.K., Serzhant A.A., Golovin V.V., Systerov S.V. Electromechanical surface hardening of tubing steels // Metal Science and Heat Treatment. 2017. T. 59. № 3–4. C. 173–175. DOI: 10.1007/s11041-017-0123-z

[8] Fedorova L., Fedorov S., Sadovnikov A., Ivanova Y., Voronina M. Abrasive wear of hilongbotnhardfacings // В сборнике: IOP ConferenceSeries: Materials Science and Engineering Сер. "International Conference on Mechanical Engineering and Applied Composite Materials" 2018. С. 012038. DOI: 10.1088/1757-899X/307/1/012038

[9] Fedorova L.V., Fedorov S.K., Ivanova Y.S., Voronina M.V. Increase of wear resistance of the drill pipe thread connection by electromechanical surface hardening // International Journal of Applied Engineering Research. 2017. T. 12. № 18. C. 7485–7489.

[10] Pakhomova, S.A. Plastic deformation hardening of iron-nickel alloys // IOP Publishing. Journal of Physics: Conference Series. 1431 (2020) 012043. DOI:10.1088/1742-596/1431/1/012043

[11] Fomina L.P. Influence of high-temperature hardening processes on the deformation of aviation engine gears // Metal technology [Tekhnologiyametallov]. 2018, No. 6, pp 28–31.

[12] Fakhurtdinov R S, Pakhomova S A andRyzhova M Y On the problems of modernizing equipment for vacuum carburization2017 // Journal of Machinery Manufacture and Reliability) 46№ 2 pp 187–192, DOI: 10.3103/S1052618816060066

[13] Marinin E.A., Chirkov A.M., Gavrilov G.N., Fetisov G.P., Chernyshov D.A., KurganovaYu.A. Experimental evaluation of the methods of laser cementation of low-alloy tool steels // Russian metallurgy [*Metally*]. 2018, No 13, pp 73–77.DOI: 10.1134/S0036029518130153

[14] Pakhomova S A, Fakhurtdinov R S, Tsinkolenko O A, Zolotov B S. The influence of carburization technology on performance properties of high-loaded gear wheels // IOP Conf. Series: Materials Science and Engineering. 747 (2020) 012126. DOI:10.1088/1757-899X/747/1/012126

[15] Kalashnikov A.S., MorgunovYu.A., Kalashnikov P.A. Advantages of gas vacuum carburizationand gas quenching under high pressure in processing a hypoid gear of the shaft type // Handbook. Engineering magazine with an app [*Spravochnik. Inzhenernyyzhurnal s prilozheniyem*]. 2018, No 8 (257), pp 3–7.

[16] LivshitsM.Yu., DerevyanovM.Yu. Yakubovich E.A. Optimal control of an object with distributed parameters on the example of vacuum carburization// Mathematical methods in engineering and technology [*Matematicheskiyemetody v tekhnikeitekhnologiyakh*]. 2017, vol. 12, pp 24–30.

[17] Kuksenova L.I., Gerasimov S.A., Alekseeva M.S., Gromov V.I. Influence of vacuum chemical heat treatment on wear resistance of VKS7 and VKS10 steels // Aviation materials and technologies [*Aviatsionnyyematerialyitekhnologii*]. 2018, No 1 (50), pp 3–8.

[18] Fomina L. P. Influence of chemical heat treatment on the endurance of teeth when bending / / Metal Technology [Tekhnologiyametallov]. 2014, No 1, pp 11–14.
[19] Semenov M.Yu. Management of the structure of carburization layers of heat-resistant steels. Part 1. // Metal Science and heat treatment. 2013, No 5, pp 31–38.
[20] Kula P., Olejnik J., Kowalewski J. New vacuum carburizing technology // Heat treatment progress. 2001. V. 1, №. 1. P. 57–65.

IOP Conf. Series: Materials Science and Engineering **963** (2020) 012007 doi:10.1088/1757-899X/963/1/012007

Quasi-static bending of thin-walled beams with different fillers under extreme load for solving the problems of crashworthiness

R Goncharov[1,2] and V Zuzov[1]

[1]Bauman Moscow State Technical University, 5 Second Baumanskaya Street, Moscow, 105005, Russian Federation

[2]E-mail: goncharov.roman@bmstu.ru

Abstract. In this paper, there are issues related to the choice of elemental modelling of thin-walled beams with various types of fillers for quasi-static bending loading, which causes washing of beams and filler, as applied to the optimal design. The objects of research are thin-walled beams made of aluminium AMG6 contains fillers made of aluminium foam, epoxy resin and composite (consisting of Poraver balls and epoxy resin). One of the best fillers for solving the problems of crashworthiness is requiring expensive technological equipment for filling. Epoxy resin is a fragile material, but inexpensive, affordable and technologically "convenient". In addition to the influence of the filler on the mechanical properties of beams, the effects of the constrained behaviour of the material of the filler under the bending loading of beams are investigated with Zwick Z100 (for hollow thin-walled beams made of aluminium and the same beams with fillers).The paper presents the results of numerical calculations and the results of the experiment (the error of calculations does not exceed 8%).The most rational is the ratio between the indicators of energy intensity - the mass is aluminium-foam (with an increase in the mass of the sample by 2,1 times the mechanical properties reach 10 of times). The remaining fillers give significantly worse results for the same parameters. To assess the effectiveness of the use and creation materials for the refinement of the cabins and car bodies according to the crashworthiness.

1. Introduction

The requirements to ensure both active and passive safety of people in the car are the most important now. Vehicle structures that perceive external impact during an accident, including shock, should be lightweight, durable and capable of absorbing energy so that living space around people is preserved [1–13]. During the collision, frame-type elements of cabs and car bodies, made primarily by welding thin-walled profiles in a closed loop, lose their bearing capacity under the influence of bending loads [14–16]. A similar character of deformation is obtained during three-point bending of a thin-walled beam, therefore, this article addresses issues related to modelling the behaviour of beams under extreme transverse quasistatic loading (causing beam and filler crushing) for subsequent interpretation of the results obtained on more complex cab and body models as applied to crashworthiness.

Thin-walled beams in the process of such bending are usually very significantly deformed at the point of contact up to the formation of a plastic joint. To achieve optimal ratios of mechanical properties during bending and structural mass, it is advisable to use a filler in the form of a foamy material, for example, aluminium-foam. Chen [17] performed an optimization to minimize the weight of the structure, which is filled with a porous material. He showed that aluminium-foam is a good energy-absorbing material. Chen [18] also studied the behaviour of foam-filled profiles during bending and proved that aluminium-foam increases the energy-absorbing properties of the structure. Hanssen [19] experimentally investigated the bending behaviour of foam-filled profiles and found that the foam filler significantly changes the local patterns of beam deformation. In articles [20, 21], issues related to the application and optimization of the properties of aluminium-foam for the refinement of elements of cabs and car bodies in order to meet the requirements for passive safety are considered. It is shown that the use of filler allows you to achieve the specified requirements with a minimum increase in mass.

The main problem limiting the massive use of foam-filled construction of cabs and bodies is related to production issues. Modern technologies for the manufacture and processing of aluminium-foam are complex and expensive, which leads to an increase in the cost of production, therefore, in this work, we also consider alternative options for fillers. One of them is epoxy resin — a cheap and affordable material and due to its excellent infusion properties, it can be "pumped" under pressure into the frame elements of cabins, which greatly simplifies and reduces the cost of production and assembly technology.

Particular attention is paid to the numerical modelling of the mechanical properties of fillers using the finite element method (FEM) and the LS-DYNA program (including the development of rational finite element models and material modelling). Two types of materials are compared: cellular elastic-plastic (aluminium-foam), which allows large deformations to fracture and brittle (epoxy resin), which has high strength in a limited range of deformations.

To assess the errors of solutions, corresponding bench experiments were carried out.

Comparison of materials with different mechanical properties is necessary to determine the limits of their applicability of cars crashworthiness and to predict the required properties of promising fillers.

2. Mathematical model

2.1. Modelling of filled aluminium beams in LS-DYNA

To study and compare the mechanical properties of cellular elastoplastic and brittle materials, the problem was posed of quasistatic loading of an aluminium beam with a filler. This problem was solved in an explicit statement by the finite element method in the LS-DYNA program. FEM of beam with filler is shown in Figure 1. The proposed type of element is eight nodal, constant stress solid and the base size of the element is 2 mm, which is typical for high-precision models [22]. This type and size of the element is best suited for solving the problems of three-point beams bending with filler [23], including multivariate and optimization [24]. The beam material (aluminium alloy AMg6) was modelled by the MAT_PLASTIC_KINEMATIC "map" with isotropic hardening (material parameters are summarized in Table 1). The diameter of the beam is 40 mm, the wall thickness is 3 mm, the length is 270 mm, the distance between the supports for three-point bending is 200 mm.

The solution to the problem in the LS-DYNA was carried out for the following types of beams:

–hollow beam;

–beam filled with epoxy resin;

–beam filled with aluminium-foam.

Aluminium-foam in LS-DYNA is described by the model proposed by Deshpande and Fleck MAT_DESHPANDE_FLECK_FOAM [25]. This model allows to simulate the mechanical properties of the foam material without directly modelling the pores, which significantly reduces the time spent on the preparation of the calculation and computer time counting.

Three variants of aluminium-foam with densities of 50 kg/m^3, 430 kg/m^3 and 800 kg/m^3, respectively, were compared in [26, 27]. The mechanical properties of the epoxy resin were set according to the data presented in Table 1 and based on research [28]. Material model MAT_PLASTICITY_POLYMER with the condition of destruction upon reaching a strain of 8%.

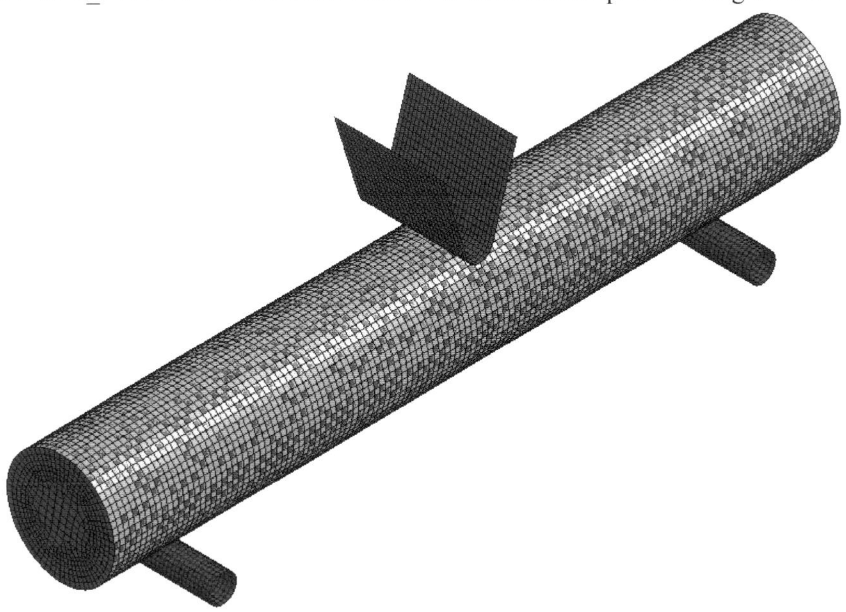

Fig. 1.FEM of filled beam and a loading device

Table 1. Mechanical properties of the materials

Parameters	AluminumAMg6	Epoxy EL with hardener EL 152 MLR	Poraver
ρ, rg/m^3	2460	1100	390
E, GPa	71	3,2	–
$\sigma_{0,2}$extension, MPa	250	–	–
$\sigma_в$extension, MPa	385	74	–
$\sigma_в$bending, MPa	–	110	–
$\sigma_в$compression, MPa	–	100	–
δ_5, %	11	8	–

This article simulates two "extreme" types of materials in which elastoplastic and brittle properties are pronounced. Other types of materials (including composite) with properties not very different from those considered can also serve as a filler for frame-type elements of cabs and car bodies. For adequate modelling of such materials, it is necessary to know the dependences of stresses on deformations for a specific load (tension, compression, bending). Such dependencies are best obtained by the experimental method [29].

The calculation time was limited to 0,1 second at a loading speed of 0,5 m/s in LS-DYNA. The dependence of strain-rate was not taken into account.

The obtained load curves are shown in Figure 2. The numerical result of the deformed state of the beam is shown in Figure 3.

IOP Conf. Series: Materials Science and Engineering **963** (2020) 012007 doi:10.1088/1757-899X/963/1/012007

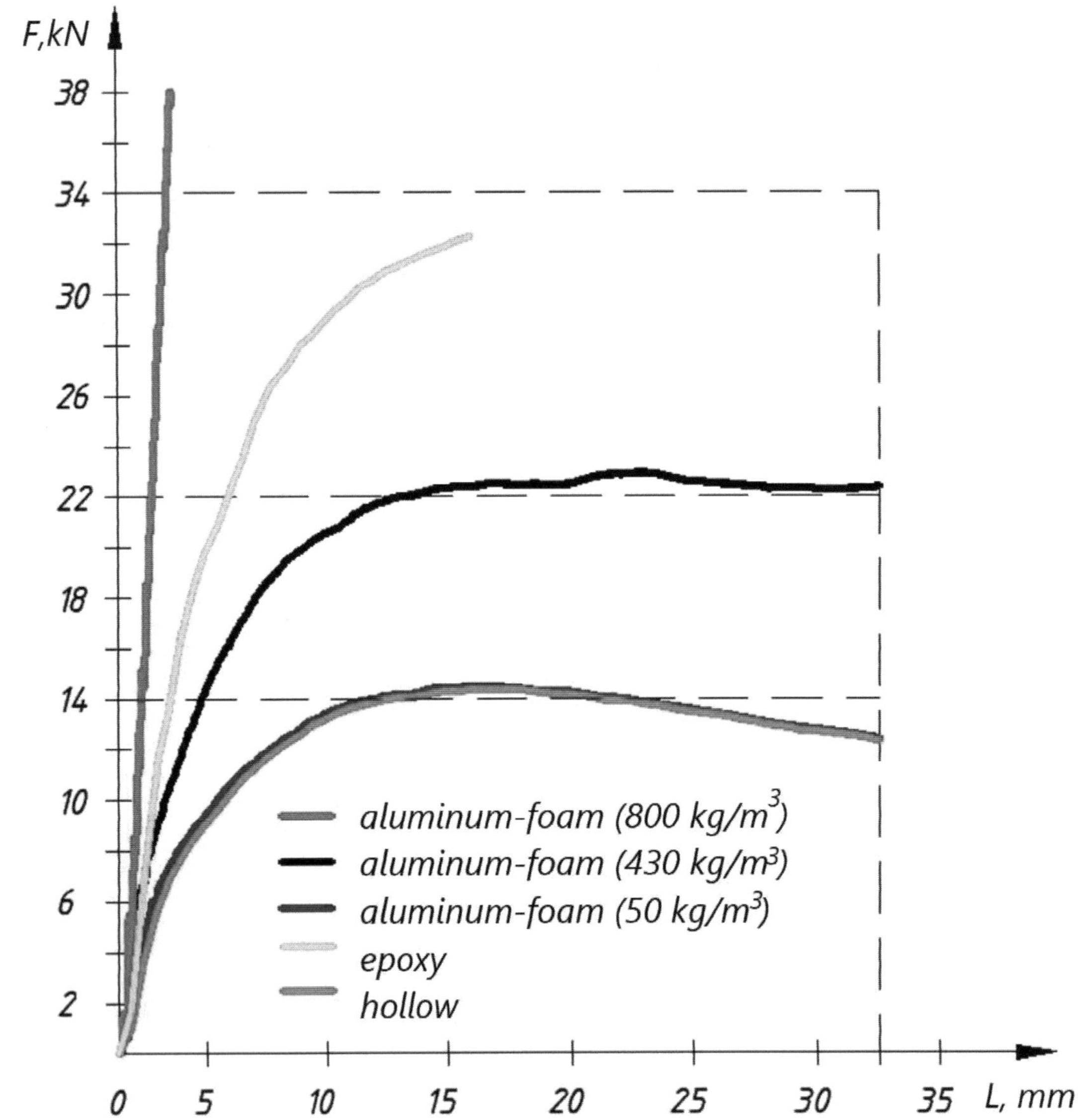

Fig. 2. Numerical impact force curves for the foam-filled beam

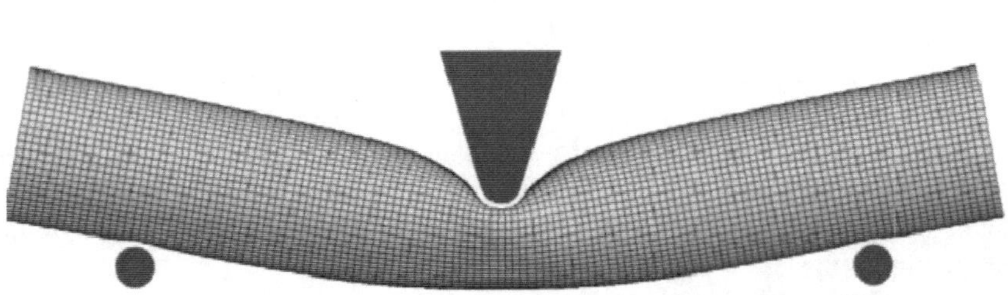

Fig. 3. Numerical deformation pattern for the filled beam

The numerical results are in Table 2.

Table 2. Results of numerical analysis

	Mass, g	Maximum force, kN
Hollow beam	250	14,1
Epoxy filled beam	520	32,3
	267 (50 kg/m^3)	14,2
Foam filled beam	396 (430 kg/m^3)	22,5
	521 (800 kg/m^3)	5140

An analysis of the results showed that brittle materials (epoxy resin) have relatively high strength in a small deformation range, in contrast to cellular elastoplastic materials (aluminium-foam), which remain operational during significant deflections and, as a result, absorb more energy during deformation, which is more important property for solving the problems of crashworthiness.

Aluminium-foam with a minimum density (50 kg/m^3) does not significantly improve the flexural rigidity of the structure. When the filler density is increased to 800 kg/m^3, the perceived force increases by 10 of times, and the mass by 2,1 times.

2.2. Verification of theoretical provisions

In order to verify, experiments were carried out for a hollow aluminium beam and a beam filled with epoxy resin. In addition, we studied the mechanical properties of beams filled with polyurethane foam and a composite material consisting of Poraverball filler and EL epoxy binder with an EL 152 MLR hardener shown in Figure 4.

In addition to the influence of the filler on the mechanical properties of beams, effects were studied in the constrained behavior of materials in the case of three-point bending. The ends were welded with rectangular aluminum plates in beams with a composite material and epoxy resin (four beams).

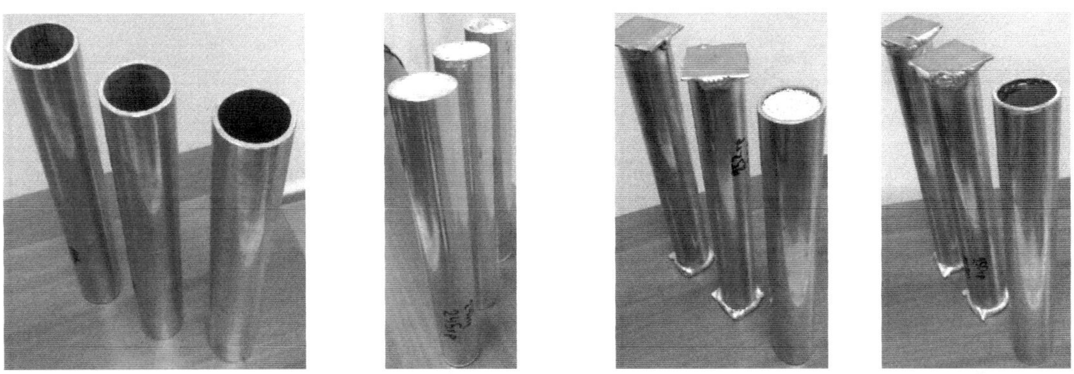

a *b* *c* *d*

Fig. 4. Test beams:
a — hollow beam; b — polyurethane foam-filled beam; c — composite material-filled beam;
d — epoxy-filled beam

IOP Conf. Series: Materials Science and Engineering **963** (2020) 012007 doi:10.1088/1757-899X/963/1/012007

Three-point bending tests were performed on a Zwick Z100 universal testing machine. The test beam and universal machine are shown in Figure 5.

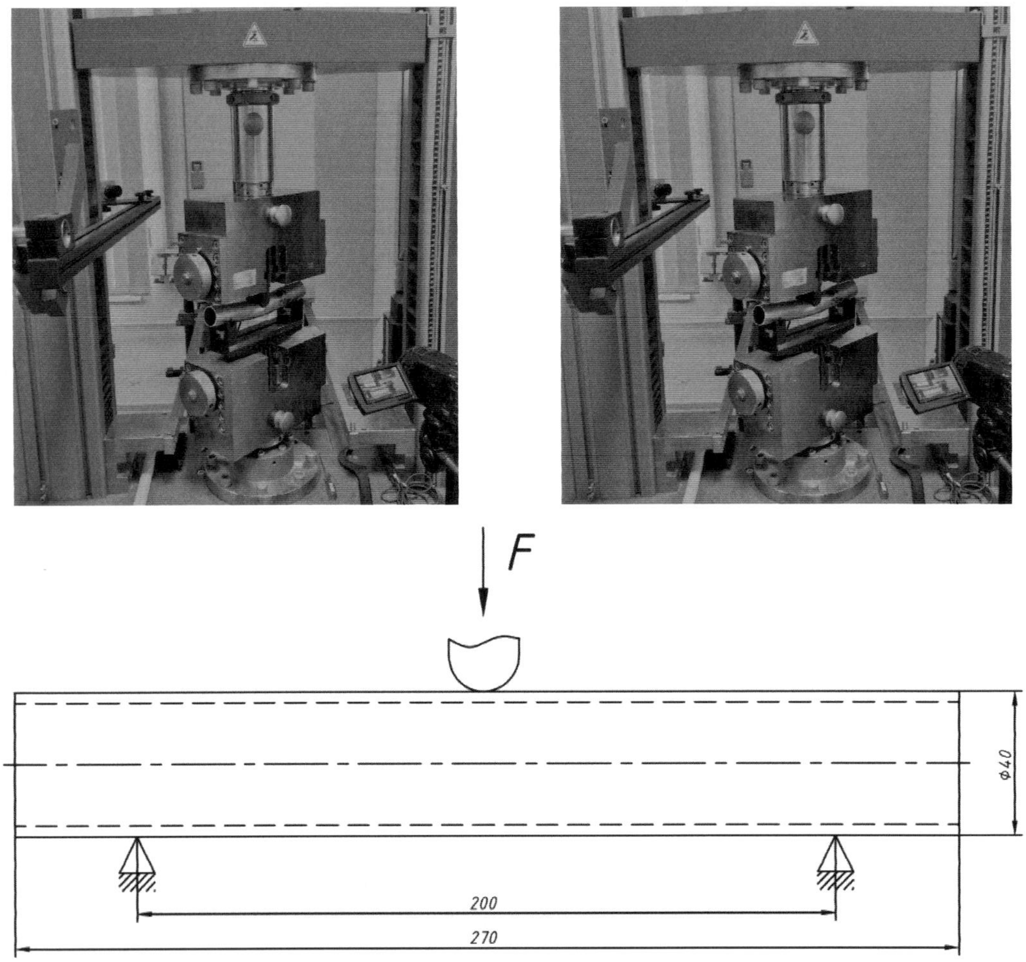

Fig. 5. Zwick Z100 testing machine and test beam

The experiment was carried out as follows:

–a gradually increasing quasistatic load was applied to the beam using an indenter with a rounded end. Load application speed 20 mm/min.

–the amount of applied force was measured depending on the movement of the beam using special sensors and software.

The experimental results are shown in the graphs of Figure 6 and Table 3, and the picture of the deformed state of the beams is shown in Figure 7.

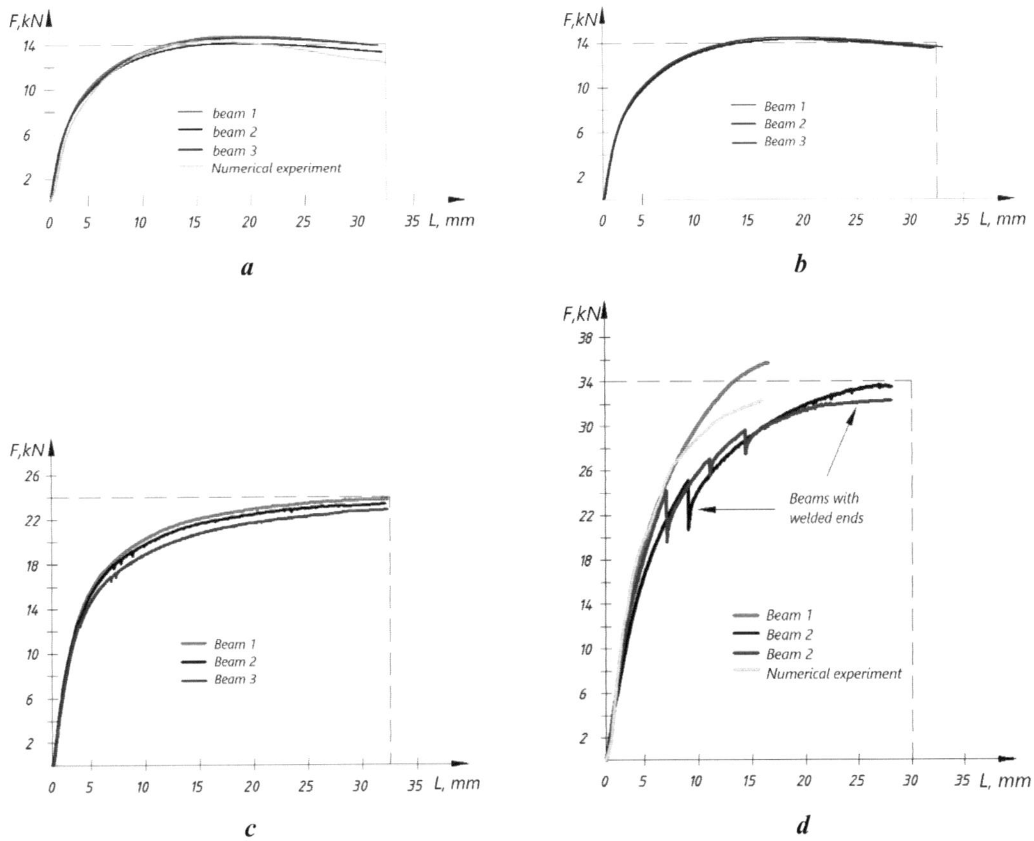

Fig. 6. Experimental and numerical impact force curves for:
a — hollow beam; b — polyurethane foam-filled beam; c — composite material-filled beam;
d — epoxy-filled beam

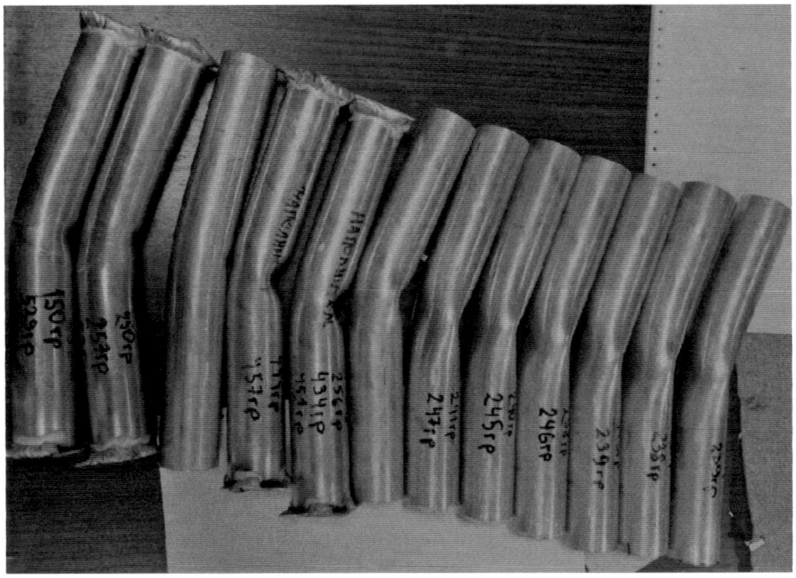

Fig. 7. Beams after full-scale experiment

Table 3. Results of the experiment

Beam type	Number	Mass, g	Maximum force, kN
	1	239	14,73
Hollow beam	2	238	14,1
	3	239	14,58
	1	241	14,27
Polyurethane foam-filled beam	2	240	14,28
	3	238	14,48
	1	413	23,79
Composite material-filled beam	2 (with welded ends)	451	23,18
	3 (with welded ends)	457	22,44
	1	514	35,35
Epoxy-filled beam	2 (with welded ends)	543	33,21
	3 (with welded ends)	545	31,42

3. Result and discussion

Analysis of the experimental results shows that the use of the proposed FEM allows to achieve acceptable results in accuracy. The error in the values of the maximum force does not exceed 8%.

In the case of a hollow beam, the loading curves for the three beams and the calculation practically coincide. The polyurethane foam-filled beam does not increase the stiffness, the loading curves in this case are comparable with the loading curves for a hollow beam.

As follows from the graphs (Figure 6 a, b) for displacements greater than 20 mm, the load acting on the beams decreases and, therefore, in both cases, the structure loses its bearing capacity.

On the other hand, the composite material as a filler allows to increase the stiffness of the beam by 1,6 times. The use of plugs at the ends of the beam does not introduce significant changes in the mechanical properties of the structure. The mass of the composite material filler is 200 g. The appearance of "teeth" on the graph is assumed to be associated with the destruction of Poraver balls and a short-term decrease in the rigidity of the entire model until neighbouring balls come into interaction and the structure begins to work again as a whole. The refinement of the frame elements of the cabs with such fillers is technological, but the effect of filling is not as significant as in the case of aluminium-foam.

Epoxy filler gives a more substantial increase in stiffness 2,23 times. In the process of deformation, the brittle resin is destroyed, as evidenced by the characteristic drops in strength on the graph. A specimen without plugs collapsed with more force than the other two, but with less strain. This is due to the fact that cracks appeared in the epoxy resin, it lost its bearing capacity and, as a result, began to extrude through the free ends of the beam. In all cases, the samples collapsed with smaller deformations (Figure 8) than in the case of using foam, therefore, despite the greater strength of the brittle material, its energy intensity is lower due to smaller deformations. Therefore, the use of brittle materials in the field of crashworthiness is not advisable (in addition, the mass of epoxy is relatively large — 275 g).

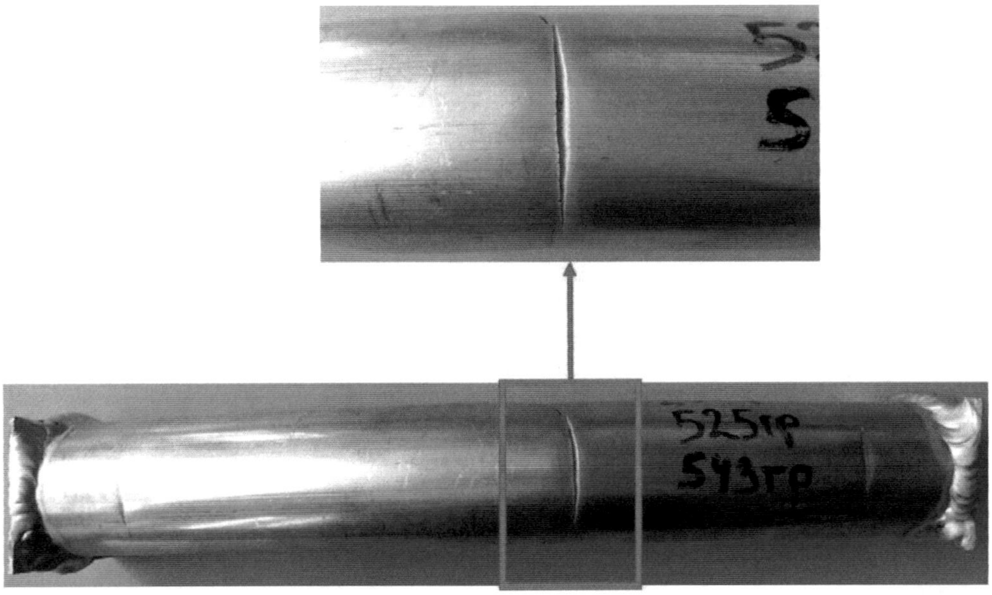

Fig. 8. The destroyed beam with epoxy filler

4. Conclusion

Based on the research, the following conclusion can be made:

1. The developed finite-element models make it possible to effectively predict the elastic-plastic behavior of a thin-walled beam (aluminum alloy AMg6) with a filler (aluminum-foam, EL epoxy resin with hardener 152 MLR, composite material) at the limit of three-point bending (error does not exceed 8% compared with the experiment0).

2. The use of aluminum-foam gives the best results in the ratio of energy capacity — mass. The maximum perceived force during bending is 10 times greater than when using other types of fillers.

3. The use of epoxy resin increased the bending stiffness of the structure 2,23 times compared with a hollow beam, while the mass increased 2,1 times.

4. The use of a composite consisting of Poraver balls and a binder in the form of an epoxy resin made it possible to increase the flexural rigidity of the structure by 1,6 times, while the mass increased by 1,8 times.

The use of brittle materials does not have a significant effect from the point of view of crashworthiness, therefore, they are not recommended for solving issues related to ensuring crashworthiness of automobiles, in contrast to elastoplastic foams (which lose their bearing capacity under significantly large displacement).

References

[1] Gorelov V A and Komissarov A I 2016 *Procedia Engineering* **150** 1322–28
[2] Gorelov V A, Komissarov A I and Miroshnichenko A V 2015 *Procedia Engineering* **129** 300–7
[3] Keller A V, Gorelov V A and Anchukov V V 2015 *Procedia Engineering* **129** 280–7
[4] Kotiev G O, Padalkin B V, Kartashov A B and Diakov A S 2017 *ARPN J. of Eng. and Appl. Sci.* **12** 1064–71
[5] Volskaya N S, Zhileykin M M and Zakharov A Y 2018 *IOP Conf. Series: Materials Science and Engineering* **315** 012028
[6] Vdovin D and Chichekin I 2016 *Procedia Engineering* **150** 1276–79
[7] Klubnichkin E E, Klubnichkin V E and Kotiev G O 2018 *IOP Conf. Series: Materials Science and Engineering* **386** 012025
[8] Zhileykin M M, Kotiev G O and Nagatsev M V 2018 *IOP Conf. Series: Materials Science and Engineering* **315** 012031
[9] Sarach E B, Kotiev G O and Beketov S A 2018 *MATEC Web of Conf.* **224** 04009
[10] Skotnikov G I, Jileykin M M and Komissarov A I 2018 *IOP Conf. Series: Materials Science and Engineering* **315** 012027
[11] Diakov A S and Kotiev G O 2018 *MATEC Web of Conf.* **224** 02096
[12] Kotiev G O, Butarovich D O and Kositsyn B B 2018 *IOP Conf. Series: Materials Science and Engineering* **315** 012014
[13] Evseev K B, Kartashov A B, Dashtiev I Z and Pozdeev A V 2018 *MATEC Web of Conf.* **224** 02039
[14] Shaban B A and Zuzov V N 2013 *Science and education* **8** 91–108
[15] Shaban B A and Zuzov V N 2013 *Science and education* **11** 95–106
[16] Shaban B A and Zuzov V N 2012 *Science and education* **11** 81–104
[17] Chen W, Wierzbicki T and Santosa S 2002 *ActaMechanica* **153** 183–206
[18] Chen W 2001 *International Journal of Solids and Structures* **38** 7919–44
[19] Hanssen A G, Hopperstad O S and Langseth M 2000 *ActaMechanica* **142** 13–31
[20] Goncharov R B and Zuzov V N 2018 *Works NSTU im. R.E. Alekseeva* **3** 130–136
[21] Goncharov R B and Zuzov V N 2018 *News of Moscow State Technical University MAMI* **4** 92–102
[22] Shaban B A and Zuzov V N 2013 *Science and education* **3** 129–156
[23] LiuweiGuo and Jilin Yu 2011 *International Journal of Impact Engineering* **38** 85–94
[24] Goncharov R B and Zuzov V N 2018 *News of Moscow State Technical University MAMI* **2** 2–9
[25] Deshpande V S and Fleck N A 2000 *J. Mech. Phys. Solids* **48** 1253–83
[26] Zhi Xiao, Jianguang Fang, Guangyong Sun and Qing Li 2015 *Int. J. Advances in engineering software* **85** 81–95
[27] Zhaokai Li, Qiang Yu, Xuan Zhao, Man Yu, Peilong Shi and Cilei Yan 2017 *Adv. Mech. Eng.* **9** 1–21
[28] Shebanov S M and Gayeva N B 2010 *Defense complex - to the scientific and technical progress of Russia* **3** 104–110
[29] Arutyunyan G A and Kartashov A B 2018 *Proceedings of higher educational institutions. Engineering* **1** 45–54

IOP Conf. Series: Materials Science and Engineering **963** (2020) 012008 doi:10.1088/1757-899X/963/1/012008

Improving wear resistance of drill pipe sub thread by using final electromechanical surface hardening

S Fedorov[1,2], V Zaripov[1], Y Ivanova[1] and A Yakovleva[1]

[1]Bauman Moscow State Technical University, 5 Second Baumanskaya Street, Moscow, 105005, Russian Federation

[2]E-mail: momd@yandex.ru

Abstract. Improving wear resistance of drill pipe sub thread is a vital task, and finding effective solution for it would much assist oil and gas companies to enhance well construction process. The report presents the results of comparative wear resistance tests for P147/147 and P133/133 subs of the top drive manufactured by OOO «NIGMASH» in Bashkortostan. The tests were performed by OOO «Tatburneft» holding company, a division of PAO «TATNEFT». Both, series and test subs were made from 40X2H4A steel. This report shows the results of metallographic studies and tests for improvement of the parts wear resistance, including the method of final electromechanical hardening (EMH) of the box and pin threads as compared with the conventional method of bulk heat treatment.The researches has developed the equipment, tools and technology for hardening of tool joint thread of drill pipe subs. The performed tests and studies aimed at enhancing efficiency of oil and gas well drilling. Their results are of much interest for oilfield operators, drilling and service contractors, as well as drilling subs manufacturers and repair companies.

Introduction

Drilling of oil and gas wells is a time- and cost-consuming process, and application of new technologies helps in making this process more efficient. Typical drill string includes such elements, as drill bit, kelly, steel drill pipe and lightweight aluminum drill pipe with steel tool joints, subs, crossovers, centralizers, various design stabilizers and reamers, and drill collars.All the elements are made of high-quality alloy steel at specialized metallurgical plants and machine works. Russian oilfield drilling, service and production sector currently imports in large quantities a wide range of oilfield tools from world leading manufacturers from China, United States, Canada and Germany.

Drill string elements are connected with one another using thread connections. The most widely used is the tool joint thread connections with thread pitches of 5.08mm and 6.35mm. External (pin end) and internal (box end) threads are made using special CNC machines with special threading carbide plates or thread chasers. Thread is cut on steel billets with the hardness of 30–36 HRC after bulk heat treatment

Drilling subsare used to connect drill string elements of different diameter or with different type threads, as well as to connect various fishing and auxiliary tools to drill pipe.

The lower sub (lower kelly cock) connects kelly with the upper drill pipe in the drill string, and the upper sub (upper kelly cock) connects kelly with the rotating swivel stem. Improving wear resistance of drill pipe sub thread is a vital task, and finding effective solution for it would much assist oil and gas companies to enhance well construction process. The most typical defects encountered on subs include OD wear and damage of internal and external threads. These defects frequently result in the forced interruption of the hole making process and rig downtime which lead to substantial economic losses.

The problem with damaged threads can only be resolved by increasing thread surface hardness [1] while maintaining high physical and mechanical properties inside the thread. The currently used technology of bulk heat treatment of billets and the following thread cutting does not ensure the required wear resistance of a thread surface. Instead of the planned 600 make-up and break-out cycles for drilling subs, 400 to 500 cycles is achieved at best. The required physical and mechanical properties of subs connection threads can be achieved by using combined methods of thread surface treatment involving concentrated energy streams (flows). These methods include such as thermochenical treatment (nitrogen hardening and carbonitriding), thermomechanical treatment, high induction hardening and laser hardening.

During high induction hardening (HIH) of medium-carbon structural steel the heat-up temperature rises to 930–950 °C. Also, heating up, heat exposure and colling down are applied in sequence while the billet is moved along the induction unit. Spray water colling is used for colling down the surface layer heated up by eddy-current. Using the HIH hardening method for tool joint tapered thread is highly problematic from the technological point of view. Furthermore, this method does not ensure the required hardening depth or the optimum physical and mechanical properties of the thread surface.

During the laser surface hardening the temperature in the treatment zone gets as high as 1000 °C—1100 °C and higher. Laser surface hardening is a non-contact hardening method featuring certain specific processes [2]. Laser surface hardening method ensures high level of the surface layer hardness, which negatively affect performance characteristics of threads of the drill string elements subjected to alternate and bending loads. Laser surface hardening of thread multiplies the risk of fatigue failure of a thread root.

Nitrogen hardening and carbonitriding are both efficient methods of hardening the drill string elements with a connection thread. Yet, the methods can only be applied in certain specialized conditions and require high standard production practices. Hence, cost of subs production significantly increases. Nitrogen hardening and carbonitriding increases hardness of not only connection threads, but also the outside diameter of subs. This causes such negative effects as wear of tong dies, galling of the sub OD, and problematic breaking out of drill pipe.

Among the methods of machines and tools surface hardening which uses the concentrated energy streams [3–5] electromechanical surface hardening technique is highly efficient[6, 7]. Theoretical studies and practical research for electromechanical hardening of connection threads [8–20] resulted in development of the electromechanical surface hardening method that can be used for improving service life of drilling subs connection threads.

The studies aimed at improving performance characteristics and extending service life of drilling subs by research, development and implementation of the final electromechanical hardening (ESH) of external and internal connection threads.

The following studies had to be performed to achieve the set objective:

1. Identify the main reasons for low service life of drill string subs and the criteria for their rejection.
2. Run comprehensive analysis of the existing methods and techniques of drilling subs external and internal connection thread surface hardening, and determine their interrelation with the operating requirements.
3. Carry out test work to support and confirm the applicability of electromechanical surface hardening technique for hardening of pin and box threads of drill string subs manufactured using pipe threading machines.
4. Run study and research to determine the effect of the thermomechanical interaction of the contact roller on alteration of geometrical and accuracy characteristics, structure and mechanical properties of the connection thread surface layer.
5. Run bench and field tests of the subs after pin and box connection threads electromechanical surface hardening.

6. Develop engineering and design documentation for the equipment, tools, accessories and process technique for electromechanical surface hardening of drill string subs external and internal connection threads.

7. Use the developed equipment and technology in subs production to harden connection threads. Run technical and economical assessment of the results.

Experimental Result

While being used in actual well drilling the subs connection threads is subjected to wear, as well as bending and torsional stresses. Uncontrolled reversal loads applied during making up and breaking out of thread connections cause such change of thread geometry as thread galling (Fig.1). For the best result to be achieved thread must feature high hardness while also maintaining sufficient modulus of resilience inside thread (Fig.2).Thread root must have a proper thread bottom shape and metal fibers should be located along the section of weakness. Thread flanks and bottom should have optimum height and microgeometry shape.

Fig. 1. Wear of the element connection thread in the formof turns of thread galling

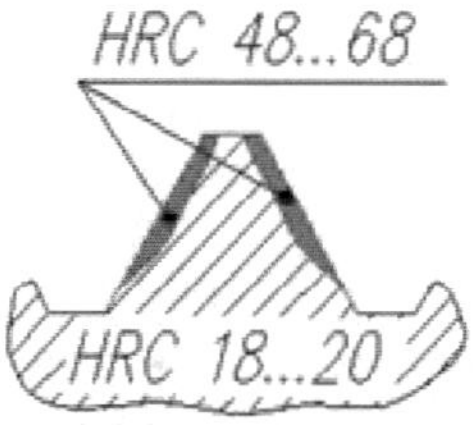

Fig. 2. Performance requirements for drill pipe sub thread

When selecting the method of the drill sting subs connection threads surface hardening the following advantages of the EMSH technique should be taken into account:

zerodecarbonisation and oxidation of the surface due to the fact that the thermomechanical cycle of "heating up — heat exposure — deformation — cooling down" takes place only in the contact zone and lasts two decimal places of a second;

ability to harden limited areas without thermomechanical effect on other surfaces of a drill string sub;

individual approach towards each specific surface with account to loading pattern and operational conditions;

ability to surface harden hollow and long elements and their individual surfaces without altering geometrical parameters of the hardened surfaces and with minimum thermal action level;

the hardened surfaces feature high quality, homogeneity of the structure and uniformity of mechanical characteristics along the section and length of the surface;

using the developed equipment and improving the tools for EMSH would allow to produce products of various configuration and dimension type;

ecological cleanness and electrical safety of the processes.

When EMSH method is used, the factor that most affects formation of the structure and, consequentially, the surface layer properties — is the temperature in the "tool - surface" contact zone generated by electric current.At optimum surface hardening mode, as a result of electric power transformation into heat energy, the contact zone instantly heated up to 1000 °C – 1100 °C. Heat removal from the high temperature zone goes through the below metal layers of the sub billet. Emulsion or water solutions with corrosion inhibitors cooling is used to preserve heat balance and prevent thread distortion.

For EMSH the sub is set in the chuck of a thread cutting machine and is rotated by rotational force D_r. The roller assembly is pushed at a certain fixed force to the sub surface, rotated about its axis and is imparted with a feed movement D_s. The contact zone between the roller and the surface been hardened the sub is been heated up to 1000...1100 °C and the surface layer cooled down fast. Cooling rate of the heated up surface layer reaches 2600 °C/sec, which is much faster than oil or water quenching. The dimensions of the heated up surface zone depend on such surface hardening technology factors as roller pressing force to the sub, roller shape and dimensions, roller feed rate, hardening material hardness, surface hardening mode, thermal conductFigivity of a drill pipe sub material, and heat transfer of the roller assembly.

Upgraded for electromechanical hardening 1M63 machine lathe of Alexandrovsky Drilling Equipment Plant was used to apply EMSH technique for hardening of drill string subs pin and box connection threads.

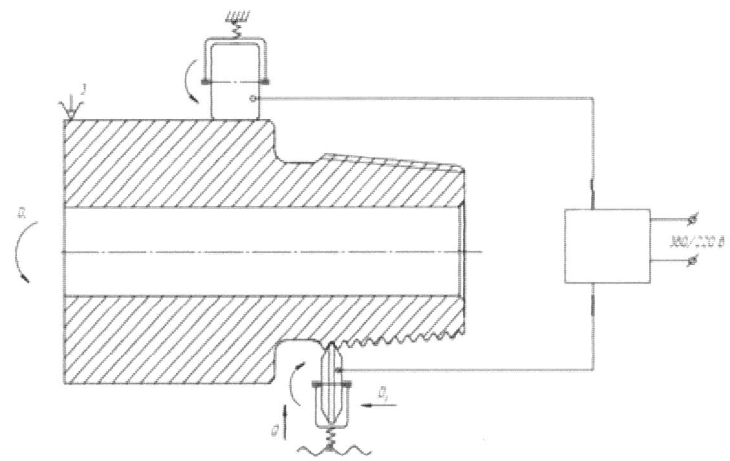

Fig. 3. Schematic diagram of the electromechanical surface hardening of drill pipe sub pin connection thread

Fig. 4. Electromechanical surface hardening of drill pipe sub pin connection thread

Fig. 5. Final electromechanical surface hardening of the drill pipe sub box thread

Same as when using other methods of intensive heating up, during EMSH the drilling sub surface layer structure must reach austenite state. Transition of austenite into martensite is related to carbon diffusion. As final EMSH is one of methods of surface contact hardening using concentrated stream of electrical power, it has certain specific features. The ESH surface hardening cycle including heating up, exposure, deformation and cooling down takes place in centisecond in the closed zone of thermomechanical contact between the EMSH tool and thread. The temperature in the hardening zone gets as high as 1000…1100 °C and higher. The heated up surface layer is cooled down by means of heat removal by the below layers of metal.

The following formula is used to calculate the amount of heat $Q_т$ released in the contact zone between EMSH tool and thread and absorbed in the high temperature zone of the thread surface being hardened during the period т:

$$Q_T = 0,24\ \eta\ I_2\ U_2\ \tau\ k_д k_и,(1)$$

η — coefficient correcting for electric power dissipation in the secondary circuit; I_2 — secondary circuit amperage, A; U_2 —secondary circuit voltage, B; τ — duration of EMSH process, c; $k_д$, $k_и$— coefficients correcting for heat removal to the sub and the tool correspondingly.

Alternatively, based on heat-balance equation

$$Q_T = H\delta B\gamma\ C\ T_e,(2)$$

g = H δBγ — high temperature mass (kg);C — specific heat, J/kg°C;T_e — temperature of metal crystalline transformation,°C;B — high temperature zone width (m); δ — hardening depth (mm); γ — density (kg/m^3

Equating Q_T into equations (1) and (2) gives us formula for calculating hardening depth based on experimentally-confirmed EMSH modes and sub material characteristics:

$$\delta = (0,24\eta I_2 U_2 \tau k_д k_и) / H B\gamma\ C\ T_e(3)$$

Regression equation of tests for secondary circuit current density (X_1) and EMSH tool pressing force (X_2) effect on hardening depth, in coded coefficients values takes the following form (4) Fig. 6.

$$\delta = 0,2115 + 0,0422\ X_1 + 0,0213X_2 — 0,0004\ X_1^2 + 3,8189\ 10^{-7}\ X_1\ X_2 — 0,0002\ X_2^2(4)$$

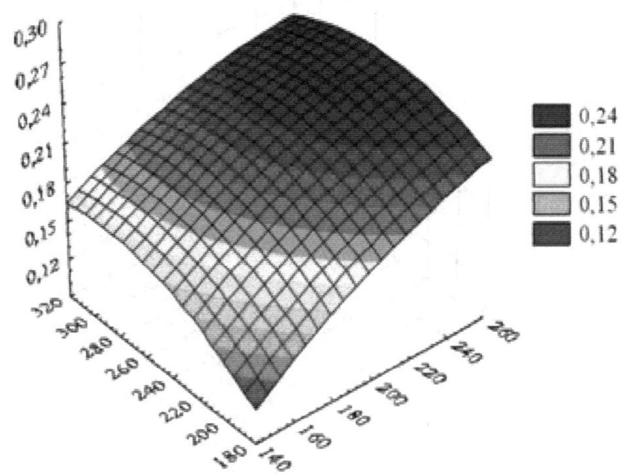

Fig. 6. Graphic presentation of yield surface of interaction between current density X_1 and tool pressing force X_2 in coefficients natural values

The metallographical test results (Fig.7) indicate gradient structure formed in the hardened layer of thread flanks. Not only the metal structure undergoes changes, but also high dispersed structure of martensite is formed in the surface layer.

Top drive drilling subs P147/147 and P133/133 (Fig. 8) manufactured by NIGMASH LLC in Bashkiria region were field tested by Tatburneft, a division of TATNEFT Public Joint-Stock

Company. The results of the field test confirmed high wear resistance of the threads hardened using EMSH technique. The test results demonstrated more than double increase of make-ups and break-outs of thread connections which was described in the statement of drilling subs connection threads wear resistance field test dated September 7, 2019. Series manufactured subs showed up to 500 load cycles. The two reasons for the subs rejection included wear of external thread and wear of the subs outside diameter.

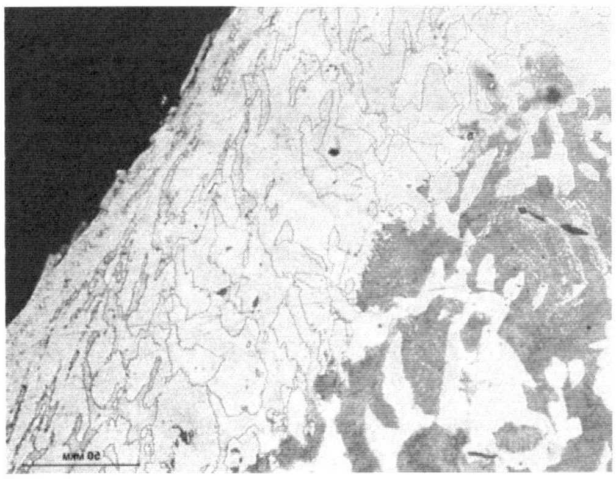

Fig. 7. Microstructure of thread flanks after final electromechanical surface hardening

Fig. 8. Top drive subs P147/147 and P133/133 after final electromechanical surface hardening

Conclusions

After EMSH technique was applied to external and internal threads of the series made subs they showed 985 cycles of connection make-up and break-out and were rejected due to sub OD wear. No wear were found on the subs connection threads.

The results of these studies and field tests allowed to develop recommendations for using final electromechanical surface hardening method for drilling subs external and internal connection threads hardening on pipe-threading machines.

1. The main reason of the low service life of drill string subs is wear of the external and internal connection threads due to low hardness (30...36 HRC) and insufficient strength of threads after bulk heat treatment.
2. Theoretical and practical studies for hardening of contact surfaces of machines and equipment by using concentrated energy streams demonstrated high efficiency of the EMSH surface hardening method.
3. The studies allowed to develop the technology and equipment for improving wear resistance of drill string subs external and internal connection threads by using final electromechanical surface hardening on pipe threading machines.
4. The research engineers developed the formula for calculating the depth of the hardened layer after the final EMSH hardening. They also determined the effect of the electrical current density in the secondary circuit and of the pressing force in the thermomechanical contact zone between the roller and the thread flank on the hardening depth.
5. Metallographical studies of the structure proved high reproducibility (over 92%) of the theoretical and experimental results related to the connection threads flank surface hardening depth depending on the used mode of EMSH method.
6. The final EMSH technique allowed to increase hardness of the connection thread surface layer from 30–36 HRC to 52–56 HRC to the depth of 0.24 mm and form fine-grained martensite structure to the depth of up to 0.1 mm.

7. Field tests of top drive subs P147/147 and P133/133 made by NIGMASH LLC in Bashkortostan region were performed by Tatburneftcompany, a division of TATNEFT. The subs used in the field test showed 985 cycles of connection threads make-up and break-out and were rejected only because of worn sub OD caused by pipe tong dies.

8. The performed studies and tests results proved efficiency of the final EMSH technology for hardening of the drilling subs connection threads. The method can also be used for improving wear resistance of drill pipe and tubing connection threads, as well as ESP housing and various other oilfield equipment.

References

[1] A.I. Yakushev, R.H. Mustayev, R.R.Mavlyutov, Improvement of strength and reliability of thread connections.Moscow, Mashinostroenie, 1979.

[2] A.G. Grigoryants, I.N. Shiganov, A.I. Misyurov, Process technologies of laser hardenin. MoscowPublishing House of N.E.Bauman MGTU University, 2006.

[3] V.S. Kraposhin Parameters Selection for Carbon Steels Laser Heating-Up to Achieve Required Hardening Depth.MITOM. 1986, №9.32–35.

[4] V.S. Kraposhin Engineering Formulas for Calculation of Metal Surface Heating Depth with High Concentration Energy Sources. MITOM. 1999, №7.31–36.

[5] V.S. Kraposhin Heating Mode Selection for Heating Using Surface Heating Source to Achieve Required Hardening Depth and Structural Condition. Moscow, Publishing House of N.E.Bauman MGTU University, 2002.

[6] B.M. Askinasi Strength Improvement and Rebuilding of Machine Parts by Using Electro-Mechanical Hardening. B.M. Askinasi. Moscow, Mashinostroenie, 1989.

[7] V.P. Bagmutov, S.N.Parshev, N.G. Dudkina, I.N. Zakharov, Electro-Mechanical Hardening: Process Technology and Physics, Characteristics, Implementation.Novosibirsk: Nauka Publishing, 2003.

[8] L.V. Fedorova, S.K. Fedorov, Y.N. Kuramshin, M.A. Artemyev, Electromechanical Hardening of Tubing Connection Threads Allows Improving Their Strength and Durability. BureniyeiNeft (2006) №1, 10–12.

[9] L.V. Fedorova, S.K. Fedorov, Technology for Improving Operational Characteristics of Thread Connections by Using Electro-Mechanical Surface Hardening.Machine Building Technologies and Problems. Papers of Visiting Machine Building Session of Principal Council, Ministry of Education of Russia. Ulyanovsk, UGTU (2003) 97–101.

[10] L.V. Fedorova, S.K. Fedorov, A.N. Semyonov, Improvement of Fast Wearing Parts Durability. World Technologies. (2008) №06, 30–32.

[11] L.V. Fedorova, V.F. Karpenkov, S.K. Fedorov, Y.S. Alekseyeva, E.V. Nagnibedova,Technology and Equipment for Improving Strength and Rebuilding of Parts Using Electromechanical Surface Hardening.Machines and Equipment for Agriculture. (2009)
№2,34–35.

[12] S.K. Fedorov, L.V. Fedorova, V.T. Sarayev, F.K. Klyuev,Implementing Electromechanical Surface Hardening Technology in Repair Workshop of Syzran Refinery. Scientific and Technical Bulletin of Rosneft Oil Company (2010) No.4, 44–47.

[13] S.K. Fedorov, L.V.Fedorova. Electro-Mechanical Hardening. RITM (2012), No. 2(70).14–16.

[14] L.V.Fedorova, S.K. Fedorov, Y.S.Ivanova, A.M. Lompas,Technology of Parts Wear Resistance Improvement by Using Electromechanical Surface Hardening. Universities New Bulletin. Mashinostroenie Publishing. (2017),№9(690).85–92.

[15] J.S. Alekseeva, L.V. Fedorova, S.K. Fedorov, I.N. Kapustin, Improving the quality of the surface layer of steel parts. Proceeding of 5-th International Mechanical Engineering Forum (IMEF). (2012) Prague, Czech Republic, 65–74.

[16] Fedorov S.K., Fedorova L.V., Ivanova Y.S., Voronina M.V. Increase of Wear Resistance of the Drill Pipe Thread Connection by Electromechanical Surface Hardening,International Journal of Applied Engineering Research ISSN 0973-4562 Volume 12, Number 18 (2017) pp. 7485–7489 © Research India Publications.

[17] L.V. Fedorova, S.K. Fedorov, A.A. Serzhant, V.V. Golovin, S.V. Systerov, Electromechanical surface hardening of tubing steels. Metal Science and Heat Treatment. 59(3–4) (2017). 173–175.

[18] S.K. Fedorov, L.V. Fedorova, V.S. Zharennikov, M.V. Pesin, Y.P. Smol'skiy Method of Thread Forming on Parts./Russian Patent No.2482942. (2011). Bulletin №15.

[19] S.K. Fedorov, L.V. Fedorova, V.S. Zharennikov, M.V. Pesin, Y.P. Smol'skiy Method of Thread Forming on Parts. Russian Patent No.2486994. (2011). Bulletin №19.

[20] S.K. Fedorov, L.V. Fedorova, Y.S.Ivanova, A.A. Yakovleva, M.V. Vlasov, V.N. Zaripov, M.A. Lashukov Tooling Unit for Thread Electromechanical Hardening. Russian Patent (2018). No.186927.

IOP Conf. Series: Materials Science and Engineering **963** (2020) 012009 doi:10.1088/1757-899X/963/1/012009

Towards consistent high quality of engineering products

L Kushnarev

Bauman Moscow State Technical University

E-mail: kushnarevl@mail.ru

Abstract. The low level of competitiveness and the constantly growing requirements of consumers for modern domestic technology indicate an urgent need to improve the quality of technological machinery and equipment produced by engineering. Ensuring a competitive world-class quality of the produced machines and equipment requires innovative development and modernization of the production organizational and technological system of mechanical engineering. Only the association of modern engineering companies is able to unit and direct the efforts of all participants in the process of creating machines towards a stable, continuous improvement in the quality of equipment. Encompassing quality control at all stages of the life cycle of a machine, the company as soon as possible and without any special financial expenses ensures the highest world level of quality of engineering products.
The goal of research is to comprehensively improve the quality of Russian-made machinery and equipment to the level of the best world analogues and higher based on the creation of an innovative production and technological system (complex), ensuring the as great as practicable level of quality of equipment at all stages of its life cycle.
The results of feasibility studies show, that a non-competitive level of quality of domestic equipment has developed and is maintained, despite the efforts of manufacturers, due to the long-term maintenance of the existing manufacturing system and production relations, requiring urgent and accelerated modernization. The main disadvantage of the existing manufacturing system in mechanical engineering is the isolation and fragmentation of enterprises and institutions working on the creation of machinery and equipment, the incompleteness of the production and technological chain, the lack of engineering and technical support of processes at all stages of the life cycle and, first of all, at the stage of operation of equipment. Therefore, in this article, we consider and propose a complex of priority organizational and technological measures providing the accelerated innovative development of the manufacturing system and improve the quality of Russian equipment above the level of the best world analogues.

Introduction

Engineering factories are going through hard times today. They are urgently faced with a huge complex of problems requiring accelerated solution: reestablishment of the high-manufacturing volumes of machinery and equipment for many machine-technological sectors of the economy; elimination of the problems of import independence and import substitution industrialization, other negative consequences of the low competitiveness of domestic engineering companies and equipment; innovative transition and development of energy-efficient machine-building production and providing machine-technological sectors of the economy with high-quality and energy-efficient equipment [1–4].

The solution to these problems involves the maximum use of the possibilities of restructurisation, integration, diversification and technological modernization of almost all types of industries and sectors of the economy - both producing and using machinery and equipment. Each of the above listed areas of innovative development of mechanical engineering and machine-technological sectors of the economy requires a progressive scientific foundation, which should, on the one hand, become the main focus of the development of the country's economy, and on the other, — provide a comprehensive innovative solution to the whole complex of problems [3–8].

The obtained results of inter-industry organizational, technological, organizational and economic analytical studies in the areas of production, usage and technical maintenance of machinery and equipment indicate technological necessity, economic efficiency and the availability of organizational capabilities, providing a packaged solution to these problems. And this is possible only with a systematic approach to the entire complex of problems, and not separately for each, since only in this case the problem of restructurisation the engineering industries and related machine-technological sectors of the economy is automatically solved [6–9].

A comprehensive and systematic approach to a detailed analysis of production and economic conditions and the relationship between manufacturers and consumers of machinery and equipment indicate that the target guidelines and criteria should be ***comprehensive support for improving the quality of engineering products***, not lower than the best foreign analogues. Only *improving the quality* of engineering products can provide a solution to all the above listed problems of manufacture [10–14].

During the entire life cycle, the quality of engineering products is formed by 75% at the stage of design and construction of machinery and equipment, by 20% — during manufacturing, and only 5% is ensured by proper operation — production (when used for its intended purpose) and technical (while ensuring operational integrity) (Fig. 1) [8, 9, 13].

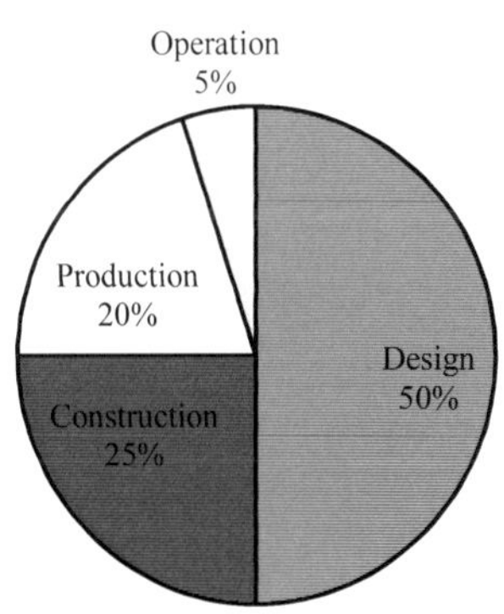

Fig.1.formation of quality at various stagesofthemachinelifecycle

At the same time it should be noted that for a high level of quality at the design and construction stage, it is necessary not only to have complete and reliable information about the quality indicators of machines and equipment, but also to have feedback with the structures that service their operation. And this just does not exist in modern system of designing, manufacturing and using machinery and equipment. Each element of this system decides to the detriment of others its own private and narrow company interests, which ultimately result not in a positive but a negative national economic effect. The products manufactured with huge costs of all types of production material and technical resources are noncompetitive, with an unacceptable price-quality ratio, which requires huge production resources to ensure operability in the form of creating a whole system of powerful technical service enterprises.

All this will continue until the manufacturers of machines and equipment will participate in ensuring the operation processes of their machines and equipment and, first of all, in ensuring their operational integrity during the warranty period of operation, which must be at least not less than presented by technical requirements and established by Russian standards.

Suggested Solutions

Therefore, at present, it is necessary in conditions of intense but healthy competition in the technology marketing area to ensure the transition to a full-size proprietary method of technical service of machinery and equipment (Fig. 2). And the need to have their own systems of proprietary technical service should relate, first of all, to foreign manufacturers and suppliers of machinery, machinery and equipment, and not only to Russian manufacturers. The first step in this direction should be done the control by Russian State Technical Authority over the quality of imported machines and equipment in completes, for the presence of visual defects, for time between failures, etc. and if necessary, prevent the sale of obviously faulty equipment to Russian consumers.

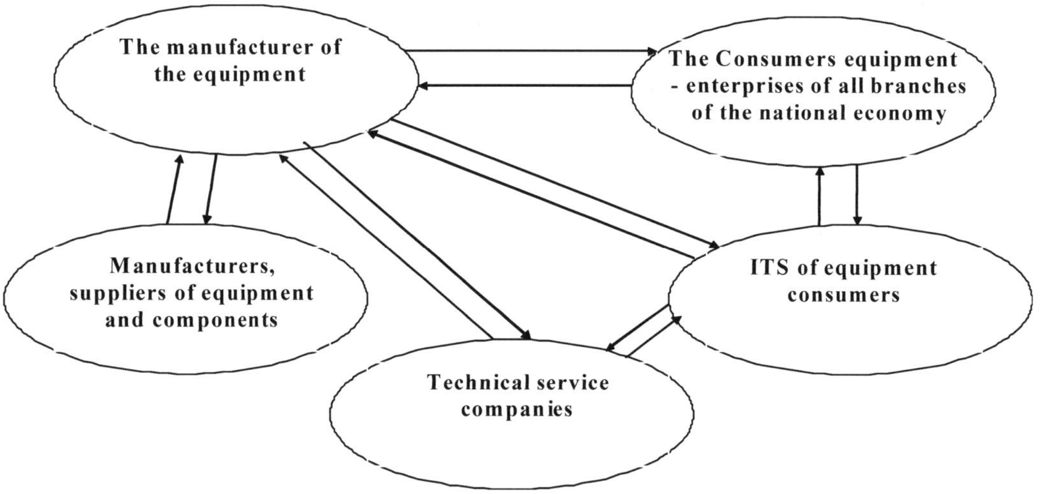

Fig. 2. The organization of the firm and technical support

In the case of failures during the warranty period of operation of machinery and equipment, Russian State Technical Authority services must involve (or provide assistance) and control the elimination of the consequences of machine failures by the forces and means of the manufacturer's warranty service. Besides, the consumer of engineering products has the right to compensation for losses (lost profits) caused by machine downtime for technical reasons. Such claims may be presented either to manufacturers of equipment, or to their technical representatives — dealer structures, depending on who is responsible for the supply of low-quality equipment to the consumer under the supply agreement.

Since in the process of supplying facilities, machinery and equipment to consumers, during the warranty and post-warranty periods of operation, production engineering measures and servicing operations are required, instead of purely intermediary structures — the "dealers" there should be engineering and technical structures — technical representatives of manufacturing. They should have the necessary facilities and finances prepared by experienced personnel for engineering and technical support and monitoring the proper operation of machinery and equipment at the consumer; do the information acquisition on the reliability of the machine, components and assemblies, operating conditions, requirements by consumers to the technical characteristics and capabilities of the machines, on reliability indicators, etc. Special attention should be paid to the time between failures, appearing failures, research and establishing the causes of their occurrence, the complexity and working time of eliminating them, and not only during the warranty period of operation, but also the after-warranty period.

Regarding the contract for the supply of machinery and equipment, — it should be wholly and completely referred at ensuring the main strategic goal of the branded technical service — comprehensive improvement of the quality of engineering products, ensuring the required level of reliability and efficiency of machinery and equipment delivered to the Russian consumer. Objectives to ensure these goals should be rationally distributed in accordance with organizational, technological and organizational and economic capabilities between the manufacturer of machines and equipment and their engineering and technical representatives. Of course, the agreement having the status of the main legal document, should be reflected the rules and requirements for the level of operation of machinery and equipment.

The main stimulating factor in ensuring consumer rights and the proper construction of relationships in the system of corporate technical service is a specially developed mechanism of organizational and economic relations. It is referred at implementing the above listed goals and objectives of corporate technical service to realistically increase the competitiveness of manufactured machines and equipment, increase their efficiency and ensure the priority of consumer rights, first of all, to purchase high-quality and reliable equipment, and not to constant servicing under warranty and post-warranty periods of operation [14–18].

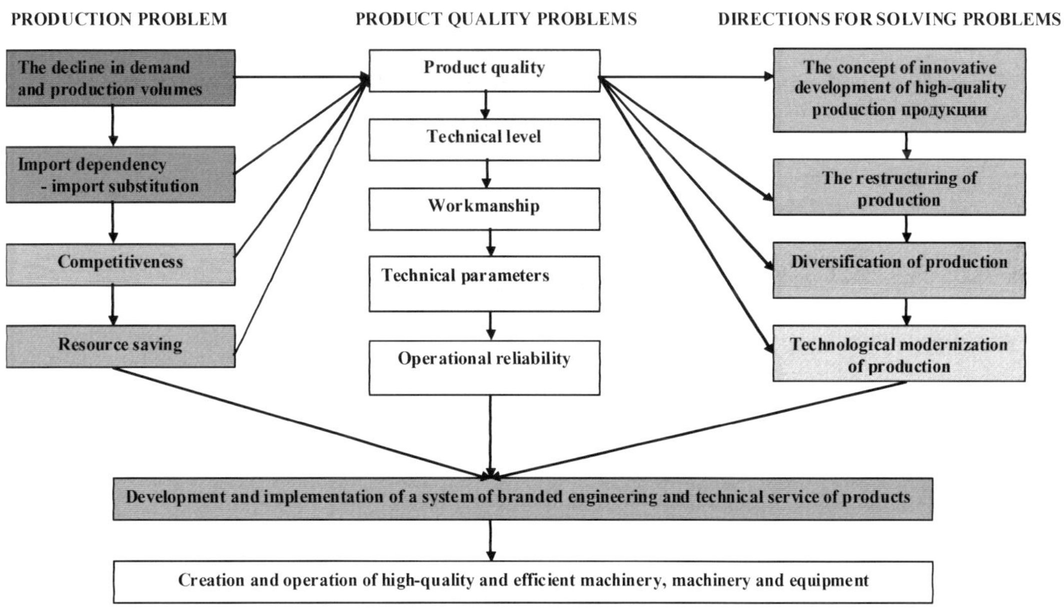

Fig. 3. Improving the quality of technology through the creation of innovative industrial and technological systems (firms)

For the practical implementation of full-scale corporate technical service in the regions in the conditions of unhealthy competition, with an unprecedented level of monopolism of equipment manufacturers and their collusion with trading organizations, with complete disregard for the rights of consumers of machinery, facilities and equipment, management and control of the renovation of machinery and equipment should be taken by territorial oversight organizations (Russian Federal Consumer Rights Protection and Human Health Control Service, Russian State Technical Authority, etc.). To improve the mechanism of relations between manufacturers and consumers of equipment the «Regulation on the organization of mandatory engineering and technical support service (provision or service) by manufacturers and suppliers to the region of their machinery, machines and equipment throughout their entire life cycle» should be developed. Moreover, they eliminate the consequences of failures that arose not through the fault of the consumer, they eliminate at their own expense, using their own financial resources and means, with the compensatory damages caused to the consumer by simply delivering low-quality equipment. In the future, on this situation, it is advisable to pass two legislative acts, the first - on the mandatory engineering and technical support of engineering products throughout its entire life cycle, the second - on the protection of the rights of consumers of industrial products [18–21].

Conclusion

It is advisable to entrust the formation of technical and technological policies to specialized services of departments of manufacturers and consumers of equipment with the involvement of scientific and educational institutions actively working to solve this problem (project) — the improving the quality of manufacturing and reliability of equipment based on the development and implementation of a proprietary method of technical service.

Currently, a significant scientific groundwork has been created on this issue, the main propositions of which can be put into practice in a short time, both on a state scale and within the framework of regions and individual companies manufacturing machinery and equipment. The final result of the research is a system developed and implemented in real production to improve the quality of manufacturing and reliability of equipment based on the proprietary method of technical service.

The economic efficiency from the introduction of research results has a multidirectional synergistic character and can amount to more than 2.0–2.5 trillion rubles for the national economic complex of the country. A significant part of the economic benefits is the decrease of the need for equipment, material and technical resources for its operation, the decrease of the need for repair, technical facilities and equipment, qualified servicing personnel, the reduction in product losses and its quality during machine and equipment downtime due to technical reasons and etc. Estimated efficiency is so high that, in aggregate, it will allow to reduce the cost of agricultural products, mechanized work and services by 35-50%. And this will positively affect the efficiency of production and economic activity of companies of many machine-technological sectors of the economy.

References

[1] Adler Yu.P. A new direction in statistical quality control — is Taguchi methods. M .: Knowledge; 1988.
[2] Kiselev M.I., Pronyakin V.I., Tulekbaeva A.K. Technical diagnostics functioning machines and mechanisms // IOP Conference Series: Materials Science and Engineering. 2018. V. 312(1). 012012.
[3] Aleshin N.P. et al. Applying nondestructive testing to quality control of additive manufactured parts // Russian Journal of Nondestructive Testing. 2016. V. 52(10). P. 600–609.
[4] Gubaydulina R.H., Gruby S.V., Davlatov G.D. Analysis of the Lifecycle of Mechanical Engineering Products // IOP Conference Series: Materials Science and Engineering. 2016. V. 142(1). 12060. DOI: 10.1088/1757-899X/142/1/012060
[5] Aleshin N.P. et al. Using nondestructive testing methods for in-production quality control of additive manufactured parts // Russian Journal of Nondestructive Testing. 2016. V. 52(9). P. 532–537. DOI: 10.1134/S1061830916090023
[6] Vasiliev A.S. Controlled Forming of Machine Components Operating Characteristics // Procedia Engineering. 2016. V. 150. P. 975–979.
[7] Nikolaev A.B. et al. Tools of production and logistics support life cycle of high-tech products // Life Science Journal 11 (10 SPEC. ISSUE). 2014. P. 238–242.

[8] Fedyukin V.K. Upravleniyekachestvomprotsessov [Process Quality Management]. St. Petersburg: Peter. 2005.
[9] OgozdinV.Yu. About the process approach in quality management // Standards and quality. 2010.Vol. 11. P. 36–38.
[10] Chernoivanov V.I., Severnyy A.E., Kushnarev L.I. and others. Problems of technical service in the agro-industrial complex of Russia (monograph). M.: GOSNITI, 2000 — 309 p.
[11] The preventive maintenance system of servicing and repair of machinery and equipment. M.: Mechanical Engineering, 2001 .- 369 p.
[12] An integrated system of servicing and repair of machinery in agriculture. M .: GOSNITI, 1985. — 143 p.
[13] Strategic management of organizational and economic stability of the company: Logistic-oriented business design / A.D. Kanchaveli, A.A. Kolobov, I.N. Omel'chenko and others. Edited by A.A. Kolobova, I.N. Omel'chenko. M: BMSTU. — 2001. — 600 p.
[14] Kushnarev L.I. Chepurina E.L., Chepurin A.V., Kushnarev S.L. The quality and reliability of domestic equipment is the basis of its competitiveness // Tractors and agricultural machinery. — 2015. — No. 11. P. 35–37.
[15] Kushnarev L.I. and others. Modernization of the technical service system of the agro-industrial complex. Scientific publication. M .: MESH. — 2017.
[16] Chepurina E.L. Sevast'yanova D.L. Improving the reliability of machinery and equipment in stock farming // Scientific and informational support for the innovative development of the agro-industrial complex. Materials of the IX International scientific-practical conference "InformAgro-2017". M .: "Rosinformagroteh". — 2017.
[17] Kushnarev S.L., Chepurina E.L. Organization of corporate technical service for agricultural machinery // Tractors and agricultural machinery. — No. 10. — P. 3–4.
[18] Kushnarev L.I. Corporate technical service of machinery and equipment. Problems. Search. Solutions (monograph). Deutschland, Saarbrücken: Palmarium Academic Publishing. — 2014.
[19] Kushnarev S.L. The principles of organization of corporate services for agricultural machinery // Repair, reestablishment, modernization. — 2004. — No. 1. — P. 35–37.
[20] Chepurina E.L., Kushnarev L.I. The role and place of agricultural equipment manufacturers in the corporate technical service. / Machinery and equipment for the village. — 2013. — No. 7. — P. 38–40.
[21] Chepurina E.L. The state of the organization of technical service of stock farming machinery and equipment. / International Technical and Economic Journal. — No. 4, — 2013. The Spectrum LLC. — P. 61–67.

Algorithms for dynamic stabilization of rear-wheel drive two-axis vehicles with a plug-in rear axle

M Zhileykin[1,2] and A Eranosyan[1]

[1]Bauman Moscow State Technical University, 5 Second Baumanskaya Street, Moscow, 105005, Russian Federation

[2]E-mail: jileykin_m@mail.ru

Abstract. Currently, two-axis front-wheel drive vehicles are widely used systems that provide increased stability and manageability by connecting the rear axle and redistributing torque between the driving axles and between the driving wheels of the rear axle. To a lesser extent, drive schemes for rear-wheel drive two-axis cars with a plug-in front axle are common, and, accordingly, the issues of traffic stabilization of such cars are not fully studied. The purpose of this work is to increase the stability and controllability of rear-wheel drive two-axis machines with a plug-in front axle by redistributing torques between the driving wheels. The analysis of the effectiveness of the 4x4 vehicle traffic stabilization system by redistributing torques between the driving axles and the wheels carried out by simulation methods allowed us to establish that the following stabilization methods are the most effective for cars with a plug-in front axle:
• a 4x4 vehicle with a plug-in front axle that redistributes torques between the driving wheels of the rear axle (the effectiveness of these stabilization methods is approximately the same):
• a 4x4 vehicle with a plug-in front axle and an automatically locked front-wheel differential;
• 4x4 vehicle with a plug-in front axle with individual braking system for individual wheels.

Introduction

The car control stabilization system is designed to maintain the stability and manageability of the car by detecting and eliminating a critical situation in advance. The system allows you to keep the car within the path set by the driver in various driving modes (acceleration, braking, driving in a straight line, in turns, and when rolling freely). Currently, two-axis front-wheel drive vehicles are widely used systems that provide increased stability and manageability by connecting the rear axle and redistributing torque between the driving axles and between the driving wheels of the rear axle [1–10].

To a lesser extent, drive schemes for rear-wheel drive two-axis cars with a plug-in front axle are common, and, accordingly, the issues of traffic stabilization of such cars are not fully studied [11].

The purpose of this work is to increase the stability and controllability of rear-wheel drive two-axis machines with a plug-in front axle by redistributing torques between the driving wheels.

The following schemes of torque control on the driving wheels of the car are considered:

1) rear-wheel drive car without torque control on the driving wheels (base for comparison);

2) a 4x4 car with a plug-in front axle without redistributing torques between the driving wheels of the same axle;

3) a 4x4 vehicle with a plug-in front axle with a reallocation of torques between the driving wheels of the rear;

4) a 4x4 vehicle with a plug-in front axle and an automatically locked front-wheel differential;
5) 4x4 vehicle with plug-in front axle with individual braking system for individual wheels.

The control algorithm for items 2 and 3 is given in the work [11].

Algorithm for redistributing torques between the driving wheels of the axle due to individual braking of individual wheels

The operation of the system is based on the recognition of the rear axle skidding or demolition of the front axle of the car. To recognize the type of skid, use the results of [12], where the parameter is used as a diagnostic sign of the onset of front or rear axle drift $\delta_V = \left\| \boldsymbol{V}_{C1} \right| - \left| \boldsymbol{V}_{C2} \right\|$, representing the difference in the estimation of linear speeds of the center of mass of the car, first using the linear speed of the center of the front axle (vector $\boldsymbol{V}_{C1}$), and then — the linear speed of the center of the rear axle (vector $\boldsymbol{V}_{C2}$). In the event of a skid the rear axle is braked the outside front wheel. In case of demolition of the front axle, the inner rear wheel is braked.

The formula for determining the braking torque on each wheel is as

$$M_{mi} = h_{ESPi} T_{\max}, i = 1,\ldots,N,$$

where $T_{\max}$ — maximum braking torque developed by the wheel brake mechanism; N — number of car wheels; $h_{ESPi} = [0\ldots1]$ — the degree of reduction of the effective braking torque on the i-th wheel due to the operation of the algorithm to counteract the demolition of the front axle during braking.

Algorithm for determining the value $h_{ESPi}, i = 1,\ldots,N$ taking into account the sign rule adopted in the simulation, it should be as follows.

If $\Theta_1 > 0°$ (turn left) & $\delta_V > 0$ (demolition of the front axle), that $h_{ESP1} = h_{ESP3} = h_{ESP4} = 0$; $h_{ESP2} = C_1 |\delta_V| + C_2 \dot{\delta}_V$.

If $\Theta_1 > 0°$ (turn left) & $\delta_V < 0$ (the drift of a back axis), that $h_{ESP1} = h_{ESP2} = h_{ESP4} = 0$; $h_{ESP3} = C_1 |\delta_V| + C_2 \dot{\delta}_V$.

If $\Theta_1 < 0°$ (turn right) & $\delta_V > 0$ (demolition of the front axle), that $h_{ESP1} = h_{ESP2} = h_{ESP3} = 0$; $h_{ESP4} = C_1 |\delta_V| + C_2 \dot{\delta}_V$.

If $\Theta_1 < 0°$ (turn right) & $\delta_V < 0$ (the drift of a back axis), that $h_{ESP4} = h_{ESP2} = h_{ESP3} = 0$; $h_{ESP1} = C_1 |\delta_V| + C_2 \dot{\delta}_V$.

In the above formulas C_1 and C_2 — regulator gain factors that can be adjusted individually for each vehicle.

Creating an additional moment of turning resistance by locking the inter-wheel differential

Front axle skidding

To increase the car's turnability, it is necessary to bring more torque to the wheels of the rear axle. However, it does not make sense to block the differential of the rear axle, since it will create a moment of resistance to turn

$$M_{c2} = \frac{B_к}{2}\left(R_{x2} + fR_{z2}\right);$$

where$B_к$ — wheel track, R_{x2}, R_{z2} — the total longitudinal and vertical reactions on the wheels of the rear axle in the projection of the X and Z axes associated with the center of mass of the car coordinate system respectively, f— rolling resistance coefficient.

The moment of resistance to the turn is harmful in this case, since it is directed against the angular speed of the turn.

Simultaneously with the redistribution of torque between the axles of the car, it is necessary to apply more torque to the external driving wheel of the rear axle in relation to the direction of rotation, which requires a controlled asymmetric differential.

The drift of a back axis
To reduce the turnability property, it is necessary to reduce the torque on the rear driving axle, transferring most of it to the front axle. At the same time, blocking the front axle differential will create a moment of resistance, which is also a moment of counter-rotation, which helps to stabilize the movement of the car

$$M_{c1} = \frac{B_K}{2}\left(R_{x1} + fR_{z1}\right);$$

whereR_{x1}, R_{z1} — total reactions on the wheels of the front axle in the projection of the X and Z axes associated with the center of mass of the car coordinate system respectively.

The algorithm for locking the front axle differential when the rear axle skids is as follows.

If $\delta_V < 0$ (the drift of a back axis), that the front axle differential is locked. Otherwise, it is unblocked.

Performance criteria for the 4x4 vehicle stabilization system with differential transmission due to individual braking of individual wheels
As criteria for the effectiveness of the stabilization system of a 4x4 vehicle with a differential transmission due to individual braking of individual wheels, you can take the following.

1) Type of vehicle trajectory (no rear axle skid or front axle drift).
2) The case when $\delta_V = 0$, corresponds to the vehicle's neutral turnability (ideal value). Therefore, the diagnostic parameterδ_V it can be used to formulate a criterion for the efficiency KE operation of motion stabilization algorithms:

$$KE = \sqrt{\frac{1}{n_r - 1}\sum_{i=1}^{n_r}\delta_{Vi}^2}, \tag{1}$$

where n_r — the number of points in the implementation process δ_{Vi}.

Investigation of the effectiveness of the proposed algorithm for the redistribution of torques between the driving axles and wheels by simulation methods
Testing the efficiency of the algorithm for redistributing torques between the driving axles and wheels was carried out on the example of a 4x4 car with a total weight of 2400 kg with a plug-in front axle and a redistribution of torque between the wheels of the rear axle. Was carried out a theoretical study of the motion of the car with the help of mathematical simulation. Features of the mathematical model of motion are considered in [13–20].

We study the movement on the "ice with snow" support base (with the coefficient of interaction of the engine with the support base at full slip of 0.35). Note that the term "support base" refers only to a solid non-deformable support surface. The front wheels of the car are steerable. The steering angle changes from zero to the set value within 3 seconds and then remains unchanged.

The mode of entering a turn and driving in a turn with a fixed radius is simulated. We studied the movement of the car at the entrance to the turn and the movement in the turn, with the speed V=25 km/h, which the driver tries to maintain constant during the movement.

The trajectory of the rear wheel drive vehicle on ice when making a turn is shown in Fig. 1.

Analysis of the driving paths of a car with a rear drive axle (Fig. 1) shows that when making a maneuver on the support surface "ice with snow" the car loses control (develops a skid of the rear axle).

The path of movement on ice of a 4x4 vehicle with a plug-in front axle without redistributing torques between the driving wheels of one axis when making a turn is shown in Fig. 2.

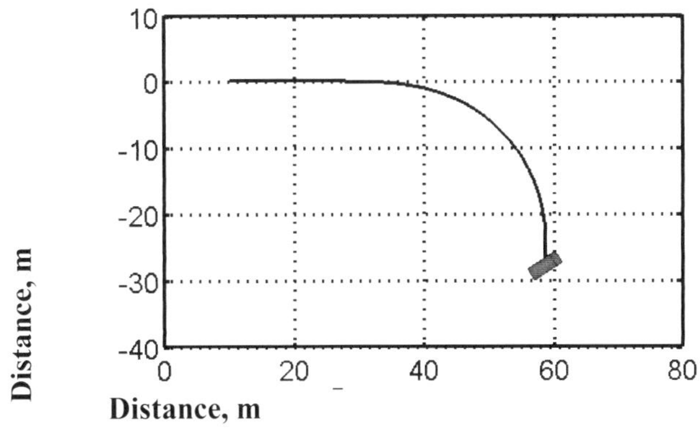

Fig. 1. The trajectory of the rear-wheel drive car in a turn on ice

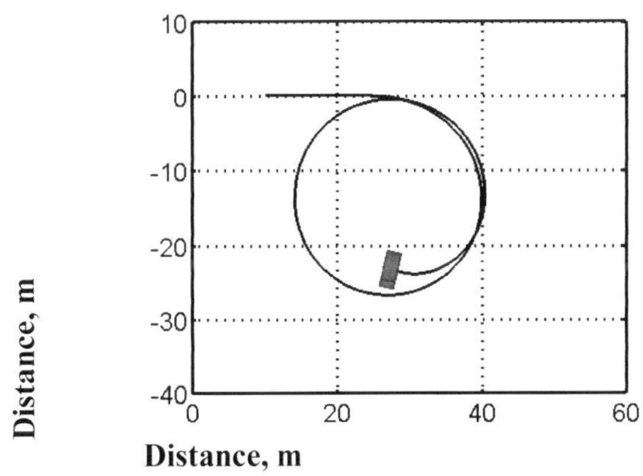

Fig. 2. The trajectory of movement on the ice of a 4x4 car with a plug-in front axle without redistributing torques between the driving wheels of the same axis when making a turn

Analysis of the driving paths of a 4x4 vehicle with a plug-in front axle without redistributing torques between the driving wheels of the same axle (Fig. 2) shows that when performing a maneuver on the "ice with snow" support surface, the car loses control (the rear axle skid develops), although much later than in a rear-wheel drive car.

The trajectory of movement on the ice of a 4x4 vehicle with a plug in front axle with the redistribution of torques between the driving wheels of the rear axle when making a turn is shown in Fig. 3.

When driving a car with a plug-in front axle and redistributing torque between the wheels of the rear axle in the same driving conditions (Fig. 3) the car remains manageable, small deviations from the trajectory can be compensated by steering.

The path of movement on ice of a 4x4 vehicle with a plug in front axle and with an automatically locked front wheel differential when making a turn is shown in Fig. 4.

When driving a vehicle with a plug-in front axle and an automatically locked front-wheel differential under the same driving conditions (Fig. 4) the car remains manageable, small deviations from the trajectory can be compensated by steering.

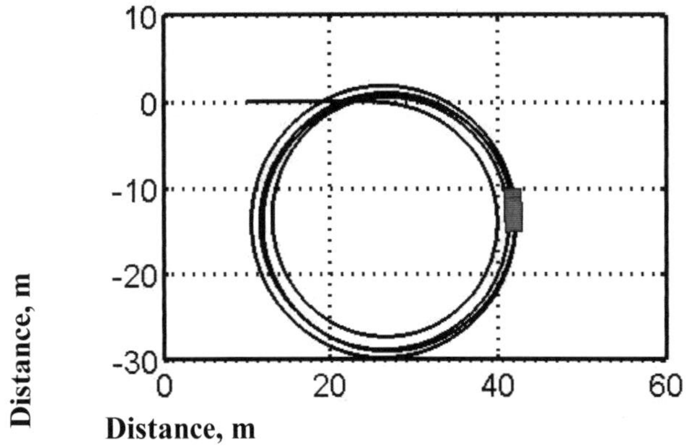

Fig. 3. The trajectory of a 4x4 vehicle with a plug-in front axle and the redistribution of torque between the wheels of the rear axle on the ice in a turn

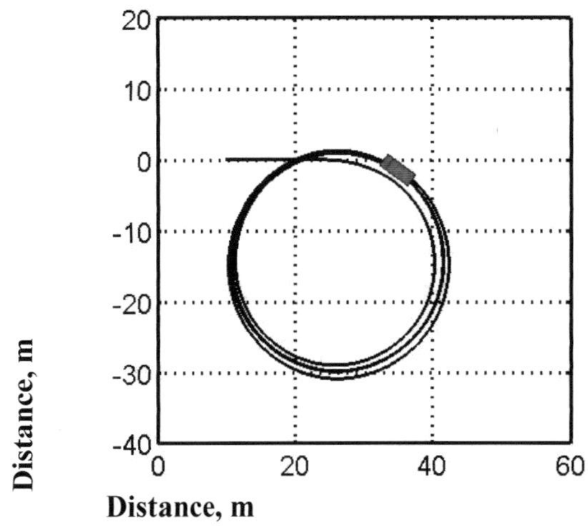

Fig. 4. The trajectory of a 4x4 vehicle with a plug-in front axle and an automatically locked front-wheel differential on ice in a turn

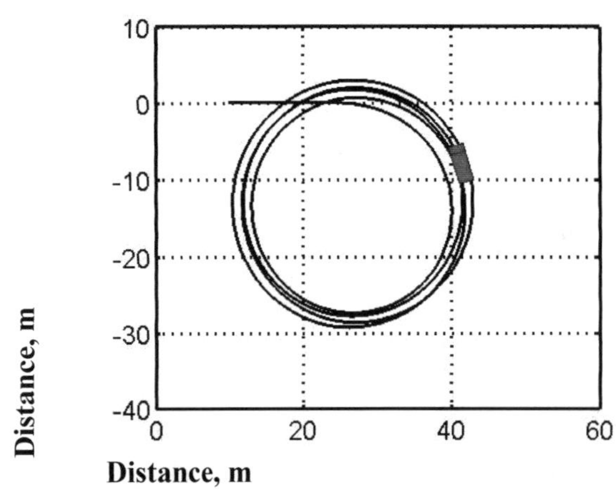

Fig. 5. The trajectory of a 4x4 vehicle with a plug-in front axle and individual braking system for individual wheels on the ice in a turn

The path of movement on the ice of a 4x4 vehicle with a plug-in front axle and with a system of individual braking of individual wheels when making a turn is shown in Fig. 5.

When driving a car with a plug-in front axle and individual braking system for individual wheels in the same driving conditions (Fig. 5) the car remains manageable, small deviations from the trajectory can be compensated by steering.

Table 1 shows the values of the *KE* performance criterion for traffic stabilization algorithms for the vehicles with a plug in front axle discussed above.

Table 1. Values of the ke performance criterion for traffic stabilization algorithms for vehicles with a plug-in front axle

№	Method of motion stabilization	*KE*, m/s
1	Rear-wheel drive car without torque control on the driving wheels (base for comparison)	104,3
2	A 4x4 car with a plug-in front axle without redistributing torques between the driving wheels of the same axle	4,18
3	A 4x4 vehicle with a plug-in front axle that redistributes torques between the driving wheels of the rear axle	0,52
4	A 4x4 vehicle with a plug-in front axle and an automatically locked front-wheel differential	0,48
5	4x4 vehicle with a plug-in front axle with individual braking system for individual wheels	0,55

The analysis of the effectiveness of the 4x4 vehicle traffic stabilization system due to the redistribution of torques between the driving axles and the wheels allowed us to establish that the following stabilization methods are most effective for cars with a plug-in front axle:

1) a 4x4 vehicle with a plug-in front axle with a reallocation of torques between the driving wheels of the rear axle;
2) a 4x4 vehicle with a plug-in front axle and an automatically locked front-wheel differential;
3) 4x4 vehicle with plug-in front axle with individual braking system for individual wheels.

The effectiveness of these stabilization methods is approximately the same.

Conclusions

The analysis of the effectiveness of the 4x4 vehicle traffic stabilization system by redistributing torques between the driving axles and the wheels carried out by simulation methods allowed us to establish that the following stabilization methods are the most effective for cars with a plug-in front axle:

- a 4x4 vehicle with a plug-in front axle with a reallocation of torques between the driving wheels of the rear axle;
- a 4x4 vehicle with a plug-in front axle and an automatically locked front-wheel differential;
- 4x4 vehicle with plug-in front axle with individual braking system for individual wheels.

The effectiveness of these stabilization methods is approximately the same.

References

[1] Diez, D.R., Velenis, E., Tavernini, D., Smith, E.N., Siampis, E., Soltani, A. Front/Rear Axle Torque Vectoring Control for Electric Vehicles (2019) Journal of Dynamic Systems, Measurement and Control, Transactions of the ASME, 141.

[2] KarogalI., AyalewB. Independent Torque Distribution Strategies for Vehicle Stability Control, in World Congress of the Society of Automotive Engineers. 2009, SAE, Inc.: Detroit, MI.

[3] Osborn R.P. Shim T. Independent Control of All-Wheel-Drive Torque Distribution. SAE paper number 2004-01-2052, 2004.

[4] Wu, Z., Xie, B., Chi, R., Du, Y., Mao, E. Active modulation of torque distribution for dual-motor front- and rear-axle drive type electric vehicle based on slip ratio (2018) Nongye Gongcheng Xuebao/Transactions of the Chinese Society of Agricultural Engineering, 34 (15), pp. 66–76.

[5] Zhao, B., Xu, N., Chen, H., Guo, K., Huang, Y. Stability control of electric vehicles with in-wheel motors by considering tire slip energy (2019) Mechanical Systems and Signal Processing, 118, pp. 340–359.

[6] Xu, W., Chen, H., Zhao, H., Ren, B. Torque optimization control for electric vehicles with four in-wheel motors equipped with regenerative braking system (2019) Mechatronics, 57, pp. 95–108.

[7] Song, Y., Shu, H., Chen, X., Luo, S. Direct-yaw-moment control of four-wheel-drive electrical vehicle based on lateral tyre-road forces and sideslip angle observer (2019) IET Intelligent Transport Systems, 13 (2), pp. 356–366.

[8] Zhao, X., Guo, G. Braking torque distribution for hybrid electric vehicles based on nonlinear disturbance observer (2019) Proceedings of the Institution of Mechanical Engineers, Part D: Journal of Automobile Engineering.

[9] J. Jung, S. Taehyun, J. Gertsch. Vehicle full-state estimation and prediction system using state observers. IEEE Transactions onVehicularTechnology,vol.58,no.8,pp.4078–4087,2009.

[10] M. Hiemer, A. von Vietinghoff, U. Kiencke, and T. Matsunaga, "Determination of the vehicle body slip angle with non-linear observer strategies," in Proc. SAE World Congress, Detroit, MI, 2005, paper no. 2005-01-0400.

[11] Zhileykin M.M., Eranosyan A.V. Improving Stability and Controllability of Two-Axial Vehicles with a Connectable Front Axle by Redistributing Torque between the Axles (2018) BMSTU Journal of Mechanical Engineering, 4 (697), p.p. 35–41.

[12] Ankinovich G.G., Verzhbitskiy A.N., Zhileykin M.M. A Method of Determining Motion Parameters of Two-Axle Wheeled Vehicles to Ensure the Operation of the Dynamic Stability System. (2017) BMSTU Journal of Mechanical Engineering, 4 (685), p.p. 11–20.

[13] Belousov, B., Ksenevich, T.I., Vantsevich, V., Komissarov, D. 8×8 platform for studing terrain mobility and traction performance of unmanned articulated ground vehicles with steered wheels (2013) SAE Technical Papers, 9.

[14] Gorelov, V.A., Komissarov, A.I., Miroshnichenko, A.V. 8×8 wheeled vehicle modeling in a multibody dynamics simulation software (2015) Procedia Engineering, 129, pp. 300–307.

[15] Keller, A.V., Gorelov, V.A., Vdovin, D.S., Taranenko, P.A., Anchukov, V.V. Mathematical model of all-terrain truck (2015) Proceedings of the ECCOMAS Thematic Conference on Multibody Dynamics 2015, Multibody Dynamics 2015, pp. 1285–1296.

[16] Gorelov, V.A., Komissarov, A.I. Mathematical model of the straight-line rolling tire — Rigid terrain irregularities interaction (2016) Procedia Engineering, 150, pp. 1322–1328.

[17] Vol'skaya, N.S., Zhileykin, M.M., Zakharov, A.Y. Mathematical model of rolling an elastic wheel over deformable support base (2018) IOP Conference Series: Materials Science and Engineering, 315 (1), article № 012028.

[18] Wong, J.Y. Theory of Ground Vehicles / J.Y. Wong. — New York: Wiley IEEE, 2001. — 560 p.

[19] H. Fennel, R. Gutwein, A. Kohl, M. Latamik, G. RollDas modulare Regler- und Regelkonzept beim ESP von ITT Automotive 7. Aachener Kolloquium Fahrzeug- und Motortechnik, 5.–7. Oktober, Aachen (1998), pp. 409–431

[20] K. LangwiederMit ESP schwere Unfalle vermeiden oder mildern ESP-Workshop, November 10, Boxberg, Germany (1999).

A multi-body dynamic model of the snowmobile to define acting loads in the its parts

D Vdovin[1], Y Levenkov[1,2], I Chichekin[1] and A Fominykh[1]

[1]Bauman Moscow State Technical University, 5 Second Baumanskaya Street, Moscow, 105005, Russian Federation

[2]E- mail: levenkov_yy@bmstu.ru

Abstract. In this paper, the snowmobile multi-body dynamic (MBD) model is presented. Mass reduction of the snowmobile parts is the important problem in the process of creating a new vehicle. It is important to determine loads to create rational designs of parts. There are various modeling methods for determining loads. More convenient method is using the MBD model. Research objective: development of the snowmobile MBD model, taking into account the main features of its design. It is possible to perform optimization and strength calculations of snowmobile parts using defined loads at early stages of the development.

Introduction

The snowmobile is the main vehicle in snow-covered regions such as northwest of Russia, Siberia, Nordic countries, Canada etc. It's designed to be operated on snow and ice and does not require a road or trail, but most are driven on open terrain or trails. For some regions, the snowmobile is the main vehicle for transporting people and goods.

It's known that mass reduction of the snowmobile parts significantly improves the driving performancesuch as the ride comfort, average speed, fuel efficiency, weights of cargo. Definition of acting loads is the main problem that occurs in the early stages of the development of the vehicle. Development of the snowmobile model has features that are due to the design [1–16].

The presented MBD snowmobile model takes into account main features of the design (front and rear suspensions, steering). Static and quasi-dynamic loading modes are considered. The load calculation is performed using the MBD program [17–26].

Mathematical model of the snowmobile

Dynamic model of the snowmobile «TAYGA PATRUL 551» is shown in Figure **.1**. This model was performed in the MBD program. Mathematical model of the snowmobile includes: front and rear suspensions, steering, frame. Also inertia mass and mass parameters of its parts are taken into account.

In the model, the following assumptions are made:

– parts of the snowmobile are rigid;
– calculations are performed for gross snowmobile weight;
– joints haven't internal friction;
– the liquid occupies the entire volume of the fuel tank;
– center of gravity of the snowmobile is constant;
–spring characteristics are linear;

–contact surface is rigid;
– shock absorber characteristics are taken into account.

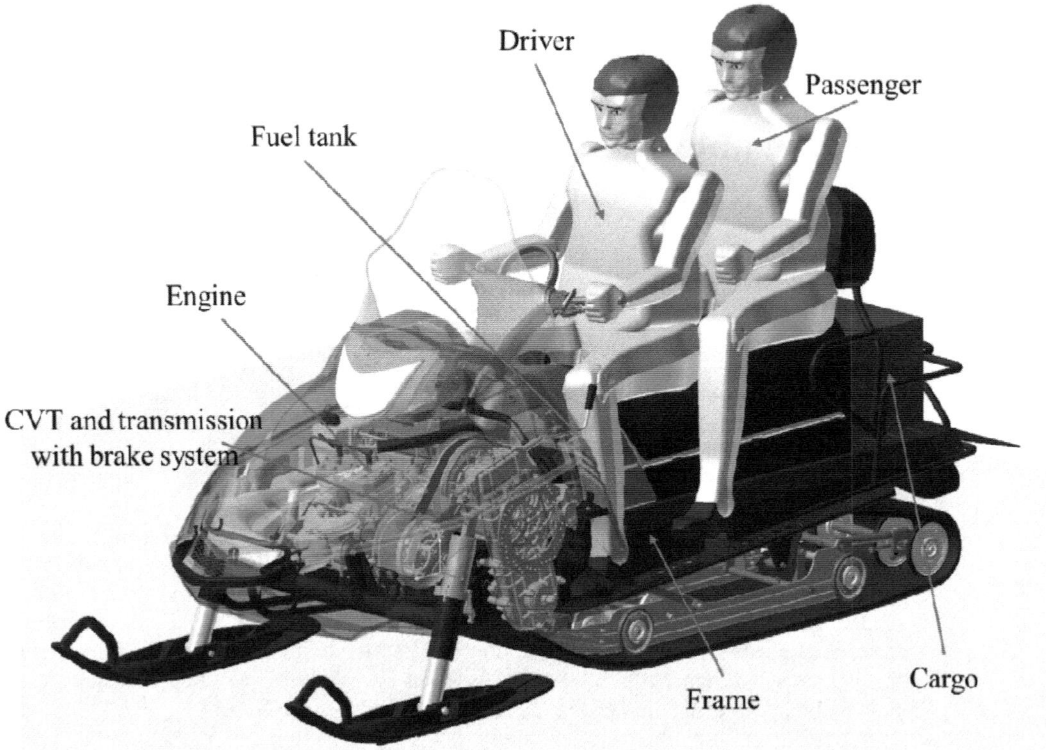

Figure.1.MDB model of the snowmobile

	Snowmobile specifications:
Gross vehicle weight	600 kg
	Gross vehicle weight distribution:
Frame weight	61 kg
Fuel tank weight	41 kg
CVT and transmission with brake system	38 kg
Driver weight	80 kg
Passenger weight	80 kg
Cargo weight	85 kg
	Front suspension:
Type	Dependent suspension with a spring strut
Spring type	Coil spring
Damper	Hydraulic shock absorber
	Rear suspension:
Type	Multi-link suspension with parallel rail designs
Spring type	Coil spring
Damper	Hydraulic shock absorber

The front suspension model is shown in Figure 2. The suspension model includes: the spring strut model that was made via spring-damper elements, the interaction of the ski and surface was modeled via contact surface-to-surface. Rotation of the spring strut relative to its axis was modeled via the revolute joint. Axial movement was modeled via the inline joint.The ski and spring strut connection is made using a bushing joint.

The rear suspension model is shown in Figure 3 and 4. All elements of the rear suspension are interconnected using revolute joints. The upper and lower arms are connected to the snowmobile frame using bushing joints. This method of attachment allows you to automatically transfer the loads at the attachment nodes of the FEM of the frame. Rollers are connected to track using revolute joints. The interaction of the track and surface was modeled using contact surface-to-surface.

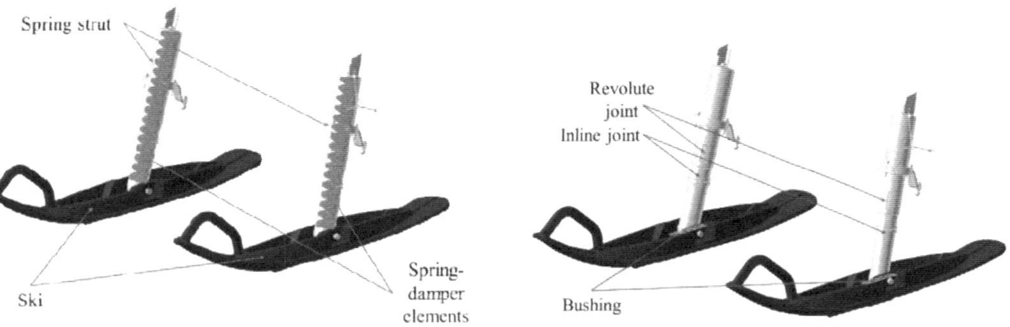

Figure.2. General elements of the front suspention

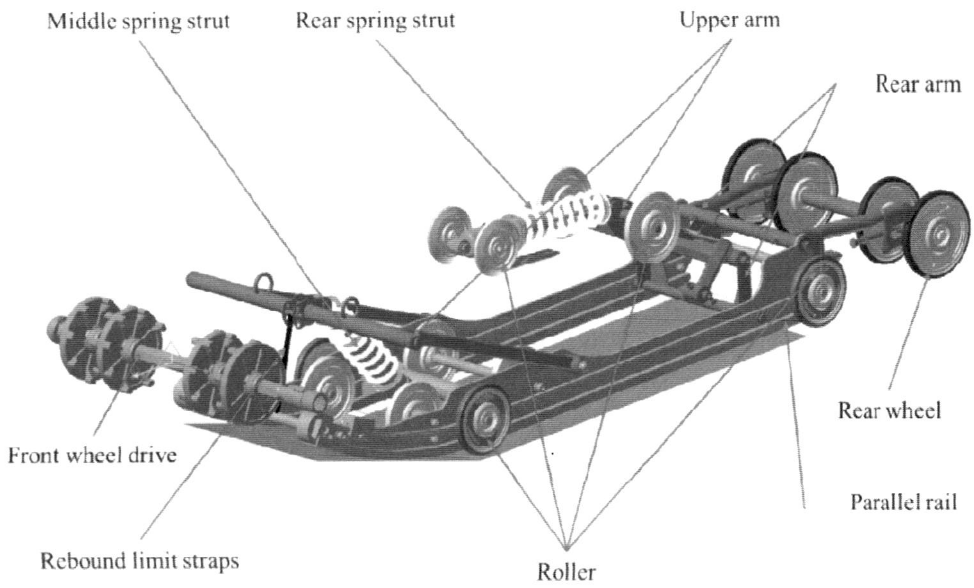

Figure 3. General elements of the rear suspention

The steering model is shown in Figure 5. The steering shaft is motionless connected to the bipod, and is connected to the frame using a rotation joint. The connection of the steering bipod and the suspension struts is made using the distance joint.

Powertrain model consists of a power unit and transmission. The power unit includes an engine, a CVT (continuously variable transmission) and a transfer case. The CVT, transfer case gear ratios and engine's torque-speed curve are defined in the model by means of an external DLL file. The engine torque is calculated as a function of the transmission rotation speed and accelerator pedal rate and is applied to the front and rear final drives, modelled as rigid bodies with gear coupling ratios. Final drives models also include a locking differential. Torque-speed curves for the power unit and CVT unit [27].

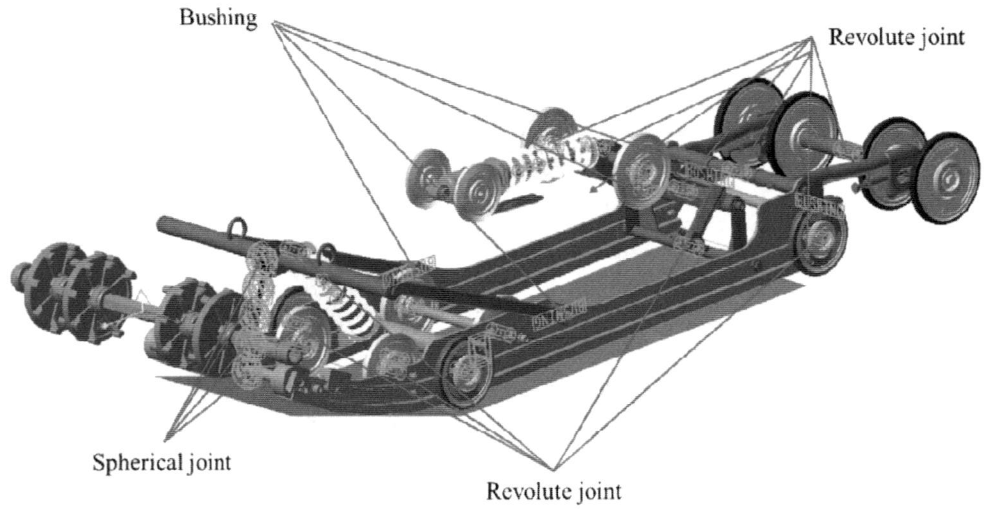

Figure4. Componentsofthemathematicalmodelofthe suspension

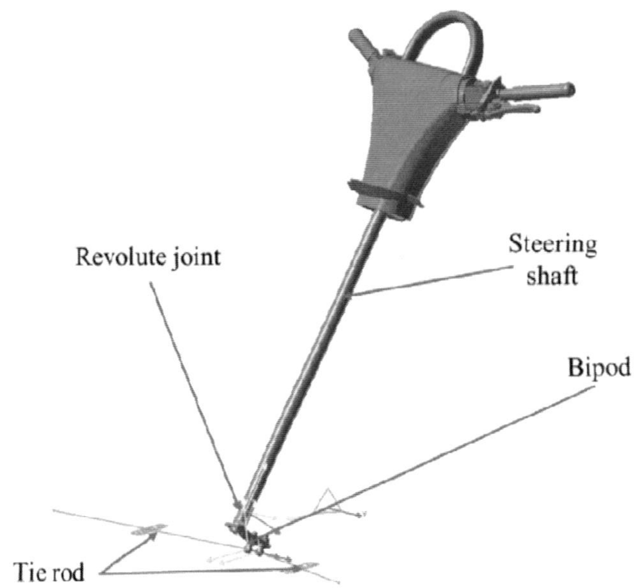

Figure5. Componentsofthemathematicalmodelofthe steering

The frame model is shown in Figure 6. The considered elements of the snowmobile model are connected to the frame.

The developed MBD model allows to simulate statics and dynamics loading cases. In this article is presented two cases: the static equilibrium on a flat road (a reference load event); the snowmobile jump (dynamic event).

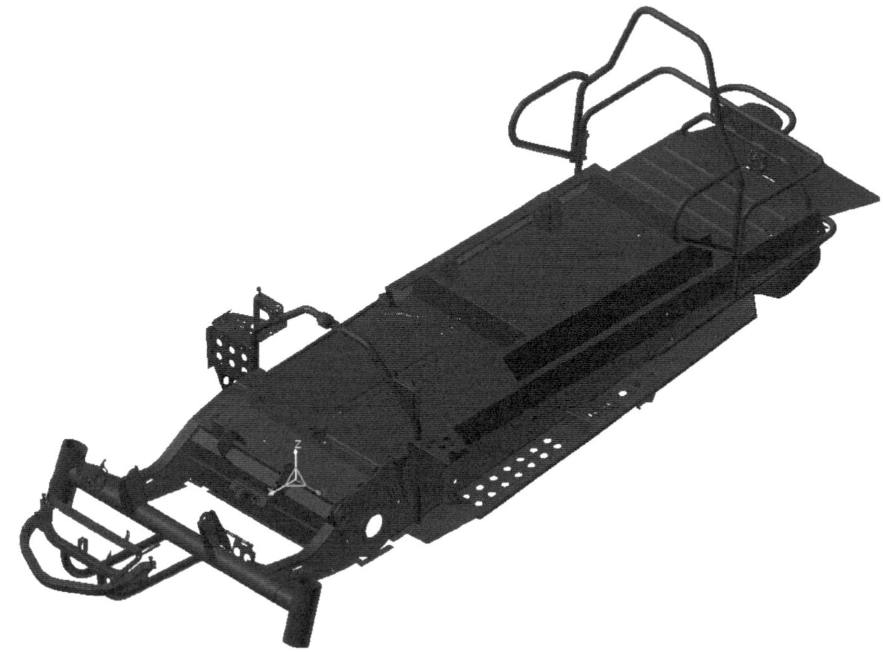

Figure6. The snowmobile frame

Estimation of the results

The static equilibrium on a flat road (Figure 7) and snowmobile jump simulations were performed (Figure 9).

As a result of simulation event produces dynamic the force time history in the suspension, frame, steering joints, etc. As an example, force vs. time diagrams in the strut revolute joint of the front left suspension is shown Figure 8 and 10. Analysis of the force time histories allows to determine the maximum loads to perform strength analysis of the snowmobile parts. Often, one event can produce several load cases.

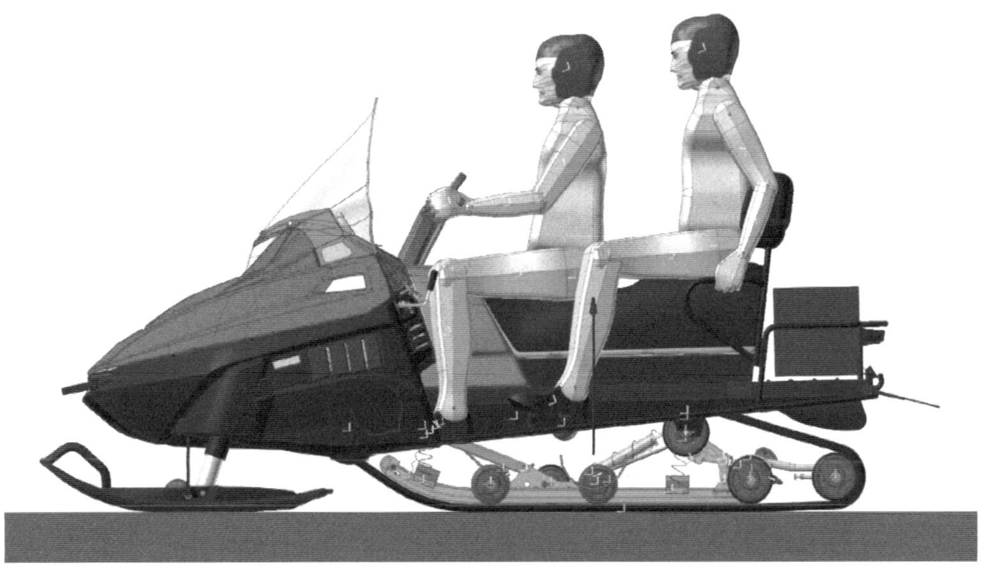

Figure7. The static equilibrium on a flat road simulation

Besides the maximum loads that determine ultimate strength of the snowmobile parts, it is also possible to simulate regular vehicle rides. These time histories can be used for further fatigue calculations of the snowmobile parts.

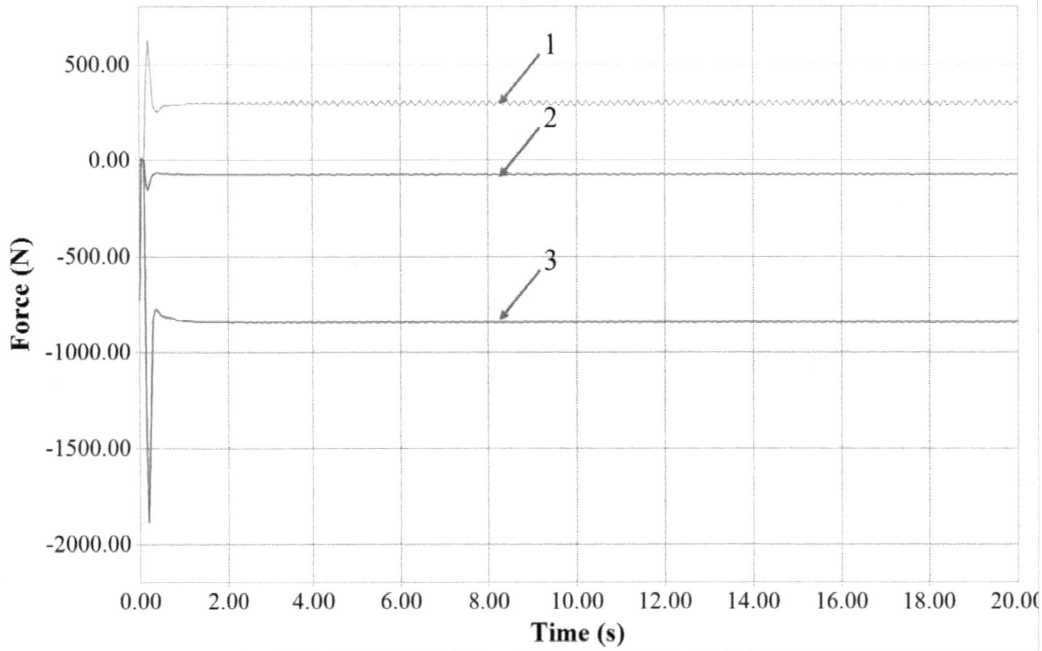

Figure8. Force magnitude vs time in the strut revolut joint of the front left suspension for the static event: — Fx — traction force, — Fy — lateral force, –Fz — vertical reaction

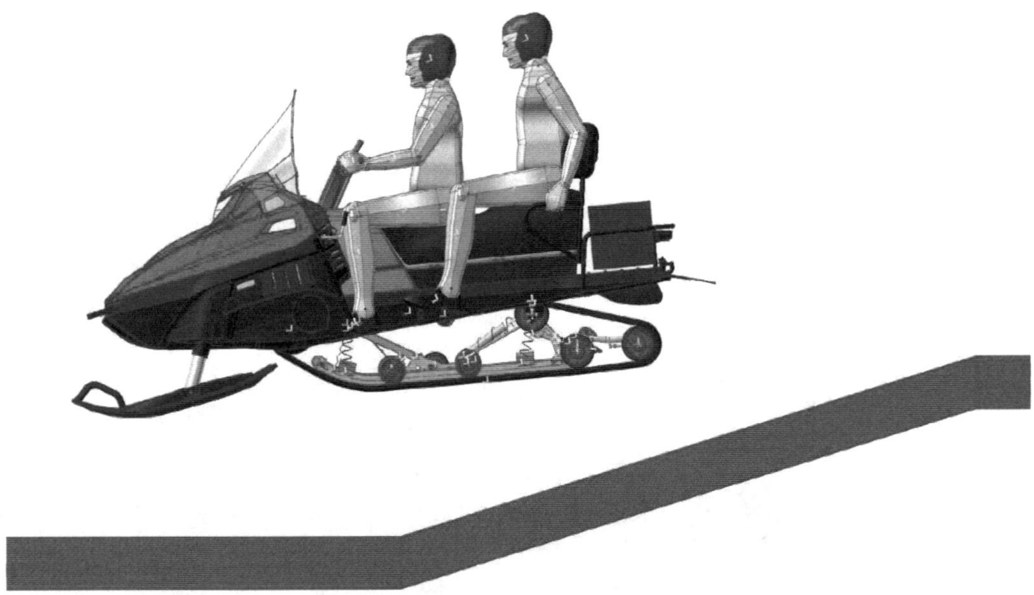

Figure9. The snowmobile jump event simulation at a**time = 13.55 s**

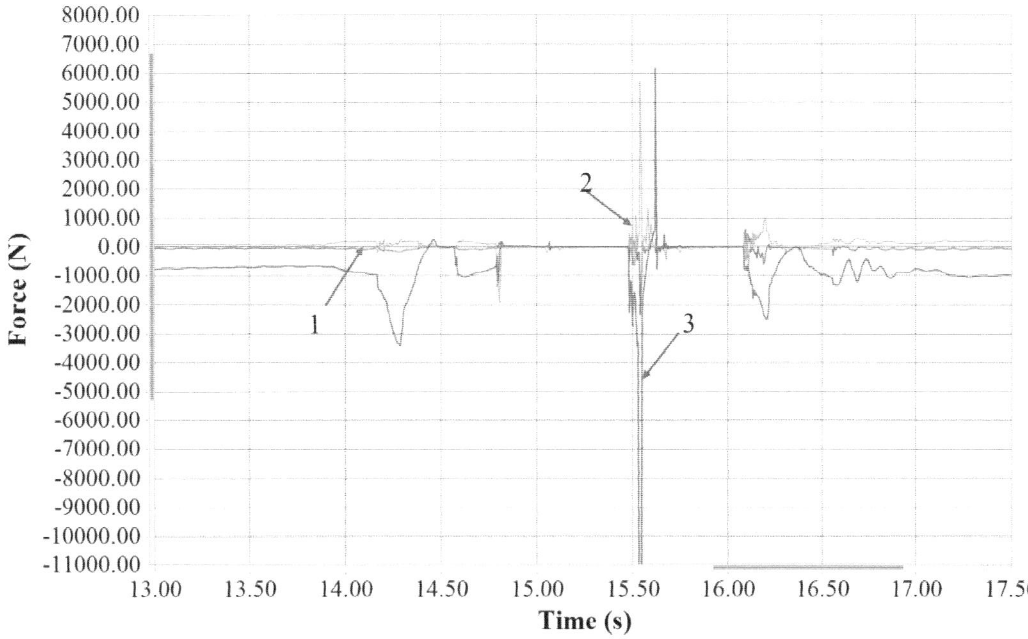

Figure10. Force magnitude vs time in the strut revolut joint of the front left suspension for the snowmobile jump event: — Fx — traction force, — Fy — lateral force, –Fz — vertical reaction

Conclusions

1. Analysis of the time histories of the forces in the snowmobile parts allows detection of the quasi-static load cases with maximum forces during the event.
2. This maximum loads can be transferred to finite element analysis software for the strength estimation of the parts.
3. Calculated loads can also be used in design optimization process, like topology optimization, at early stages of the snowmobile development.
4. The time histories of the forces can be used for fatigue life analysis of the snowmobile parts.

Acknowledgments

It should be noted that this work was carried out at the Bauman Moscow State Technical University, with financial support from the government in the face of the Russian Ministry of Education under the project: №14.577.21.0272. (Identification number: RFMEFI57717X0272).

References

[1] Hashemi E, Pirani M, Khajepour A, Kasaiezadeh A 2016. *Vehicle System Dynamics***54** 1736–61
[2] Wei Y, Liu Y, Li X, Oertel C 2016.*Vehicle System Dynamics***54** 463–73
[3] Taheri M, Ahmadian M 2016.*Vehicle System Dynamics***54** 653–66
[4] Rodriguez J, Freeman P T, Wagner J, Pidgeon P, Alexander K 2016.*Int. J. of Automotive Technol.***17** 71–81
[5] Xia X, Xiong L, Sun K, Yu Z P 2016. *Int. J. of Automotive Technol.***17** 991–1002
[6] Diakov A S, Kotiev G O 2018. *MATEC Web of Conf.***224** 02096
[7] Evseev K B, Kartashov A B, Dashtiev I Z and Pozdeev A V 2018. *MATEC Web of Conf.* **224** 02039
[8] Kotiev G O, Padalkin B V, Kartashov A B, Diakov A S 2017. *ARPN J. of Eng. and Appl. Sci.* **12** 1064–71.
[9] Sarach E B, Kotiev G O, Beketov S A 2018. *MATEC Web of Conf.***224** 04009

[10] Klubnichkin V E,Diakov A S, Klubnichkin E E, Zakharov A Y, Vakhidov U Sh, Suchenina A S and Basmanov I V 2019.J. of Phys.: Conf. Series **1177** 012048
[11] Volskaya N S, Zhileykin M M and Zakharov A Y 2018. *IOP Conf. Series: Materials Science and Engineering***315** 012028
[12] Klubnichkin E E, Klubnichkin V E, Kotiev G O 2018. *IOP Conf. Series: Materials Science and Engineering***386** 012025
[13] Zhileykin M M,Kotiev G O, Nagatsev M V 2018. *IOP Conf. Series: Materials Science and Engineering***315**012031
[14] Kotiev G O, Butarovich D O, Kositsyn B B 2018. *IOP Conf. Series: Materials Science and Engineering***315** 012014
[15] Skotnikov G I, Jileykin M M and Komissarov A.I 2018. *IOP Conf. Series: Materials Science and Engineering***315** 012027
[16] Ejsmont J, Taryma S, Ronowski G, Swieczko-Zurek B 2016. *Int. J. of Automotive Technol.* **17** 237
[17] Vdovin D, Chichekin I 2016. *Procedia Engineering***150** 1276–79.
[18] Vdovin D S, Chichekin I V, Levenkov Y Y 2018.*Trudy NAMI***1** 36
[19] Vdovin, D S, Chichekin I V, Levenkov Y Y, Shabolin, M L 2019. *IOP Conf. Series: Materials Science and Engineering* **534(1)** 012024
[20] Vdovin D, Chichekin I and Ryakhovsky O 2019. *IOP Conf. Series: Materials Science and Engineering* **589(1)** 012025
[21] Volskaya N S, Chichekin I V 2019. *IOP Conf. Series: Materials Science and Engineering* **534(1)** 012022
[22] Lee Y.-L, BarkeyM E, Kang H.-T 2011 *Metal fatigue analysis handbook. Practical problem-solving techniques for computer-aided engineering* (Oxford, Elsevier) p 581
[23] Gorelov V A, Komissarov A I 2016. *Procedia Engineering***150** 1322–28.
[24] Gorelov V A, Komissarov A I, Miroshnichenko A V 2015. *Procedia Engineering***129** 300–7
[25] Levenkov Y Y, Vol'skaya N S, Rusanov O A 2019. *IOP Conf. Series: Materials Science and Engineering* **534(1)** 012023
[26] Keller A V, Gorelov V A, Anchukov V V 2015. *Procedia Engineering***129** 280–7.
[27] Vdovin D, Levenkov Y and Chichekin V 2019. *IOP Conf. Series: Materials Science and Engineering* **589(1)** 012026

Mathematical modeling of autonomous single-stage electro-hydraulic servo drive

N Sosnovsky[1,2] and Nguyen Van Hoa[1]

[1]Bauman Moscow State Technical University, 5 Second Baumanskaya Street, Moscow, 105005, Russian Federation

[2]E- mail: sosn@bmstu.ru

Annotation. The study is devoted to mathematical modeling of an autonomous single-stage electro-hydraulic servo drive. Two structural schemes of the drive are considered — simplified and more complex. The drive transient conditions of these two circuits are given.

Introduction

Currently, hydraulic steering machines are widely used in various technical sectors, including those used in aviation and space-rocket hardware. This circumstance is due to several advantages of hydraulic drives in comparison with electric and gas:

- high level of working fluid pressure (up to 56 MPa and more),
- low compressibility of the working fluid,
- significant energy intensity (specific power),
- high speed (dynamic properties),
- relatively small mass and size of hydraulic units.

The disadvantages of hydraulic drives include the complexity of installation, operation and the high cost of a hydropower source, the influence of temperature on the characteristics, technological limitations.

Steering machines according to the type of energy used are divided into manual, gas, electric and hydraulic. Various combinations of these basic types are often used, for example electro-hydraulic. [1]

A steering engine is an electro-hydraulic unit consisting of a hydraulic motor, a hydrodistributor and an electro-hydraulic power steering and designed to transmit, amplify and convert an electrical signal containing control information into the mechanical movement of the rudders or controls. [2]

According to the type of control signal, the hydraulic power actuators are divided into amplifiers with a mechanical and electrical control signal. There are various types of feedbacks in electro-hydraulic servo mechanisms. Rigid feedback on the output coordinate is widely used, which has high tracking accuracy and stability against oscillations.

According to the design of the control element, the hydraulic power actuators is divided into amplifiers with throttling valves of the spool type; type nozzle-flapper; with jet tube; rotary (crane) type; with needle throttle. The most widespread are spool-type hydraulic power steering with a throttle nozzle, and to a lesser extent, with a jet tube.

According to the number of cascades of amplification, the hydraulic power amplifiers is divided into single, double and multi-stage. Single-stage power steering is used when the output requires a relatively small power. In them, the control signal acts on the movable element of the control device. If you want to create more power at the output, then the power steering is included in the hydraulic system with high pressure and flow rate of the working fluid. This leads to an increase in the design dimensions of the power actuator and the power of the control signal, as well as to a decrease in the sensitivity of the power steering. In this case, it is more advantageous to use two- or multi-stage hydraulic power amplifiers in which the power steering of the first amplification stage with a low flow rate and liquid pressure controls the operation of the power steering of the second stage with a higher flow rate and liquid pressure. Thanks to the use of two- and multi-stage amplifiers, it becomes possible to increase the output power with a low-power control signal. [3]

At present, analog electrohydraulic steering machines are widely used as executive bodies of aircraft motion control systems.

A single-stage steering machine is one of the simplest and most common steering machines and is often used to control the direction of the engine relative to two axles [4].

The steering machine consists of a casing, with an electric motor mounted on it and a pump located inside the casing, a control unit, a power mechanism with a feedback sensor. The output rod of the power mechanism moves the controls in accordance with the commands of the control system.

The operation principle of the steering machine (SM) will be considered according to the electro-hydraulic circuit diagram (Fig. 1).

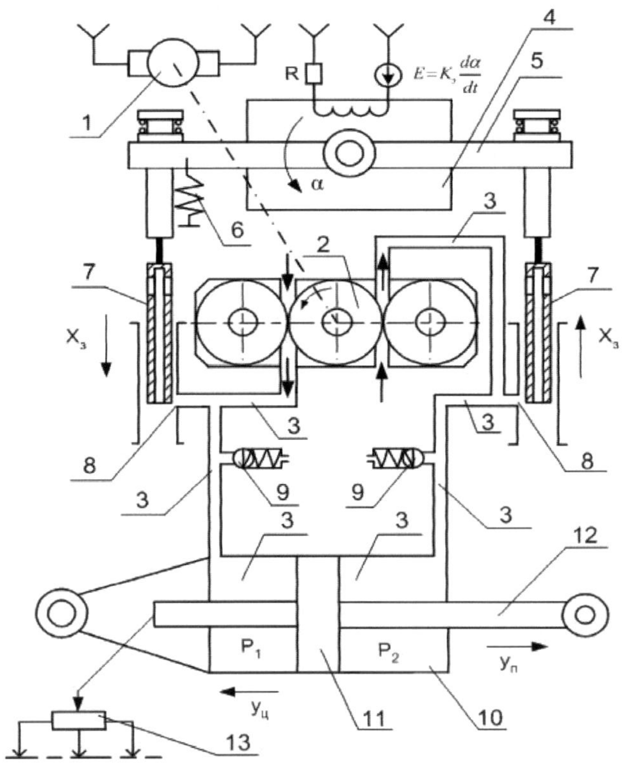

Fig. 1. Schematic diagram of the drive

The electric motor 1 drives the three-gear pump 2, which creates flows of the working fluid directed to the spool plungers 7. In the absence of a signal supplied from the electronic power steering, the liquid flows through the windows opened by the spool plungers. Due to the equality of the area of the windows, the pressure difference in the cavities of the hydraulic cylinder 10 is zero and the piston 11 together with the rod 12 are stationary. In the presence of a signal in the form of a voltage at the ends of the winding of an electromechanical converter (EMC), the rocker 5 rotates clockwise or counterclockwise, depending on the polarity of the signal. The rotation of the rocking chair causes the movement of the spool plungers, which increase the opening of one window and reduce the opening of another. Accordingly, the pressure in one cavity of the hydraulic cylinder decreases, and in the other it increases. Under the action of the force created by the pressure difference in the hydraulic cylinder, the piston 11 moves until the feedback signal from the sensor 13 does not compensate for the input signal. Safety valves 9 installed on pressure lines of the pump limit the excessive pressure increase in the hydraulic cylinder. Elements 3 to 8 form a single-stage hydraulic power steering (HPS) [5].

1. The mathematical model of the hydraulic drive

In the modeling process, the following assumptions were made [5 - 7].

- Since the range of working fluid pressure changes in the cavities of the SM is small ($< 50\ kgF / cm^2$), the dependences of the density, viscosity and bulk modulus of the working fluid on pressure are not taken into account, and their values are assumed to be constant.
- Due to the short duration of the studied processes, the temperature of the working fluid is assumed to be constant.
- Since there is a temperature compensator in the drain cavity of the SM case, the pressure in the drain cavity of the SM is assumed to be constant and equal to atmospheric
- Since the Reynolds number of working fluid flows in the extension orifices of the SM spool hydraulic power steering exceeds 2000 (turbulent flow regime), the flow coefficient of the extension orifices is assumed to be constant.
- Since the length of the connecting pipelines of the SM is not large, their hydraulic resistance is assumed to be small and not taken into account.
- The working edges of the spool plungers and throttle orifices of the sleeves spool hydraulic power steering are ideally sharp; the design of the hydraulic power steering is assumed to be symmetrical, and the dimensions of the extension orifices are exactly the same.
- The resulting non-stationary component of the hydrodynamic force due to the symmetrical design of the hydraulic power steering is assumed to be negligible and not taken into account.
- We neglect the hydrodynamic and hydrostatic forces acting on the spool plungers due to smallness
- Leaks of the working fluid through the safety valves of the SM are assumed to be small and are not taken into account in the mathematical model of the SM.
- The mass of the moving parts of the hydraulic cylinder is neglected in comparison with the reduced mass of the load.
- The force of internal friction in the hydraulic cylinder is taken into account in the sum of the resistance forces of the reduced load.
- The direction of movement of the body of the power hydraulic cylinder SM is back to the direction of movement of its piston.

1.1 The mathematical model of an electro-hydraulic power steering (EHPS)

The equation of the voltage balance in the electric circuit of an electromechanical converter (EMC) SM (in a polarized relay):

$$L \cdot \frac{dI_k}{dt} + K_e \cdot \frac{d\alpha}{dt} + R \cdot I_k = U_k, (1)$$

where: U_k— command voltage supplied to the EMC winding;

I_k— current flowing through the EMC winding; R— active resistance of the EMC winding;

L— inductance of the EMC winding; α— angle of rotation of the EMC roller;

K_e—coefficient of electromagnetic high-speed communication [V·s/rad].

Transform the equation according to Laplace:

$$\frac{I_k(T_1 \cdot s+1)}{1/R} = U - K_e \cdot s \cdot \alpha.$$

where $T_1 = \frac{L}{R}$.

The equation of the balance of moments reduced to the EMC shaft:

$$J \cdot \frac{d^2\alpha}{dt^2} + M^d_{cm} \cdot sign\left(\frac{d\alpha}{dt}\right) + K_{MV} \cdot \frac{d\alpha}{dt} + K_{mp} \cdot \alpha + R_\alpha \cdot F_g = \left(\frac{K_{Mi}}{K_p}\right) \cdot I_K;$$

where: J— moment of inertia of the moving parts of the EHPS reduced to the EMC shaft;
R_α— the arm of the power steering spool;
F_g— the resulting hydrodynamic force and the stationary component of the hydrodynamic force;
M^d_{cm} —the moment of dry friction of motion, reduced to the EMC;
K_{MV}— coefficient of viscous friction moment reduced to the EMC shaft;
K_{mp}— stiffness coefficient of the spring connected to the EMC shaft;
K_{Mi}— coefficient of steepness of the moment characteristic of the electromagnetic field;
K_p— coefficient of alignment of dimensions ($K_p = 0.0981$).

Applyingtheassumptions, weobtain

$$J \cdot \frac{d^2\alpha}{dt^2} + K_{mv \cdot e} \cdot \frac{d\alpha}{dt} + K_{Mn} \cdot \alpha = \left(\frac{K_{Mi}}{K_p}\right) \cdot I_K;$$

where: $K_{mv \cdot e}$ — equivalent coefficient of the moment of viscous friction.

Transform the equation according to Laplace:

$$\alpha \cdot \left(\frac{J}{K_{mp}} \cdot s^2 + \frac{K_{mv \cdot e}}{K_{mp}} \cdot s + 1\right) = \left(\frac{K_{Mi}}{K_p K_{mp}}\right) \cdot I_K; (2)$$

The equation of the relationship of the angular displacement of the EMC roller with the linear displacement of the spool plunger:

$$X_3 = R_\alpha \cdot \alpha, (3)$$

Based on equations (1–3), a structural diagram of the EHPS was constructed: [4, 5].

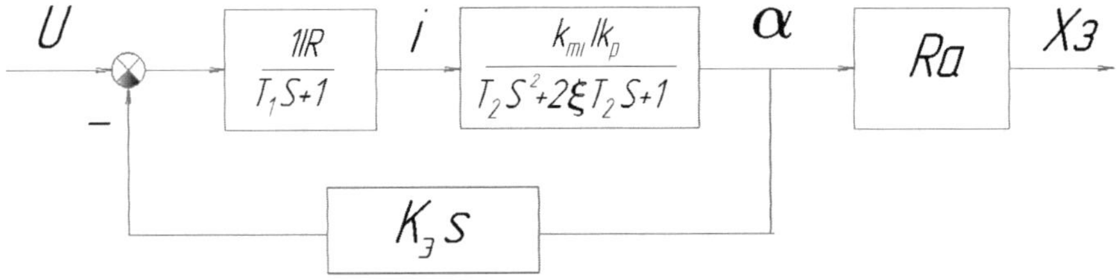

Fig.2. Block diagram of the EHPS

1.2 Linearized hydraulic drive model (simplified)

The equation of the family of flow-differential characteristics of the EHPS: [7]

$$Q = K_{QX} X_з - K_{Qp} \Delta p$$

where: $\Delta p = p_1 - p_2$— pressure difference between the cavities of the SM;
Q— low rate of the working fluid between the cavities of the SM.

The coefficients K_{QX} and K_{Qp} are determined by approximating the flow-differential characteristics.

The equation of the balance of costs when moving the piston of the power cylinder

$$Q = S_n \frac{dY_p}{dt} + \frac{V_0}{2E_c}\frac{d(\Delta p)}{dt}.$$

where: Y_p — displacement of the piston rod of the power hydraulic cylinder SM;

$E_c = \dfrac{B_g}{1 + \dfrac{V_l}{V_o} + 2 \cdot \dfrac{F_c^2 \cdot B_g}{V_o \cdot c_{op}}}$ — reduced modulus of elasticity of the hydraulic cylinder;

V_o —the effective volume of the cavity of the power hydraulic cylinder with an average piston position.

Equation of load motion

$$m_l \frac{d^2 Y_l}{dt^2} + K_{fr}\frac{dY_l}{dt} + c_l Y_l = c_{cv}(Y_h - Y_m)$$

where: c_l — positional load coefficient; K_{fr} — coefficient of viscous friction;

c_{cv} —coefficient of the place compliance where the rod of the SM power hydraulic cylinder is connected to the control object (load); m_l — equivalent load mass.

The equation of the balance of forces acting on the piston of the power cylinder SM

$$c_{cv}(Y_h - Y_m) = \Delta p S_n,$$

where $\dfrac{1}{c_\kappa} = \dfrac{1}{c_o} + \dfrac{1}{c_{cv}}$ — coefficient of the place compliance where the rod of the SM power hydraulic cylinder is connected to the control object (load);

Y_m — moving the control object (load).

From the previous equations we get:

$$T_{HD} s(T_c^2 s^2 + 2\xi_c s + 1) y_m(s) = X_з - K_l y_m(s),$$

$T_{hd} = \dfrac{F_c}{K_{Qx}}$ —hydraulic drive time constant,

$T_c = \sqrt{\dfrac{m}{c_c}}$ —cylinder time constant,

$c_c = \dfrac{2 \cdot F_c^2 \cdot E_c}{V_o\left(1 + \dfrac{2F_c^2 E_c}{V_0 c_{cv}}\right)}$ — reduced stiffness of the loaded hydraulic cylinder,

$\xi_c = \dfrac{T_{dc}}{2T_c}$ —coefficient of relative damping of the hydraulic cylinder,

$T_{dc} = \dfrac{K_{Qp} m}{F_c^2} + \dfrac{k_{mp} V_0}{2F_c^2 E_c} + \dfrac{k_{mp}}{c_{cv}}$ — time constant of relative damping,

$K_н = \dfrac{K_{Qp} c_н}{K_{Qx} F_c}$ —coefficient of internal feedback.

For small liquid leaks through the distributor and small displacements of the spool from the neutral position $K_н$, it is usually small and can be neglected

$$T_{HD} s(T_c^2 s^2 + 2\xi_c s + 1) y_m(s) = X_з \quad (4)$$

Feedback signal equation:

$$U_{oc} = K_{oc}(Y_h + Y_c).$$

We have $c_н$ □ c_{cv}, $c_н$ □ c_o,
From the equations we get:

$$U_{oc}(s) = K_{noc}\left(\frac{m}{c_K} + \frac{k_{mp}}{c_K} + 1\right)y_m(s),$$

$\frac{1}{c_K} = \frac{1}{c_O} + \frac{1}{c_{cv}}$ —flexibility of mounting the hydraulic cylinder

The value c_K is so great that it is possible to exclude the forcing link of the second order and accept $y_m = y_h$. Then we get:

$$U_{oc}(s) = K_{poc}y_m(s) \quad (5)$$

The voltage balance equation of the integrator:

$$U_e = U_{vx} - U_{oc}, \quad (6)$$

where: U_e — error signal.

Based on equations (1–6), a block diagram of the drive is constructed, shown in Fig. 3. [7, 8]

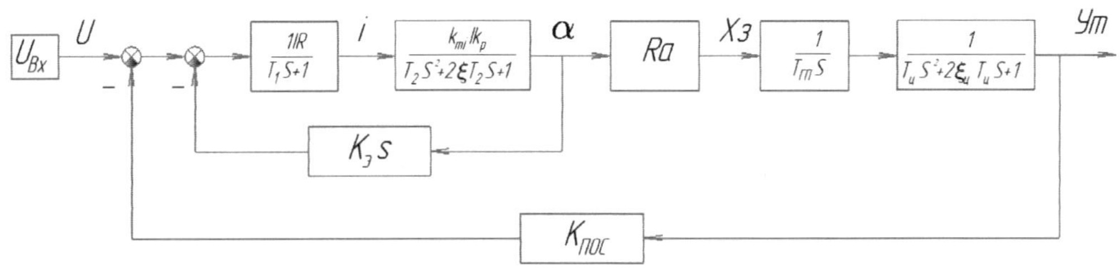

Fig.3. Block diagram of the drive

3. Sophisticated hydraulic drive model(taking into account the effect of external load and displacement of the support of the object)

Accepted assumption: the flow rate Q depends only on the displacement of the spool $X_з$, then

$$Q = K_{QX}X_з.$$

Equation of load motion

$$m_l\frac{d^2Y_l}{dt^2} = c_{cv}(Y_h - Y_m) - K_{mp}\frac{dY_l}{dt} - c_lY_l - F_l, \quad (7)$$

whereF_n — constantly acting force.

From the equation of the balance of forces acting on the piston of the power hydraulic cylinder, we obtain:

$$Y_h = \frac{\Delta pS_n}{c_{cv}} + Y_m \quad (8)$$

Substituting (8) in the equation of the balance of the flow rate we get:

$$Q = S_n\frac{dY_m}{dt} + \frac{V_0}{2E^н{}_c}\frac{d(\Delta p)}{dt}m, \quad (9)$$

where $E_c^{II} = \dfrac{B_g}{1+\dfrac{V_l}{V_o}+2\cdot\dfrac{F_c^2\cdot B_g}{V_o\cdot c_K}}$ — reduced modulus of elasticity of the hydraulic cylinder.

Based on equations (1–9), the structural diagram shown in Fig. 4 [5].

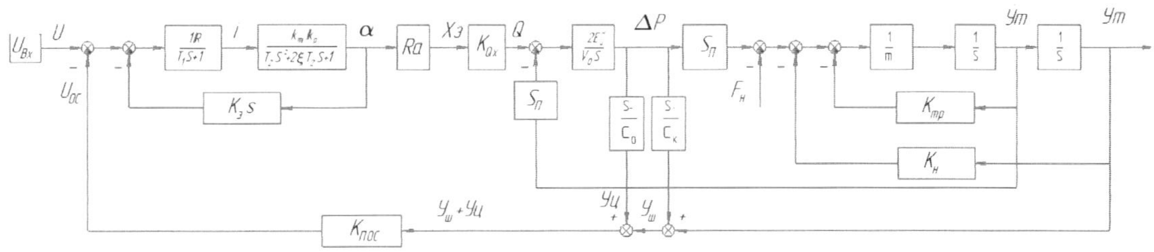

Fig.4. Block diagram of the drive

2. The results of mathematical modeling.

Using the BMSTU software complex, modeling was carried out, transient drive processes were obtained at various levels of the input signal.

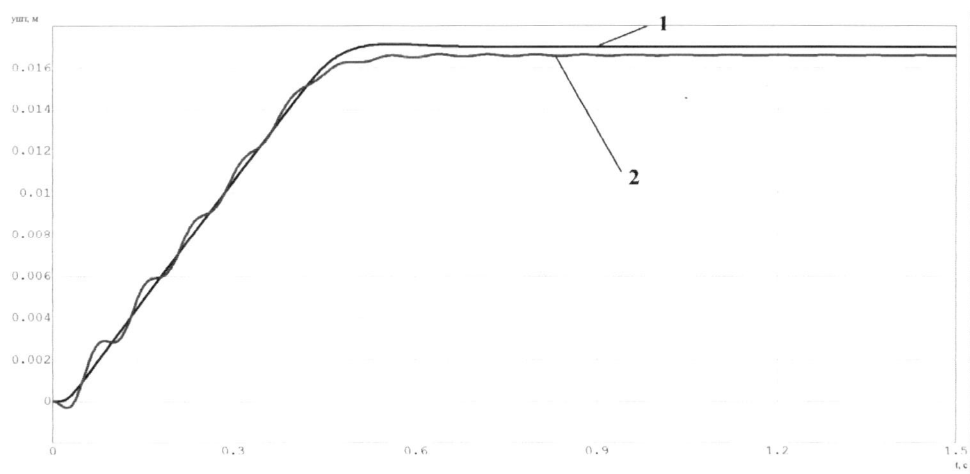

Fig.5. Transient drive. 1 - simplified model, 2 complex model (whenU_{BX}= U_{max}=12 V).

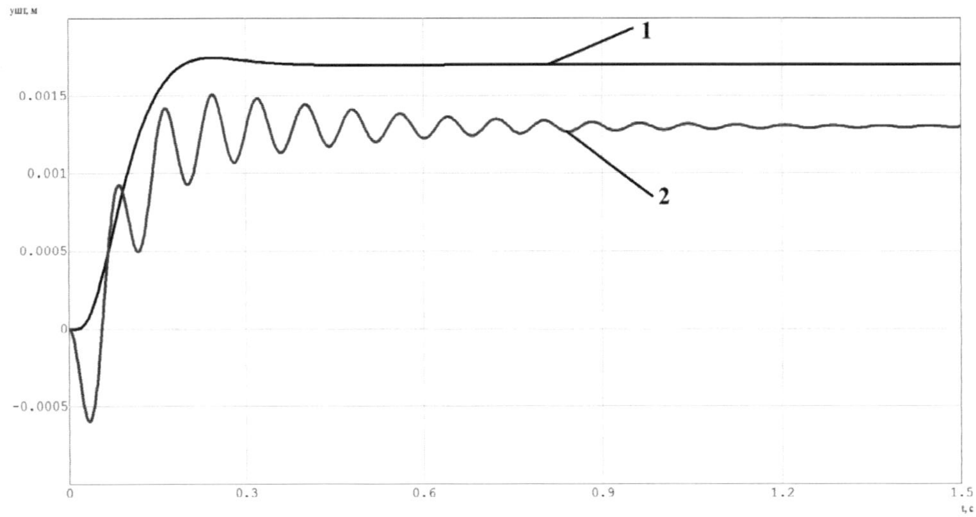

Fig.6. Transient drive. 1 - simplified model, 2 complex model
(when U_{BX}=10% Umax=1.2 V)

Conclusion

In the work, the parameters of an autonomous single-stage electro-hydraulic servo drive are calculated and a mathematical model of such a drive is created. Using the BMSTU software package, modeling was performed, and the dynamic characteristics of the drive were obtained. The results show that the mathematical model of the drive meets the requirements of the system and can be used for further research.

References

[1] Redko P.G., Ambarnikov A.V., Ermakov S.A., Karev V.I., Selivanov A.M., Trifonov O.N. Hydraulic units and flight control drives for aircraft. Moscow: Olita, 2004
[2] VasechkinYu.S., ObolenskyYu.G. Hydraulic drives of aircraft: Textbook. allowance. — M.: Publishing House of BMSTU, 2008
[3] Redko P.G. Increased uptime and improved performance of electro-hydraulic servo drives. M.: Janus-K, 2002
[4] Feodosiev V. I. Fundamentals of rocket flight technology. M .: Science. The main edition of the physical and mathematical literature, 1981
[5] Malyshev V.N., Popov D.N., Sosnovsky N.G. Identification of Autonomous Electro-Hydraulic Tracking Drive // Science and Education: Electronic Scientific and Technical Edition, http://technomag.edu.ru/doc/66269.htmI. —Informregister No. 0420700025 \ 0027. — 2007. — Issue. 6 (June)
[6] Rocket and space technology. Calculation, design, construction and testing of space systems. Series XII Issue No. 3–4. S.P.Korolev Rocket and Space Corporation Energia, 1998
[7] Popov D.N. Dynamics and regulation of hydraulic and pneumatic systems. — 2nd ed., Revised. and add. — M: Engineering, 1987.
[8] Multicriteria optimization of hydraulic systems: textbook. manual for universities / Borovin G.K., Popov D.N.; Bauman Moscow State Technical University. — M.: Publishing House of BMSTU, 2007.

Choice of finishing and strengthening treatment method for cycloidal screw surfaces for multi-product production conditions

P Akulinichev[1,2], I Zenin[1] and A Goncharov[1]

[1]Bauman Moscow State Technical University, 5 Second Baumanskaya Street, Moscow, 105005, Russian Federation

[2]E-mail: aka11996@mail.ru

Annotation. This work is devoted to the study of modern methods of finishing complex profile surfaces, in particular, considered cycloidal screw surfaces, typical for single-screw machines of volume type. The operating conditions of parts with cycloidal screw surfaces are analyzed. In the article the technical requirements provided in the process of finishing are identified, and a method and device for implementing the method of finishing in the conditions of automated multi-product production are offered.

Introduction

Improving the design and performance of machines is often associated with an increase in the proportion of parts with complex surfaces [1]–[4]. A characteristic product that includes complex-profile parts is pumping and compressor equipment, in particular, single-screw pumps. Single-screw pumps are widely used in many industries (engineering, instrumentation, oil and gas industry, food production), where they are used to solve a wide range of tasks: lifting oil and reservoir fluids from wells, in-field pumping of liquids and oil and gas mixtures, transportation of various products in the food and wine industry, dosing of liquids and applying various mixtures on the surface, 3D printing, etc. [5].

Pumps of this type are volumetric rotary hydraulic machines that provide performance and durability when pumping liquids containing mechanical impurities and do not have lubricating capabilities. These characteristics are achieved by the principle of operation and design of the working bodies (screw pair). The principle of operation of single-screw pumps is to create a pressure of the injected liquid or other substance, carried out by moving the liquid by a screw metal rotor rotating inside the stator of the appropriate shape (figure 1). The screw pair consists of an elastic lining of the cage-the stator, and a screw with a wear-resistant surface-the rotor. The working bodies of these pumps have cycloidal screw surfaces with different types of profiles.

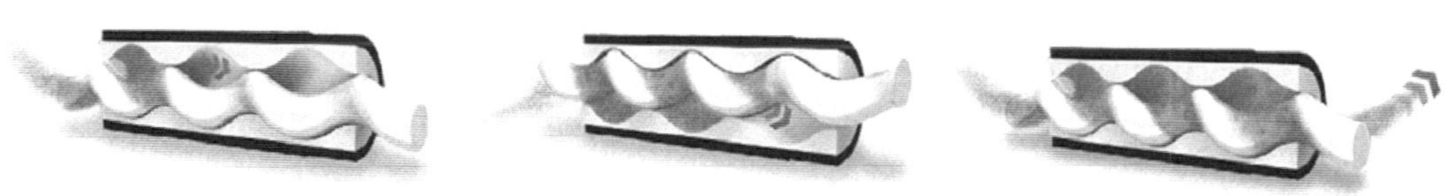

Figure 1. Principle of operation of a screw pair

The vast scope of application of such pumps makes it necessary to manufacture various versions of the screw pair geometry, the sizes of which have a wide range: from small sizes — in dispensers, to large — sized-in the oil industry. The development of methods and approaches to improve the quality of complex-profile surfaces of working bodies of single-screw pumps in conditions of multi-product production is an urgent problem [6]–[9], the solution of which is devoted to this work.

Screw pumps in operation can work in quite difficult conditions: with liquids and pastes containing abrasive particles; with chemicals; with food additives and other chemically aggressive environments. These operating conditions negatively affect the life of the pump's working parts. The most common causes of failure of the rotor of a screw pump are abrasive wear, acid exposure, fatigue wear, and pitting corrosion (figure 2) [5].

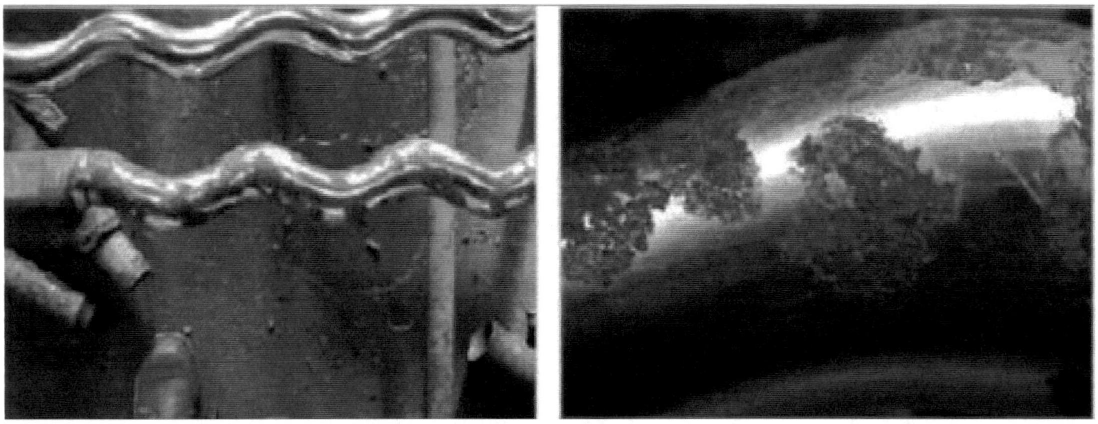

Figure 2. The rotor is exposed to corrosion

The influence of factors that lead to rotor failure during operation must be minimized. In order to reduce the chemical and corrosion wear of the rotor, it is necessary to use materials that have sufficient chemical resistance for the manufacture of the rotor. This material is stainless austenitic steel. Increasing wear resistance is realized by increasing the surface quality parameters of the cycloidal screw surface: increasing the hardness, reducing the surface roughness.

These requirements can be met by heat treatment of parts with subsequent application of finishing methods. The following methods are used as finishing methods: abrasive treatment (bound and free abrasive), surface plastic deformation (SPD), chemical and electrochemical methods [10]–[14]. The use of these methods depends on the volume of production.

Methods

Processing of rotors made of corrosion-resistant austenitic steels is possible only by PPD and magnetic abrasive polishing methods, which allow to increase the microhardness of the surface and at the same time reduce the surface roughness.

The most promising methods of SPD include combined hardening, software methods for applying additional oscillating and oscillating movements to the main process, and static-pulse processing of SPD [15], [16].

The essence of magnetic abrasive polishing is that the processed surface of the part or powder with magnetic and abrasive properties placed in a magnetic field, report a forced movement relative to each other. Metal removal is carried out as a result of the force action of the powder on the surface of the part and the specified relative movements (figure 3) [17].

The method of static-pulse processing of SPD consists in periodic pulse action on the loaded surface by a striker through a statically loaded waveguide, which allows more precisely controlling the distribution of microhardness and residual stresses at a significant depth of the hardened surface layer (figure 4) [15].

Figure 3. A device for magnetic abrasive polishing

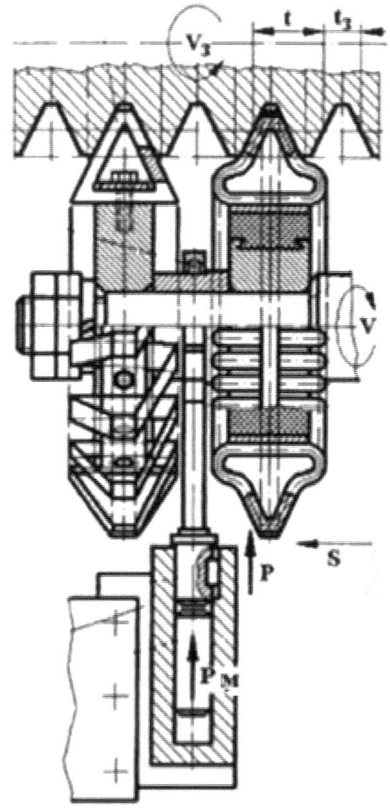

Figure 4. Device for hardening thread

These methods are based on the use of special equipment and tools for processing complex-profile surfaces, which limits their use in conditions of multi-product production. In view of this, it was proposed as a method of finishing processing to apply rolling toroidal roller of complex profile surfaces directly on metal-cutting equipment with CNC without changing the scheme of basing and fixing the workpiece. This solution provides high flexibility due to software control of the rolling process, will avoid the appearance of basing errors due to the implementation of the "one-step" processing principle, as well as reduce the labor intensity of production of parts containing complex-profile screw surfaces, due to the implementation of processing in automated mode.

Based on the above, it can be concluded that it is advisable to use software methods of PPD on CNC machines. As a tool, it is proposed to use a roller with a toroidal deforming surface [18], which will provide the possibility of rolling the cycloidal helical surface of rotors with different geometric parameters on standard CNC equipment.

Results and discussion

To implement the method of rolling a complex profile surface on a multi-purpose CNC machine, it is necessary to develop a tool that allows running on metal-cutting equipment [19] [20].

As a tool for processing, it is proposed to use a device containing a roller with a toroidal working surface, a housing that acts as a guide and accommodates an elastic element, and a tool cone for installation in the machine spindle by an automatic tool change system (figure 5). Creating the necessary pressure in the contact zone of the roller and the workpiece is carried out by deforming the elastic element inside the device body by a certain amount. The constant value of the deformation is provided by controlling the processing process by means of the CNC. Moreover, given that the stiffness of the technological system can differ significantly at different points, the value of the working deformation of the elastic element should be an order of magnitude greater than the value of the deformations of the technological system under the action of the rolling force [21].

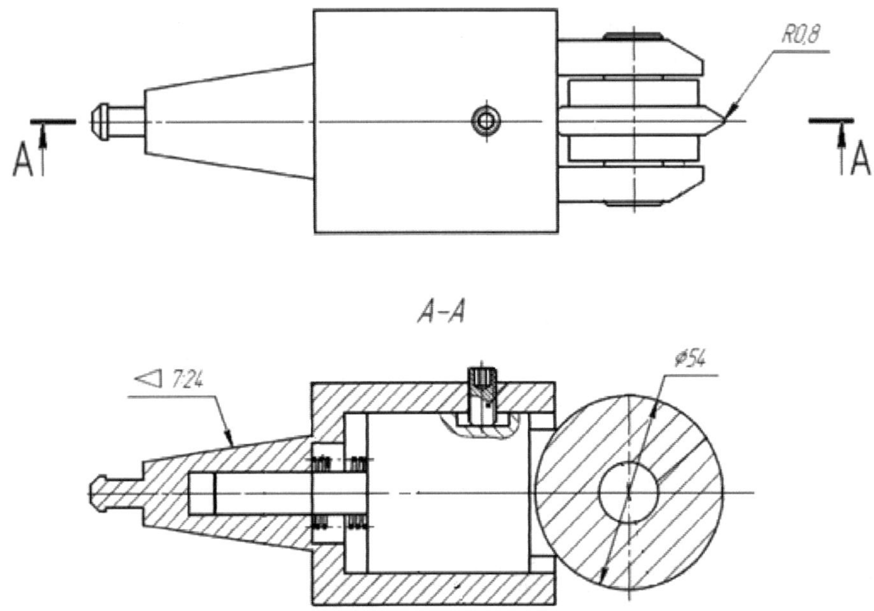

Figure 5. Sketch of the device for running-in

The use of the proposed device will allow for forming and finishing-hardening processing in automatic mode on a single operation, which becomes possible due to the expansion of technological capabilities of CNC equipment. In addition, this approach will reduce the preparatory and final time by increasing the concentration of the transition on a single operation, and the implementation of the principle of processing "in one step" allows you to avoid the occurrence of errors based on the stage of finishing processing of a complex surface.

Conclusions

As a result of the analysis of modern methods of finishing and strengthening processing of complex profile surfaces, it was found that the most progressive for processing cycloidal screw surfaces of parts made of austenitic steels in the conditions of automated multi-product production is software rolling with a toroidal roller on multi-purpose machines. The developed device for rolling on a CNC milling machine allows you to expand the technological capabilities of this class of equipment and produce shaping and finishing-hardening processing in an automated mode, which is an indisputable advantage for multi-product production.

In the future, it is planned to simulate the process of rolling in conditions of variable stiffness of the technological system, due to this, to clarify the modes of rolling and then implement the processing process on complex-profile blanks, rolling in one step after forming on a CNC milling machine.

References

[1] Vasiliev A. S., Goncharov A. A. Some aspects of problematics in designing technological complexes / / IOP Conference Series: Earth and Environmental Science. — 2018. P. 1–7.

[2] Vasiliev A. S. Directed formation of operational properties of parts in technological environments // Bulletin of the South Ural state University, 2017, no. 17, Pp. 33–40.

[3] Lizardyn P. I., Kheifets M. L., Klimenko S. A., Vasiliev A. S. Technological and operational inheritance of quality indicators in the life cycle of machine-building products / / Reports of the national Academy of Sciences of Belarus. — 2005. no. 2. Pp. 130–135.

[4] Averchenkov V. I., Vasiliev A. S., Heifets M. L. Technological heredity in forming the quality of manufactured parts / / science-Intensive technologies in mechanical engineering. — 2018. no. 10. Pp. 27–32.

[5] Enríquez-Méndez Y. M., Torres-Toledano J. G., Rojas-Sánchez F. A. Progressive cavities pump: a case of study / / Gasmexico/ — 2015. P. 1–17.

[6] Goncharov A. A., Vasiliev A. S., Gemba I. N. Modern methods of processing screw surfaces of screw pump rotors // Bulletin of the Rybinsk state aviation technological Academy named after P. A. Solovyov, 2017, no. 1, Pp. 202–208.

[7] Vasiliev A. S., Goncharov A. A. Special strategy for processing complex conical screw surfaces of working bodies of single-screw compressors / / Notes of the mining Institute. — 2019. no. 235. Pp. 60–64.

[8] Goncharov A. A., Akhmad D. N., spazhilov A. S. Expanding the technological capabilities of standard CNC equipment for processing screw surfaces of single-screw pump rotors // Guide. Engineering journal with Appendix. — 2018. no. 8. P. 8–13.

[9] Goncharov A. A., Gemba I. N. Processing of complex profile details of the type of rotation bodies on CNC machines / / Chief mechanic. — 2015. no. 8. Pp. 26–31.

[10] Blumenstein V., Krechetov A. A., Mahal M. S. Modern competitive technologies of finishing-hardening treatment of surface plastic deformation // Hardening technology and coatings. — 2012. No. 7. S. 7–11.

[11] Kirichek A.V., Soloviev D. L. New technology of processing by surface plastic deformation // Hardening technology and coatings. — 2012. No. 8. P. 36–42.

[12] Ovseenko A. N., Klauch D. N., Kushcheva M. E., Ovseenko E. S. Quality of the surface layer of parts hardened by methods of surface plastic deformation / / Strengthening technologies and coatings. — 2010. — №6. Pp. 13–18.

[13] Ivanov D. V., Ivanov V. S., Kuznetsov A. P. Cycloidal shaping of polyhedra in two-rotor systems / / Russian engineering research. 2018. no. 12. P. 949–955.

[14] Sakhvadze G. Zh., Kavtaradze R. Z., Nikabadze M. U., Kikvidze O. G. Increasing the fatigue life of stainless steel samples using laser shock wave processing technology // Problems of mechanical engineering and machine reliability — 2019. no. 4. P. 83.

[15] Kirichek A.V., Solovyov D. L., Tarasov D. E. Devices for implementing wave deformation hardening of composite surfaces / / Fundamental and applied problems of engineering and technology. — 2013. no. 3. Pp. 120–128.

[16] Russian Patent no. 2098259. Method of static-pulse processing by surface plastic deformation [Text] / A. G. Lazutkin, A.V. Kirichek, D. L. Solovyov. Byull. # 34, 1997.

[17] Maksarov V. V., Keksin A. I. Technological improvement of the quality of composite surfaces by magnetic-abrasive polishing / / Metalloobrabotka. — 2017. No. 1. Pp. 47–57.

[18] 10. Drozd M. S., Matlin M. M., Sidyakin Yu. I. Engineering calculations of elastic-plastic contact deformation. — Moscow: Mashinostroenie, 1986. 224 p.

[19] Odintsovo, L. G. Hardening and finishing of parts by surface plastic deformation: Handbook. — Moscow: Mashinostroenie, 1987, 328 p., Il.

[20] Running-in for surface treatment / / encyclopedia of mechanical engineering XXL URL: https://mash-xxl.info/info/224912/ (accessed 24.04.2019).

[21] Butakov B. I., Zubekhina A.V. Rigidity of the machine-tool-part system when rolling parts with rollers / / Technical Sciences. Bulletin of agricultural science of Primorye. — 2008. №4. Pp. 210–223.

Development of methods for calculating cutting modes when processing the internal cycloidal screw surface using the original method used in drilling

M Albov[1,2] and A Goncharov[1]

[1]Bauman Moscow State Technical University, 5 Second Baumanskaya Street, Moscow, 105005, Russian Federation

[2]E-mail: maxim.albom@yandex.ru

Annotation. In this paper, we consider the problems that arise when using screw volumetric machines when working with chemically aggressive media. To solve this problem, it was proposed to use a chemically resistant material of fluoroplast for the manufacture of screw machine clips. To implement the process of creating clips, the kinematics of methodagerotor drilling of cycloidal screw holes was proposed, and a method for calculating processing modes was developed.

Introduction

Single-screw pumps and units based on a screw pair are widely used in various industries for pumping oil-containing products, volumetric dosing of viscous substances, and even in 3D printing [1]. The working body of single-screw pumps is a screw pair: the rotor and the cage. The main surface of these parts is a complex profile screw surface, which is generally characteristic of parts of modern machines [2]–[5]. Most manufacturers of screw machines and dispensers use injection molding of elastomeric materials in a form where the hole profile is formed by using a rod made using a similar technology as the rotor of such machines [6]–[9]. The materials most common in the manufacture of screw machine clips include: nitrile butadiene rubber (NBR), ethylene propylene rubber (EPDM), fluoro rubber (FKM), polyurethane (PU). But foundry materials used in the creation of screw clips do not have sufficient chemical resistance to use them for transporting chemically aggressive media, such as: ammonia, aniline, acetone, butyl acetate, arsenic, ammonium phosphate, ethyl acetate, and others.

To solve the problem of using single-screw pumps with chemically active substances, it was proposed to use fluoroplast (PTFE) as the material of the cage, which is not susceptible to the effects of the above materials. But fluoroplast cannot be cast into a casting mold, it can only be processed by mechanical methods. There are some variants of processing a complex-profile screw surface by blade methods: stretching, "drilling", milling with a disk cutter [10]–[11], however, the implementation of these methods is practically impossible.

The pulling method has several disadvantages such as the complexity of chip removal in connection with the length of the screw holes, the complexity of manufacturing tools, as well as the kinematics of the movement of the broach, that is, the tool must not move along the straight path, and to twist in a spiral, which complicates the use of this method. Milling with a disc cutter, since the availability of the inner surface of the holder is very limited, the tool stiffness is greatly reduced by reducing the cross section of the holder, which will reduce the already not very high quality of the surface obtained as a result of milling. The method of laser drilling is based on a complex mutual kinematics of the movement of the clip and the profiled tool, as a result of which the trajectory of the cutting edge describes the full profile of the cycloidal screw surface. With this method, the greatest complexity is created by the shape of the cutting tool, as well as the joint complex kinematics of the movement of the tool and the clip.

Methods

Let's focus on the method of "drilling", as of all these methods, it is the most promising in terms of implementation in production.

Figure 1 shows the kinematics of the drilling method.The position (1) shows the shape of the intended instrument.Position (2) denotes a billet in the form of rolled products, the tool guide is a bronze bushing, shown in position (3). With the mutual complex rotational movement of the workpiece and the tool, as well as the axial feed of the tool, the internal cycloidal helical surface is formed.The proposed processing method is performed using a kinematic system with three controlled coordinates, two rotational and one translational. With this method of processing, the most important aspect for obtaining a high-quality surface of the screw cage is strict compliance with the joint kinematics of the movement of the rotor-tool and the workpiece-cage:

$$N_p = \left(2N_o + \frac{S}{h} \right), \tag{1}$$

Even a small deviation in the speed of rotation of the tool or workpiece can lead to jamming of the tool or incorrect formation of the cycloidal screw surface.

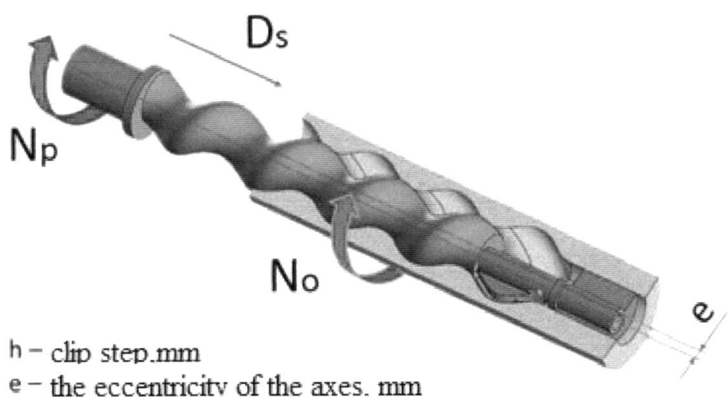

Figure 1. Cinematic mod "drilling»

For stable cutting conditions, it is necessary to choose the tool sharpening, so that the material is cut, and not plastically deformed, and choose the cutting angles: back, front. Based on the recommendations for processing PTFE, the front angle should be at least 10°. Since the shape of the" drill" corresponds to the shape of the rotor that rotates in the screw cage, the back angle can be made 0° to give additional rigidity to the tool. Selecting a suitable cutting edge for forming a complete profile of the internal screw surface, the analysis of the intersection of the tool surfaces with the workpiece and the already processed area was performed.

Drawing 2 shows a simulation written in the program "Graphing Calculator-Desmos". The section of the workpiece (POS. 1) and the tool (POS. 2) is presented. the intersection Point of the back and front surface of the tool describes the "ellipse" trajectory in the workpiece (POS.3).During the simulation, it was found that in different sections at different times, the front and rear angles change in different ranges. For the front and rear corners, the range is from -12.5 ° to +12.5°, which is unacceptable,so it was decided to choose the cutting edge so that the rear angle varies from 0 to +25°, and the front angle is separately sharpened and is in the range of 10 ° to 35°, which is acceptable for processing soft materials such as fluoroplast [12]–[14].

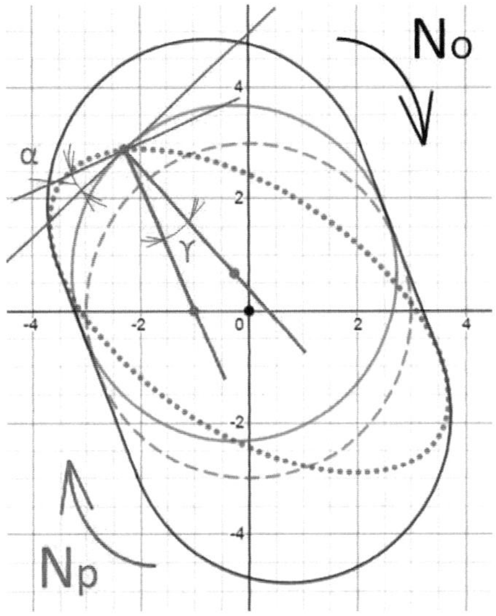

Figure 2. The trajectory of the cross section of the cutting edge and tool angles

We calculate the required cutting power and torque for eccentric drilling of plastic material[15]–[18]. When calculating the drilling capacity, it is necessary to calculate the specific material removal-Q (cm3 / min):

$$Q = \frac{s_{min} \cdot \pi \cdot D^2}{4 \cdot 1000}, \tag{2}$$

wheres_{min}— minute tool feed (mm / min);
D — drill diameter (mm).

Using this formula, it is not possible to calculate the specific material removal when drilling a cycloidal screw surface, because the cross-sectional area differs from classical drilling. Then, according to (Fig. 3), the formula for the specific removal of material for eccentric drilling after all transformations will have the form:

$$Q = \frac{s_{min} \cdot \pi \cdot (d_r^2 - d_o) + 16 \cdot e \cdot d_r \cdot s_{min}}{4 \cdot 1000}, \tag{3}$$

where s_{min} — table feed of the tool[mm/min];
d_r — rotor diameter[mm];
d_o — hole diameter in the workpiece[mm];
e — the eccentricity of the rotor-tool[mm].

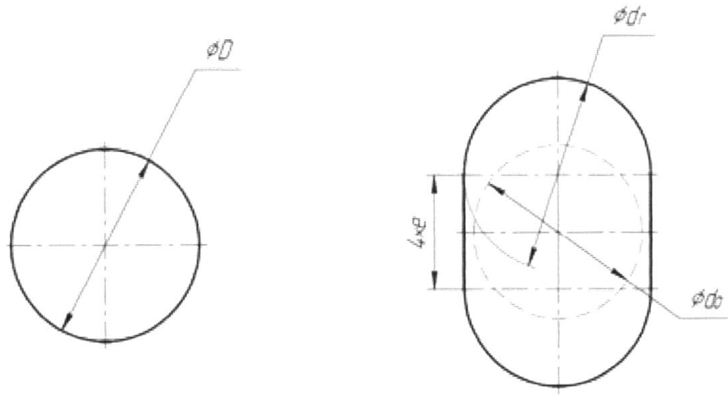

Figure 3. The cross-sectional area when drilling

The calculation of the required power P (kW) for traditional drilling is calculated as follows:

$$P = \frac{Q \cdot k_c}{60000 \cdot \eta}, \tag{4}$$

where Q — specific material consumption[cm3/min];

k_c — specific cutting force [N / mm2], table value;

η — the efficiency of the main drive of the cutting movement of equipment, dimensionless value;

Substituting the formula (3) in (4), and presenting the minute feed as a product $s_{min} = s_{rev} \cdot n$, where s_{rev} — tool feed per revolution, n — relative speed of the tool relative to the clip, $n = (N_p - N_o)$. Receive:

$$P = \frac{k_c \cdot s_{rev} \cdot n \cdot (\pi \cdot (d_r^2 - d_o) + 16 \cdot e \cdot d_r)}{\eta \cdot 240000000}, \tag{5}$$

where the coefficient 240,000,000 is used for converting units of measurement.

Then, the required torque can be calculated using the following formula:

$$M = \frac{k_c \cdot s_{об} \cdot (\pi \cdot (d_r^2 - d_o) + 16 \cdot e \cdot d_r)}{\eta \cdot 25263}. \tag{6}$$

Find the relationship between the speed at the edge of the cutting edge at the maximum radius of the eccentric drill from the angle of rotation of the rotor.

R — maximum radius of rotation of the cutting edge of the tool in this section, constant value[mm];

α — the angle of rotation of the clip-blank, the angle of rotation of the rotor, respectively, is 2 times greater, since its rotation speed is twice as large[degrees];

r — the radius of rotation of the cutting edge in the clip, a variable value, depends on the angle of rotation of the clip [mm];

V_o — speed of the rotor cutting edge in the cage [m/min];

V_p — rotor cutting edge speed [m/min];

θ — the angle between the velocity vector [degrees];

V_{rev} — speed of the rotor cutting edge relative to the cage [m/min].

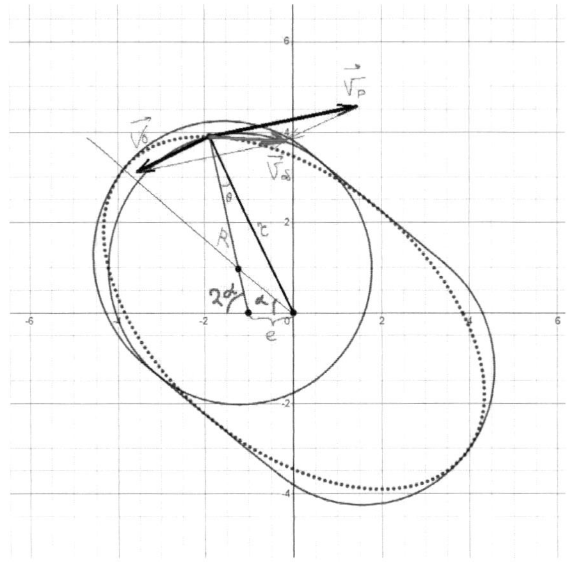

Figure 4. Calculation of cutting speed

The modulus of the sum of vectors is calculated using the formula:

$$\overrightarrow{V_{rev}} = \overrightarrow{V_o} + \overrightarrow{V_p} = \sqrt{|\overrightarrow{V_o}|^2 + |\overrightarrow{V_p}|^2 - 2 \cdot |\overrightarrow{V_o}| \cdot |\overrightarrow{V_p}| \cdot cos\theta}, \quad (7)$$

Rotor cutting edge speed:

$$|\overrightarrow{V_p}| = \frac{\pi \cdot 2R \cdot Np}{1000}, \quad (8)$$

where Np — rotation speed of the tool rotor[rpm].

Speed of the rotor cutting edge in the cage

$$|\overrightarrow{V_o}| = \frac{\pi \cdot 2r \cdot No}{1000}, \quad (9)$$

where No — the speed of rotation of the workpiece-clips [rpm].

The value " r " is unknown, so we will find the dependence of this value on α and other known values. According to Figure 5, by the cosine theorem, the dependence takes the following form:

$$r^2 = R^2 + e^2 - 2 \cdot R \cdot e \cdot cos(180 - 2\alpha), \quad (10)$$

where e — the eccentricity between the axes of rotation of the tool and the workpiece [mm].

Then, formula (9) takes the form:

$$|\overrightarrow{V_o}| = \frac{\pi \cdot 2\sqrt{R^2 + e^2 - 2 \cdot R \cdot e \cdot cos(180 - 2\alpha)} \cdot No}{1000}, \quad (11)$$

According to Figure 5, by the cosine theorem, the last component of the formula (7), namely $cos\theta$ has the form:

$$e^2 = R^2 + r^2 - 2 \cdot R \cdot r \cdot cos(\theta),$$

$$cos(\theta) = \frac{R^2 + r^2 - e^2}{2 \cdot R \cdot r} \quad (12)$$

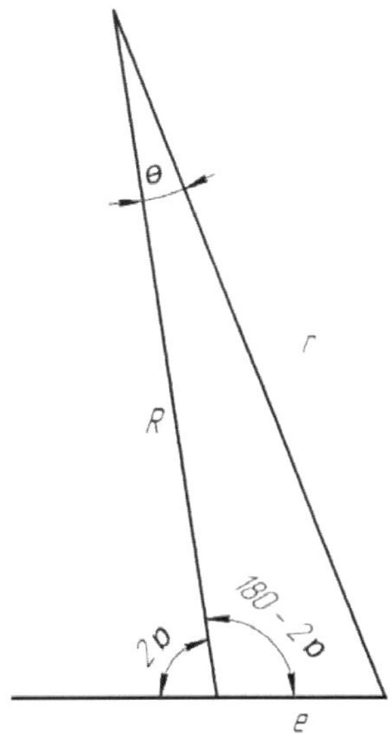

Figure 5. The scheme of calculation of angles

Substitute (12), (11), (10), (9), (8) to the formula (7):

$$V_{rev} = \frac{2\pi No}{1000}\sqrt{R^2 + e^2 + 2 \cdot R \cdot e \cdot cos(180 - 2\alpha)}. \quad (13)$$

Results and discussion
According to formula (13), the speed distribution of the cutting edge of the tool relative to the angle of rotation of the clip will have the following form, and the point of extremum and therefore the maximum cutting speed will be located at the angle $\alpha = 90$ figure 6.

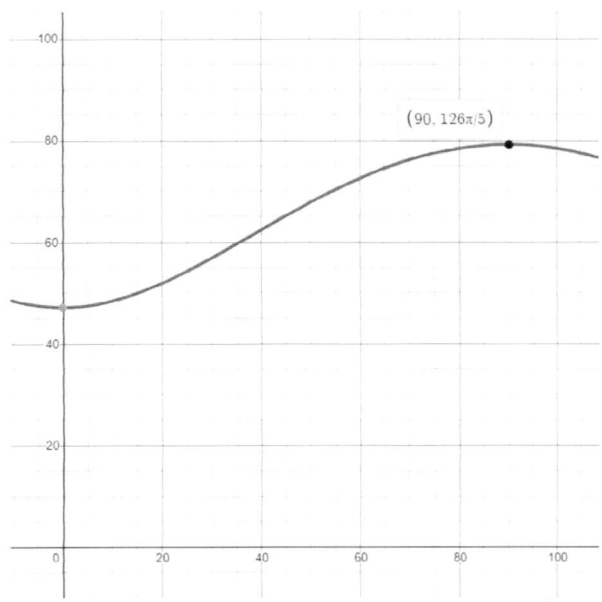

Figure 6. Distribution of the cutting speed depending on the angle of rotation of the clip

Then, the maximum and minimum cutting speed for a specific cutting edge section will be:

$$V_{rev.max} = \frac{2\pi No(R+e)}{1000}, \quad (14)$$

$$V_{rev.min} = \frac{2\pi No(R-e)}{1000}. \quad (15)$$

Based on the graph shown in Figure 6, the cutting speed will not be constant, but will vary in the range of 40% and this is considered only one section of the cutting edge, with the highest cutting speed. Such a change can both positively and negatively affect the quality of the resulting surface [19]–[20].

According to (6) it is possible to find out the torque on the tool, as well as the cutting force acting on the tool. But this is the total force that does not show the load distribution along the cutting edge, so the development of methods for determining the distribution of cutting force is a promising task.

Conclusions

As a result of analysis of modern methods of forming of internal helical surfaces of the clips are single screw pumps, it was identified that the most promising method blade processing not casting materials clips is the method of "drilling", the implementation of which involves a number of technologicheskaya, but it is feasible and enables a high quality inner cycloidal helical surfaces. The developed method of calculating the cutting moment and cutting speed requires experimental confirmation.

In the future, it is planned to develop a design of the installation based on the proposed kinematics of the movement of the tool and the workpiece, which will clarify the processing modes and geometry of the tool.

References

[1] Enríquez-Méndez Y.M., Torres-Toledano J.G., Rojas-Sánchez F.A. Progressive cavities pump: a case of study // Gasmexico/ — 2015. P. 1–17.

[2] Vasiliev A.S., Goncharov A.A. Some aspects of problematics in designing technological complexes // IOP Conference Series: Earth and Environmental Science. — 2018. P. 1–7.

[3] Vasiliev A. S. Directed formation of operational properties of parts in technological environments // Bulletin of the South Ural state University, 2017, no. 17, Pp. 33–40.

[4] Lizardsyn P. I., Heifets M. L., Klimenko S. A., Vasiliev A. S. Technological and operational inheritance of quality indicators in the life cycle of machine-building products / / Reports of the national Academy of Sciences of Belarus, 2005, no. 2, Pp. 130–135.

[5] Averchenkov V. I., Vasiliev A. S., Heifets M. L. Technological heredity in forming the quality of manufactured parts / / science-Intensive technologies in mechanical engineering. — 2018. no. 10. Pp. 27–32.

[6] Goncharov A. A., Vasiliev A. S., Gemba I. N. Modern methods of processing screw surfaces of screw pump rotors // Bulletin of the Rybinsk state aviation technological Academy named after P. A. Solovyov, 2017, no. 1, Pp. 202–208.

[7] Goncharov A. A., Gemba I. N. Processing of complex profile details of the type of rotation bodies on CNC machines / / Chief mechanic. — 2015. no. 8. Pp. 26–31.

[8] Vasiliev A. S., Goncharov A. A. Special strategy for processing complex conical screw surfaces of working bodies of single-screw compressors / / Notes of the mining Institute. — 2019. no. 235. Pp. 60–64.

[9] Goncharov A. A., Ahmad D. N., Schadilov, A. S., Expansion of technological possibilities of the standard equipment with CNC for machining of the helical surfaces of the rotor single screw pumps // Guide. Engineering journal with Appendix. — 2018. no. 8. P. 8–13.

[10] Maksimova E. sh., tyutrin Nikolay Orestovich, problems of manufacturing parts from fluoroplast-4 by mechanical processing methods. Irkutsk national research technical University / / life cycle of structural materials (from receipt to disposal) Irkutsk, April 26–28, 2017. P. 138–144.

[11] Valovsky V. M. Screw pumps for oil production, Moscow, publishing house "Oil economy", 2012., 248s.

[12] Erenkov O. Yu., Ivakhnenko A. G., Hosen R. I. New combined methods of processing polymer materials by cutting on the basis of preliminary physical, chemical and mechanical effects. Vladivostok: Dalnauka, 2007. 219 p.

[13] Gavrilova A.V. Improving the quality of blade processing of polymer materials by preliminary mechanical deformation of blanks: abstract dis. Cand. tech. sciences'. Komsomolsk n/A: 2009. Typewriting.

[14] Composite materials based on polytetrafluoroethylene. Structural modification / Yu. K. Mashkov [et al.]. — Moscow: Mashinostroenie, 2005. — 240 p.

[15] Calculation of cutting forces when processing plastic materials in a wide range of thicknesses of the cut layer / Grubiy S. V. / / Izvestiya Vuzov. Ser. "Engineering". — 2018. — № 2. — P. 3–10.

[16] Petrushin S.I., Proskokov A.V. Theory of constrained cutting: Chip formation with a developed plastic-deformation zone. Russian Engineering Research, 2010, vol. 30, no. 1, pp. 45–50.

[17] Rosenberg J. A. Cutting of materials. Kurgan, TRANS-Ural region, 2007. 294 p.

[18] Kabaldin Yu. G., Kuzmishina A. M. Quantum-mechanical modeling of deformation and fracture of the shear layer during cutting. Vestnik mashinostroeniya, 2016, no. 4, pp. 65–71.

[19] Dreval A. E., M. S. Lugansky built-up edge and the operability of the cutting tool//News of higher educational institutions. Engineering. 2012. no. 12. Pp. 3–7.

[20] Sergiev A. P., Vladimirov A. A., Shvachkin E. G. On the question of physical phenomena in the cutting zone//Federal state Autonomous educational institution of higher education " national research technological University "MISIS". RUSSIAN SCIENTIST. 2017. C.20–30.

IOP Conf. Series: Materials Science and Engineering **963** (2020) 012015 doi:10.1088/1757-899X/963/1/012015

Analysis of various design methods for unloading a centrifugal pump rotor from axial forces

K Volodin[1,2] and V Lomakin[1]

[1]Bauman Moscow State Technical University, 5 Second Baumanskaya Street, Moscow, 105005, Russian Federation

[2]E-mail: kirvol@gmail.com

Abstact. This article is devoted to a review of the main methods of unloading the rotor of a horizontal multistage pump from axial forces. Two methods are considered: a hydrofoam and an opposite arrangement of blocks of driving wheels. The drawings are given and a method for compensating axial force for each of them is described. Their advantages and disadvantages are described. General design guidelines are provided.

Introduction

Centrifugal pumps are widely used in various industries, such as mining, chemical, food, as well as in the field of public utilities. Currently, a lot of numerical research is being conducted aimed at improving the performance of pumping equipment. The main area of such research is pump optimization. However, do not forget about other equally important parameters, such as reliability. [1,2].

Optimization is performed using hydrodynamic modeling methods [3]–[7]. However, the application of these methods is sometimes difficult due to the large number of necessary calculations. In [8]–[13] other methods for modeling processes occurring in centrifugal pumps are described.

A large axial force acts on the rotor of a high-pressure horizontal multistage pump, which, in the absence or failure of unloading devices, can lead to axial displacement of the shaft with impellers in it in the axial direction towards the suction side.

Axial force in a multistage centrifugal pump and methods of balancing it

In the impeller of a one-way inlet due to its lack of symmetry relative to a plane perpendicular to the axis of rotation of the pump, an unbalanced hydraulic force arises, directed along the axis towards the entrance to the wheel. [14] This is due to the difference in pressure on the walls of the impeller (Fig. 1).

To balance this force in multistage pumps, the following discharge methods are used:

1. Hydraulic heel (hydraulic balancing device)The most common device for compensating axial force in multi-stage sectional pumps is the hydroheel. The hydrofoil contains a unloading disk 5 rigidly fixed to the shaft, a fixed support ring (pillow) 2, sequentially arranged cylindrical 1 and end 3 inductors and a chamber 4 separating these inductors (Fig. 2)

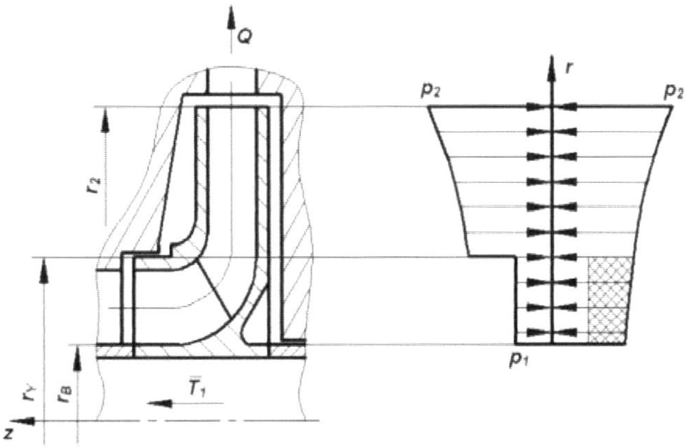

Fig. 1. Pressure distribution on the walls of the impeller

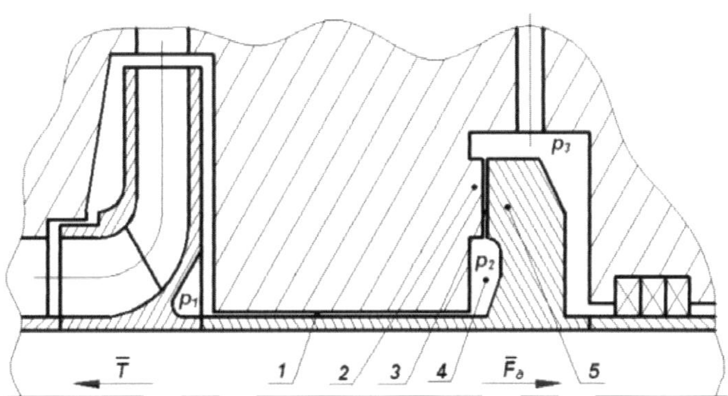

Fig. 2. Hydraulic heel

The full differential pressure at the heel is the difference between the discharge pressure and the pressure in the chamber behind the hydro-point, which is usually connected by a bypass pipe to the pump inlet. Part of the total pressure drop is throttled at the end throttle, the conductivity of which depends on the axial displacement of the rotor. For example, if under the influence of excessive axial force the rotor moves to the left, then the clearance in the end throttle decreases, which in turn will cause the pressure in the chamber to increase to a level that ensures the restoration of the equilibrium of the rotor. In addition to the axial balancing of the rotor, the hydrofoil performs an additional function of unloading the end seal of the pump from the discharge side from high pressure.

2. Opposed arrangement of groups of impellers

In using this method, the impellers are installed in groups in the opposite direction to each other (Fig. 3). The application of this method allows almost completely compensating for the axial force, as well as getting rid of the unloading unit, which increases the manufacturability of the design and increases the volumetric efficiency of the pump, since there will be no leakage through the unloading device. At the same time, it is necessary to provide for the construction of complex transfer channels, which leads to an increase in the dimensions of the pump and an increase in hydraulic losses.

In designing, it is important to choose the most optimal method of unloading from the axial force. The main design criteria in this case should be the efficiency and service life of the pump. The joint analysis of these criteria can be carried out by analogy with the methodology described in [15]. To improve the overall efficiency of the pump, you must also pay attention to the design of the guide device for each stage. In addition, to verify the results obtained, a simulation can be performed using the method described in [16]. At the same time, it is necessary to take into account the criteria for manufacturability of the guide device [17].

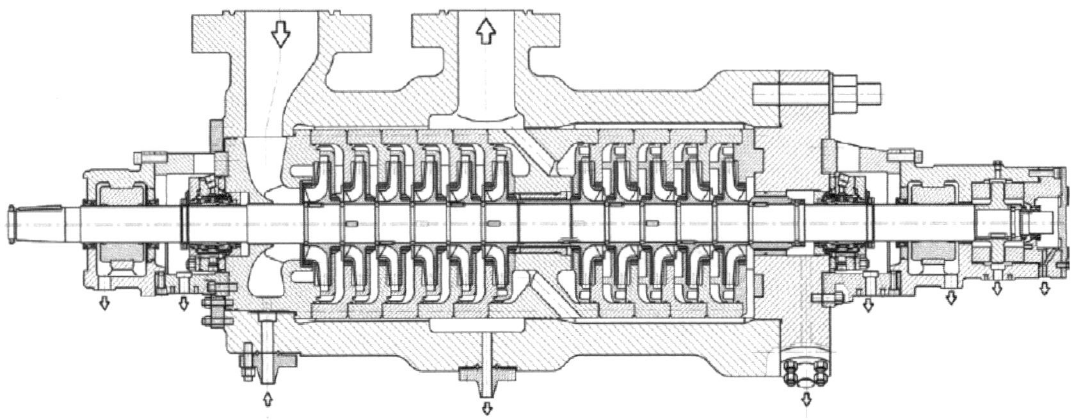

Fig. 3. The design of the centrifugal pump with the opposite arrangement of stages

Conclusion

In the article were considered the basic structural methods of unloading the rotor of a horizontal multistage sectional high-pressure pump from axial forces. According to the criteria of constructive reliability and ease of maintenance, the opposed arrangement of the groups of impellers is a more optimal constructive solution. General recommendations were given on the choice of a solution and the design of appropriate pumps.

References

[1] A Protopopov and D Bondareva 2019 IOP Conf. Ser.: Mater. Sci. Eng. 492 012002
[2] V.O. Lomakin et al. / Procedia Engineering 176 (2017) 476–482
[3] T Valiev and APetrov 2019 IOP Conf. Ser.: Mater. Sci. Eng. 492 012038
[4] V Lomakin and O Bibik 2019IOP Conf. Ser.: Mater. Sci. Eng.492 01200
[5] A Petrov et al 2019 IO OPP Conf. Ser.: Mater. Sci. Eng. 492 012036
[6] V Cheremushkin and V Lomakin 2019 IOP Conf. Ser.: Mater. Sci. Eng. 492 01203
[7] Yousefi, H., Noorollaahhi, Y., Tahani, M., Fahimi, R., &Saremian, S. (2019). Numerical simulation for obtaining optimal impeller's blade parameters of a centrifugal pump for high-viscosity fluid pumping. Sustainable Energy Technologies and Asessments, 34, 16–26. doi:10.1016/j.seta.2019.04.01
[8] A Shablovskiy and E Kutovoy 2019 IOP Conf. Ser.: Mater. Sci. Eng.492 012035
[9] Zhang, Z. -., Chen, H. -., Ma, Z., He, J. -., Liu, H., & Liu, C. (2019). Research on improvingthe dynamic performance of centrifugal pumps with twisted gap drainage blades. Journal ofFluids Engineering, Transactions of the ASME, 141(9) doi:10.1115/1.4042885
[10] Zhang, S., Li, H., & Xi, D. (2019). Investigation of the integrated model of side chamber, wear-rings clearance, and balancing holes for centrifugal pumps. Journal of Fluids Engineering, Transactions of the ASME, 141(10) doi:10.1115/1.4043059
[11] Yun, R., Zuchao, Z., Denghao, W., &Xiaojun, L. (2019). Influence of guide ring on energy loss in a multistage centrifugal pump. Journal of Fluids Engineering, Transactions of the ASME, 141(6) doi:10.1115/1.4041876
[12] Pirouzpanah, S., Patil, A., Chen, Y., & Morrison, G. (2019). Predictive erosion model for mixed flow centrifugal pump. Journal of Energy Resources Technology, Transactions of the ASME, 141(9) doi:10.1115/1.4043135
[13] Blume, M., & Skoda, RR. (2019). 3D flow simulation of a circular leading edge hydrofoil and assessment of cavitation erosion by the statistical evaluation of void collapses and cavitation structures. Wear, 428-442 29, 457–469. doi:10.1016/j.wear.2019.04.01

[14] Mikhailov A.K., Malyushenko V.V. Vane pumps. Theory, calculation and design // Moscow: "Engineering", 1977, 288 pp.AProtopopovand C Jakovich 2019 IOP Conf. Ser.: Mater. Sci. Eng. 492 012034
[15] K Abramov 2019 IOP Conf. Ser.: Mater. Sci. Eng. 589 012013
[16] Lomakin V.O., Kuleshovav M.S. and Bozh'eva S.M. 2016 Numerical Modeling of Liquid Flow in a Pump Station Power Technology and Engineering 49 324–327
[17] Ivanov A.S. and Murkin S.V. 2016 Threaded joints in drives Russian Engineering Research 36 611–619

Overview of methods for optimizing the flow of the centrifugal pump

N Gradilenko[1,2] and V Lomakin[1]

[1]Bauman Moscow State Technical University, 5 Second Baumanskaya Street, Moscow, 105005, Russian Federation

[2]E-mail: natali_msh@mail.ru

Abstact. This article is devoted to a review of the flow part of the centrifugal pumps(CFP), as well as methods for its optimization. The components of the flow part (inlet, impeller and outlet) were considered, and methods for their optimization were given. The analysis of the literature was carried out. Conclusions are drawn on the development and improvement of vane pumps.

Introduction
Pumps of the centrifugal type are widely used in various fields of industry: mining, chemical, municipal engineering, etc. [1]. So the centrifugal pumping system is designed to pump neutral cold water with a temperature of 1 to 45 ° C with a small amount of mechanical impurities, the first type of centrifugal pump is used to pump hotter water up to 105 ° C, the second type of centrifugal pump is used to pump acidic water, the third type of centrifugal pump is used to pump oil and the fourth type of centrifugal pump - for work in the oil system.

Since before the beginning of the previous century, pumps of the CFP type were produced on the basis of centralized orders of enterprises, by the development of serial production (1968–1970), there were more than 30 types, and the number of standard sizes produced exceeded 250 units. By the beginning of the 90s, the nomenclature was reduced by manufacturers to 22 types. On the one hand, the multiplicity was justified by the variety of operating parameters and conditions of water pumping, on the other — the lack of a sound scientific approach. Further actions in the field of CFP development were aimed at building a rational parametric device with a minimum number of pumps in this series and optimizing the flow part [2].

The main characteristics of vane pumps are resource and efficiency. These two parameters are often mutually exclusive: when you increase the first, you have to sacrifice the second. Thus, the optimization task is to find a compromise solution that will satisfy both parameters [3].

Optimization of the flow part
The flow part of the vane pump (Fig. 1) consists of three main elements: supply, impeller and outlet. The inlet is designed to supply liquid from the inlet pipe to the impeller. The impeller transmits the energy of the liquid. And the outlet transfers the fluid from the impeller to the pump outlet pipe or to another stage [4].

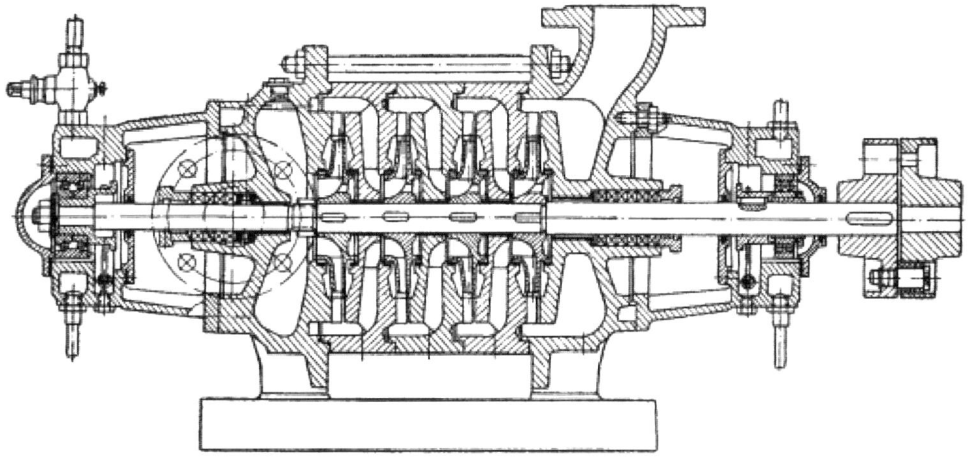

Fig. 1. Multi-stage impeller pump

Most often, the goal of optimization is to increase the efficiency of the pump, but it is equally important to improve their cavitation quality, reliability and vibro-acoustic characteristics. Optimization can be performed using two methods: an intuitive method or optimization algorithms [5].

With intuitive optimization, a number of experiments are performed under the same conditions with a change in a certain part of the pump and the effect of the change is investigated. When using optimization algorithms, calculations are performed in the program using hydrodynamic modeling methods [6–10]. The person sets the step and sequence of actions, and selects the best result. The use of optimization algorithms reduces the influence of the human factor on the research result, but it is more difficult to perform [10–15] describes other methods for modeling processes occurring in centrifugal pumps.

Methods for optimizing the supply include:

1. Shifting the maximum efficiency towards higher feed rates by increasing the area of the inlet to the taper cone. Increasing the area allows you to increase the design flow rate and reduce the head, but increases the size of the pump [16];
2. Method for optimizing the side semi-spiral branch (Fig. 2), in which the entire supply is divided into three sections: the confuser section before entering the wheel, the spiral section from the confuser to the transition section, and the section from the transition section to the inlet pipe. In the zone of the spiral section, a "tongue" is installed that optimizes the flow of the shaft and prevents the flow of liquid in the direction opposite to the rotation of the wheel. The method is described in more detail in [17]. it allows optimizing the geometric parameters of the side semi-spiral branch, but there is an uneven distribution of liquid velocities and their moments at the inlet to the impeller, especially with large pump feeds. This in turn leads to a decrease in efficiency and deterioration of cavitation qualities;
3. Optimization semi-spiral inlet, based on [17], which decreases the width of the calculated and the intermediate sections, increase the radius of the volute, and changing the shape of the section "language", thereby achieving higher uniformity of the velocity distribution of the fluid and moments and reduce the vortex formation at the entrance to the wheel, which leads to increased efficiency and better cavitation characteristics [18];
4. Semi-spiral inlet with an increased moment of fluid velocity at the inlet to the impeller, which leads to an increase in the pump efficiency [19];
5. The use of B-splines for constructing a side semi-spiral inlet when working in CAD helps to simplify the design process of the inlet [20];

6. Optimization of the cross-section area and edge structure geometry in the study of the supply geometry under non-structural conditions, in which it is shown that the performance parameters depend on both geometric conditions and operating conditions [21].

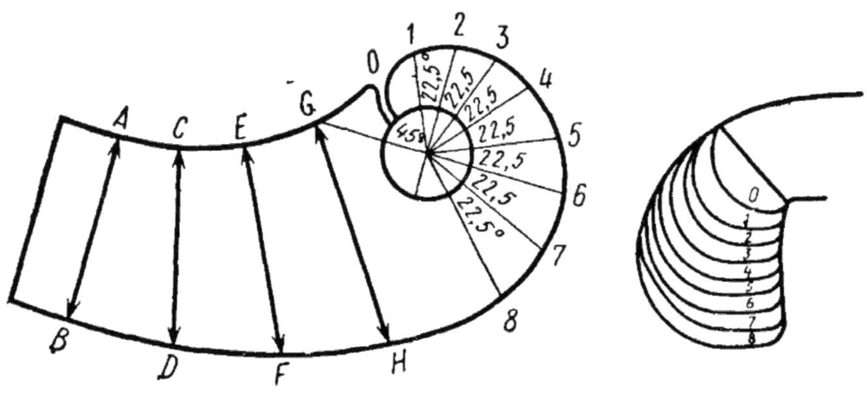

Fig. 2. Diagram of a semi-spiral inlet

Ways to optimize the impeller:

1. The impeller with slits allows for a more uniform distribution of pressure, reduce cavitation, with the correct location and configuration will not affect the efficiency of the pump, but such a wheel will be more difficult to manufacture [22];
2. Reducing the radius of the impeller leads to a reduction in the optimal flow rate, but maintaining efficiency, this allows you to trim the impellers [17];
3. Changing the shape of the impeller blades affects the efficiency of the pump, this is considered and given in the form of an algorithm in [23];
4. The angle of installation of the blades at the outlet affects the head, so when the installation angle increases, the theoretical head increases, when the installation angle is greater than 90°, the proportion of dynamic head increases intensively, and for static head falls [24];
5. The number of blades depends on the speed coefficient and the size of the wheel, with a smaller size, usually choose a smaller number of blades to reduce the flow constraint of the blades [24].

Ways to optimize the outlet and guide device:

1. Changing the diameter of the guide device, the angle and the number of guide channels affects the efficiency. Thus, increasing the number of guide channels increases the efficiency of the pump [25, 26];
2. Comparison of channel and blade-less guide devices is given in [27];
3. The graph from [28] shows that as the channel width of the guide device increases, the maximum position shifts towards larger feeds;
4. Reducing the throughput of the spiral output is achieved by inserting into the outlet section or adding the outlet tongue, this is required to improve performance [29];
5. Greater diffusivity reduces the radial force by orders of magnitude [30]

Conclusions:

Various ways of optimizing the flow part of a CFP-type pump were considered. These methods are mainly aimed at increasing the efficiency of the pump. In addition, its cavitation qualities, reliability and vibro-acoustic characteristics are improved. In the design of the inlet and outlet for optimization, it is often suggested to use diffusivity and" tongues". Optimization is recommended using algorithms, since this saves time, reduces the likelihood of human error, and avoids creating a large number of real models of poor quality, and therefore saves money.

References

[1] A Protopopov and D Bondareva 2019 IOP Conf. Ser.: Mater. Sci. Eng. 492 012002
[2] Palamarchuk T. N. Principles of construction of a unified standard-size series of centrifugal sectional pumps / T. N. Palamarchuk // Collection of scientific papers Donizht. — 2016. no. 43. — p. 31–34.
[3] A Protopopov and C Jakovich 2019 IOP Conf. Ser.: Mater. Sci. Eng. 492 012034
[4] Lomakin, A. A. Centrifugal and axial pumps / A. A. Lomakin ed. T. P. Malashkin-M., 1966. — 364s.
[5] V.O. Lomakin et al. / Procedia Engineering 176 (2017) 476–482
[6] T Valiev and APetrov 2019 IOP Conf. Ser.: Mater. Sci. Eng. 492 012038
[7] V Lomakin and O Bibik 2019 IOP Conf. Ser.: Mater. Sci. Eng.492 01200
[8] A Petrov et al 2019 IO OPP Conf. Ser.: Mater. Sci. Eng. 492 012036
[9] V Cheremushkin and V Lomakin 2019 IOP Conf. Ser.: Mater. Sci. Eng. 492 01203
[10] Yousefi, H., Noorollaahhi, Y., Tahani, M., Fahimi, R., &Saremian, S. (2019). Numerical simulation for obtaining optimal impeller's blade parameters of a centrifugal pump for high-viscosity fluid pumping. SustainableEnergyTechnologiesandAsessments, 34, 16–26. doi:10.1016/j.seta.2019.04.01
[11] A Shablovskiy and E Kutovoy 2019 IOP Conf. Ser.: Mater. Sci. Eng.492 012035
[12] Zhang, Z. -., Chen, H. -., Ma, Z., He, J. -., Liu, H., & Liu, C. (2019). Research on improving the dynamic performance of centrifugal pumps with twisted gap drainage blades. JournalofFluidsEngineering, Transactionsofthe ASME, 141(9) doi:10.1115/1.4042885
[13] Zhang, S., Li, H., & Xi, D. (2019). Investigation of the integrated model of side chamber, wear-rings clearance, and balancing holes for centrifugal pumps. JournalofFluidsEngineering, Transactionsofthe ASME, 141(10) doi:10.1115/1.4043059
[14] Yun, R., Zuchao, Z., Denghao, W., &Xiaojun, L. (2019). Influence of guide ring on energy loss in a multistage centrifugal pump. JournalofFluidsEngineering, Transactionsofthe ASME, 141(6) doi:10.1115/1.4041876
[15] Pirouzpanah, S., Patil, A., Chen, Y., & Morrison, G. (2019). Predictive erosion model for mixed flow centrifugal pump. Journal of Energy Resources Technology, Transactions of the ASME, 141(9) doi:10.1115/1.4043135
[16] Blume, M., & Skoda, RR. (2019). 3D flow simulation of a circular leading edge hydrofoil and assessment of cavitation erosion by the statistical evaluation of void collapses and cavitation structures. Wear, 428–442 29, 457–469. doi:10.1016/j.wear.2019.04.01
[17] Ovsyannikova, B. V. high-Speed blade pumps / B. V. Ovsyannikova edited by A. N. Sherstyuk-M., 1975–337c.
[18] A. K. Mikhailov, Malyushenko V. V. Vane pumps. Theory, calculation and construction./Mikhailov A. K. M.: Mashinostroenie, 1977. 289s.
[19] Patent 2013147386/06, 2013.10.24. Ways to optimize the geometric parameters of a side semi-spiral inlet of a two-way centrifugal pump // Russian Patent No. 2532870 / Kushnarev V. I., Kushnarev I. V., Obozny Yu. S.
[20] Lomakin V. O., Petrov A. I., shcherbachev P. V. Development of a side semi-spiral supply with an increased moment of fluid velocity at the inlet to the impeller// News of higher educational institutions. Engineering. 2012. Special Issue. P. 3.
[21] Artemov A.V., shcherbachev P. V., Tarasov O. I. Application of B-splines for constructing a side semi-spiral pump supply // Engineering Bulletin-M. 2014. no. 12
[22] H S Shim and K Y Kim 2016 IOP Conf. Ser.: Earth Environment. Sci. 49 032007
[23] S F Xie et al 2015 IOP Conf. Ser.: Mater. Sci. Eng. 72 032023
[24] Hu Hao , Li Xinkai and Gu Bo Hydraulic Optimization of Multiphase Pump Based on CFD and Genetic Algorithm // International Journal of Grid Distribution Computing — Vol. 8, No.6, (2015)

[25] Rudnev S. S., and Matveev I. V. Methodological guide to the course design of blade pumps Rotaprint MVTU, 1974, 70 p.
[26] E Morozova et al 2019 IOP Conf. Ser.: Mater. Sci. Eng. 589 012008
[27] N Isaev et al 2019 IOP Conf. Ser.: Mater. Sci. Eng. 589 012009
[28] K Abramov 2019 IOP Conf. Ser.: Mater. Sci. Eng. 589 012013
[29] Davydov I. V. Trudy VIGM / / issue XX, MOSGIZ-1958
[30] Troshin G. A., Petrov A. I. Methods of modification of the flow part of oil main pumps of the NM type // Engineering Bulletin. Bauman Moscow state technical University. Electronic scientific and technical journal. 2014. № 11.

Thermal processes in the electromechanical mandrel of holes in steel billets

A Morozov[1,3], G Fedotov[1], Y Ivanova[2], L Khabieva[1] and A Knyurov[1]

[1]Ulyanovsk state agrarian University named after P.A. Stolypin, avenue NovyVenets, 1, Ulyanovsk, 432017, Russian Federahion

[2]Bauman Moscow State Technical University, 5 Second Baumanskaya Street, Moscow, 105005, Russian Federation

[3]E-mail: alvi.mor@mail.ru

Abstract. The article presents findings on the distribution of temperature fields during Electromechanical mandreling (EMM) of holes in the surface of the workpiece and the tool. The design temperatures significantly exceed the phase transformation temperatures, but the melting of structural components in the surface layer of the workpiece does not occur due to high cooling rates. It is not possible to fix the maximum process temperatures experimentally due to their short duration of action.

Introduction

The level of quality and durability of both individual parts and assembly units as a whole is determined by the quality of the source material and the perfection and efficiency of the technological systems used in the manufacture of component parts.

The energy capacity of technological systems and the efficiency of energy use in them are constantly increasing with the development of production [1].

Electro-mechanical processing (EMP) refers to finishing and strengthening technologies, while the surface of the workpiece is subjected to simultaneous heating with phase transformations and force action on the surface in the austenitic state. The resulting white layer on the surface does not have a characteristic crystal structure, has a high hardness and high enough impact strength. These properties white layer under EMP acquires under the influence of ultra-fast heating, pressure and rapid cooling [2–6].

The surface quality required by the operating conditions is achieved by the correct assignment of the technological modes of EMP, their optimization. However, the purpose of technological modes for averaged (fixed) values of temperatures at various points along the section of the workpiece, significantly differ from the real ones and do not allow that to get a surface of the specified quality. Therefore, the task of modeling thermal processes in EMP is extremely important.

The use of EMP processes for processing small-diameter holes (PSD) is very relevant, since the use of known hardening technologies is problematic [7–11].

Materials and methods

For ESD of holes with a diameter of less than 40 mm, a mandrel is used as a tool, and the heat source would be bandwise (Fig. 1) [12, 13].

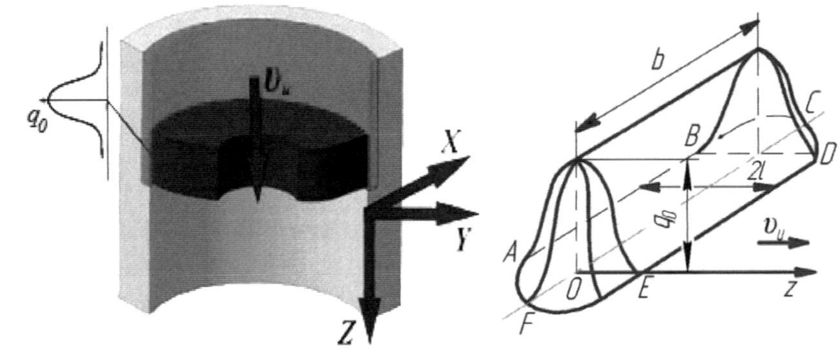

Fig. 1. Scheme for determining the temperature field of the hole to be processed

The solution to finding the temperature field when heated by a band-type high-temperature source q moving along the surface of the workpiece hole at the speed of v is found on the basis of the differential equation of thermal conductivity

$$c_i(T_i)\rho_i(T_i)\frac{\partial T_i(\vec{r},t)}{\partial t}=\frac{1}{r}\frac{\partial}{\partial r}\left(r\lambda(T_i)\frac{\partial T_i(\vec{r},t)}{\partial r}\right)+\frac{\partial}{\partial z}\left(\lambda(T_i)\frac{\partial T_i(\vec{r},t)}{\partial z}\right),\ \vec{r}\in V_i,\ i=\text{I, II},$$

whereV_i — spatial regions of an axisymmetric cylindrical structure; $\vec{r}=\vec{r}(r,\phi,z)$, $(0\le\phi\le 2\pi)$; λ_i, c_i, ρ_i — coefficients of thermal conductivity, specific heat capacity and density of structural elements.

Initial condition:

$$T_i(\vec{r},0)=T_c,$$

where: T_c — initial temperature equal to the ambient temperature 300 K.

The complexity of thermal phenomena that occur when a high-temperature moving band source is exposed to the surface of a metal material in the EMM process is caused by the interaction of a large number of factors on which they depend. These factors include: heat release as a result of electrophysical effects (Joule heat release, Peltier, Thomson, Seebeck effects, electrolytic phenomena), heat released during tool friction on the surface of the workpiece, heat released as a result of deformation of the metal surface layer, heat exchange between the tool and the surface layer, heat release as a result of phase transformations, heat transfer from the source to the workpiece, and heat exchange with the environment.

The solution of this problem was carried out by the most common finite element method in the theory of thermal conductivity.

When choosing a modeling method, should take into account not only the effectiveness of the mathematical calculation procedure itself, but also the ability of the software product itself to ensure the proper level of adequacy.

For convenience and clarity, the simulation was carried out in two stages.

At the first stage, thermal processes were modeled with A band-band high-temperature source without taking into account tool movement (quasi-static model) in the COMSOL Multiphysics V5 mathematical modeling environment.2.

In the second stage, simulation was carried out with the movement of the tool relative to the surface (quasidynamic model) using environment mathematical modeling ANSYS 14.5.

The use of two environments of mathematical modeling allowed a more accurate way to assess convergence of the results.

To form the boundary conditions of the modeled process, the calculation schemes are presented (Fig. 2).

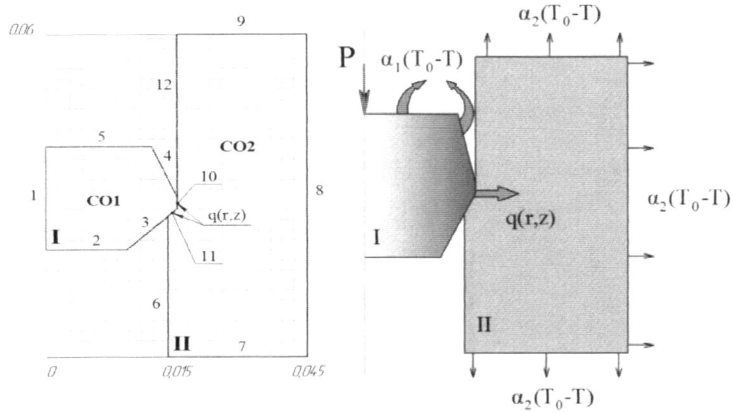

ab

Fig. 2. Calculation schemes: a — for boundary conditions on the external surfaces of the structure; b— for thermal effects from a moving source

For the center 1 of the axisymmetric structure I (figure 2 a), the following conditions apply:

$$\left.\frac{\partial T_I(\vec{r},t)}{\partial r}\right|_{r=0} = 0,\ T_I(\vec{r},t)\Big|_{r=0} \neq \infty.$$

Boundary conditions on the external surfaces of the structure:

$$5, 4 — \lambda_I \frac{\partial T_I(\vec{r}, t)}{\partial n} + \alpha_{\text{вын}}(T_I(\vec{r}, t) - T_c) = 0,$$

$$12 — \lambda_{II} \frac{\partial T_{II}(\vec{r}, t)}{\partial n} + \alpha_{\text{вын}}(T_{II}(\vec{r}, t) - T_c) = 0,$$

$$2, 3 — \lambda_I \frac{\partial T_I(\vec{r}, t)}{\partial n} + \alpha_{\text{ест}}(T_I(\vec{r}, t) - T_c) = 0,$$

$$6, 7, 8, 9 — \lambda_{II} \frac{\partial T_{II}(\vec{r}, t)}{\partial n} + \alpha_{\text{ест}}(T_{II}(\vec{r}, t) - T_c) = 0,$$

where $\frac{\partial}{\partial n}$ — derivative in the direction of the external normal to the corresponding surface, $\alpha_{\text{вын}4}$, $\alpha_{\text{вын}5}$, $\alpha_{\text{вын}12}$, $\alpha_{\text{ест}2}$, $\alpha_{\text{ест}3}$, $\alpha_{\text{ест}6}$, $\alpha_{\text{ест}7}$, $\alpha_{\text{ест}8}$, $\alpha_{\text{ест}9}$ — coefficients of natural (*αест*) and forced (*αвын*) heat exchange with the environment, W/(m² K).

In the zone of contact between the tool and the part with different thermophysical parameters, the heat flux density is continuous, that is

$$-\overrightarrow{n_I}(\lambda_I \nabla T_I) - \overrightarrow{n_{II}}(\lambda_{II} \nabla T_{II}) = q(\vec{r}),$$

where the function $q(\vec{r})$ is modeled by the normal distribution function in the following form

$$q(\vec{r}) = \left(2{,}056 \cdot 10^7\right) e^{-4{,}938 \cdot z^2}.$$

The resulting system of equations allows us to calculate the temperature field in a metal alloy for any time.

The models were built using real dimensions. When creating a finite element model, special attention was paid to splitting the area where there is a direct contact interaction between the tool and the hole surface to be processed, this is expressed in the need to significantly reduce the size of the grid elements in the area of contact between the tool and the part.

A two-dimensional idealization of the object in question was used to construct the geometric model. At the same time, the number of finite elements was about 3500.

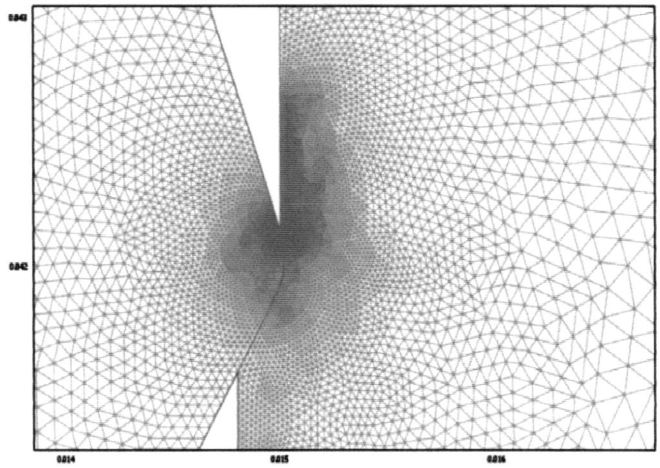

Fig. 3. General view of the finite element partition of the contact zone

The view of the end elements and the diagram of the division of the contact zone of the tool with the surface of the hole of the part is shown in figure 3.

Research Results and Discussion

As an example, consider the process of electro-mechanical mandreling (EMMP) of a sample hole made of steel 403 S 17 with a carbide mandrel made of VK8 in the following processing modes: current I = 5000 A, I = 5200 A, I = 5400 A; tension i = 0.1 mm.

In the study of heat processes was analyzed the time required to reach the maximum temperature on the surface of the workpiece as the source of electric current and the temperature change into the contacting objects under quasi-static model (EMM).

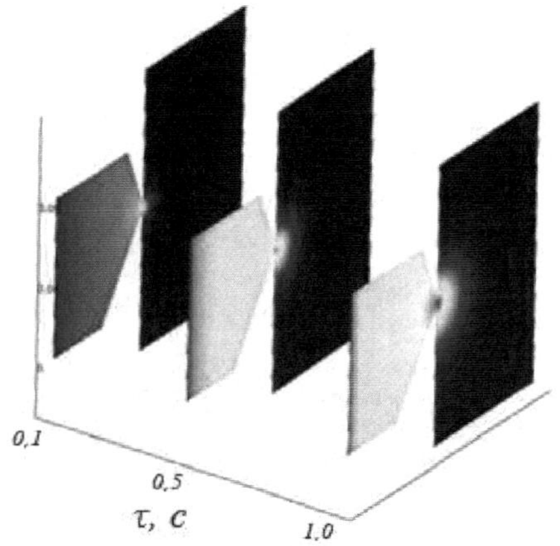

Fig. 4. Visualization of temperature field propagation as a function of time

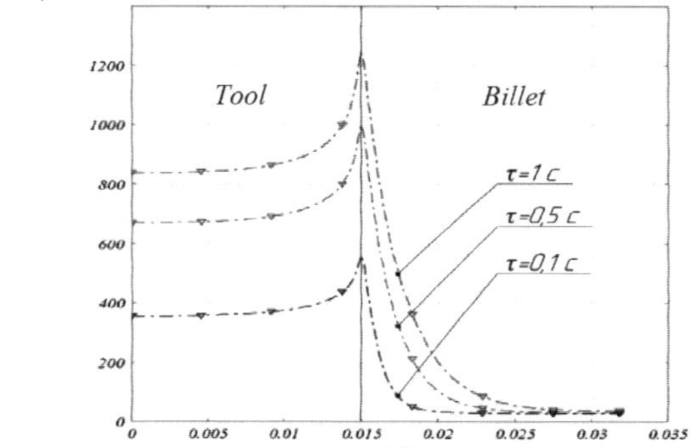

Fig. 5. Change in the temperature of the tool and workpiece at EMM from the time of action of the source at I=5400 A

Figure 4 shows the dynamics of temperature changes in the body of the workpiece and tool of the sample and tool depending on the time of action of the heat source. Based on the results of calculations, it was found that the maximum temperature on the surface of the workpiece is reached in about one second and at a current strength of I = 5400 A is 1214 °C.

The temperature change in the tool and the workpiece depending on the time of action of the heat source is shown in figure 5.

The study of the temperature along the line of contact of the tool with the surface of the workpiece revealed that its maximum value is reached at the point of contact of the lower part of the spherical surface of the mandrel with the sample (figure 6). This is due to the high contact pressures at this site, which ensures the best contact.

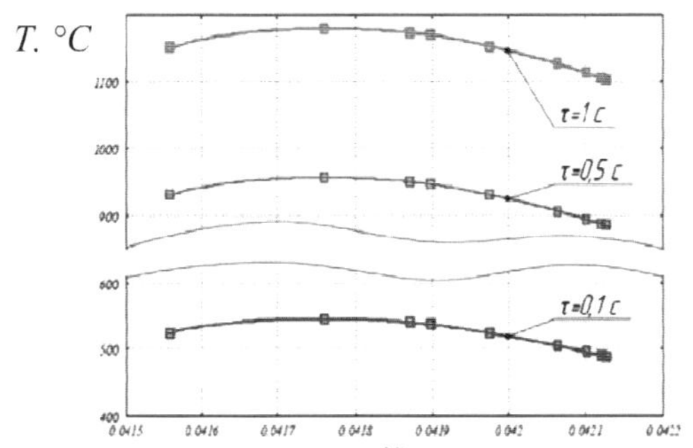

Fig. 6. Temperature change from time along the contact line of the tool with the workpiece at I=5200 A

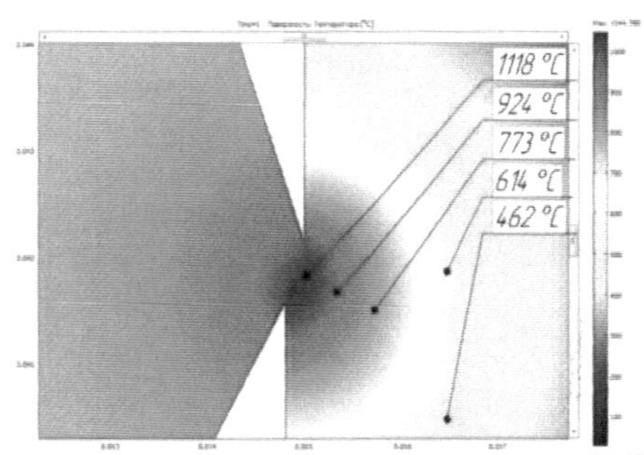

Fig. 7. Temperature field for the depth of the workpiece at I = 5200 A

As a result of calculations, it was found that with an increase in the current strength in the range from 5000 A to 5400 A, there is an increase in the surface temperature from 921 °C to 1214 °C (Fig. 8).

As can be seen from figures 7 and 8, the temperature values decrease sharply in the depth of the surface layer.

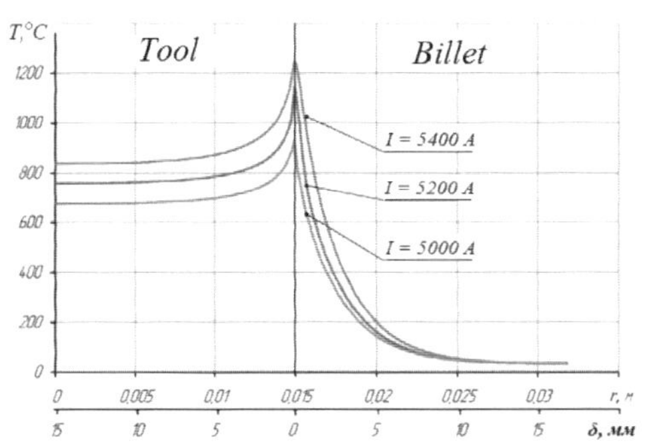

Fig. 8. Influence of electric current on the temperature field of the tool and workpiece from a constant source

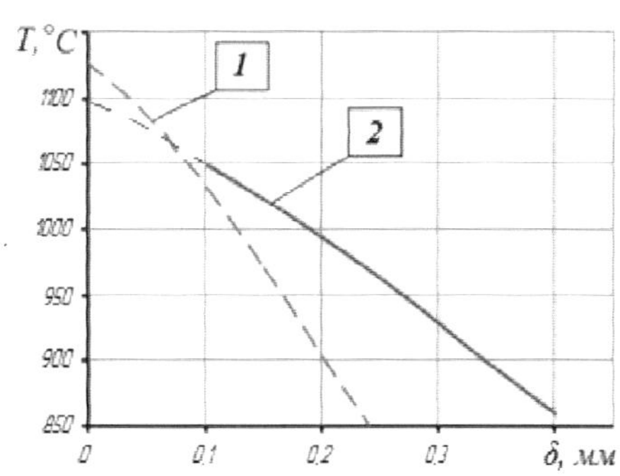

Fig. 9. The temperature field of the workpiece at I=5200A: 1-calculation; 2-experiment

Figure 9 shows the results of calculation and measurement of temperature changes in the depth of the workpiece under the same conditions.

The discrepancy of 4% can be explained by the fact that the simulation did not take into account possible current losses during its passage from the source (transformer) to the tool and workpiece, since current losses depend on a number of factors, it is almost impossible to adequately consider them in the simulation.

IOP Conf. Series: Materials Science and Engineering **963** (2020) 012017 doi:10.1088/1757-899X/963/1/012017

Conclusion

The results of solving the thermal problem are: visualization of the distribution of temperature fields, numerical values of heating-cooling rates depending on the modes of the studied EMM processes by a band-band high-temperature source.

References

[1] A. Abramov, A. Morozov, A. Koshkina, S. Petryakov, Ju. Nuretdinova, S. Bratan. The cleaning process model of diesel fuel in an electric field. Matec web of conferences. The proceedings International Conference on Modern Trends in Manufacturing Technologies and Equipment: Mechanical Engineering and Materials Science. (2019). C. 00095.

[2] B.M. Askinasi Strength Improvement and Rebuilding of Machine Parts by Using Electro-Mechanical Hardening. B.M. Askinasi. Moscow, Mashinostroenie, 1989.

[3] V.P. Bagmutov, S.N. Parshev, N.G. Dudkina, I.N. Zakharov, Electro-Mechanical Hardening: Process Technology and Physics, Characteristics, Implementation.Novosibirsk: Nauka Publishing, 2003.

[4] Fedorov S.K., Fedorova L.V., Ivanova Y.S., Voronina M.V. Increase of Wear Resistance of the Drill Pipe Thread Connection by Electromechanical Surface Hardening,International Journal of Applied Engineering Research ISSN 0973-4562 Volume 12, Number 18 (2017) pp. 7485–7489 © Research India Publications.

[5] L.V. Fedorova, S.K. Fedorov, A.A. Serzhant, V.V. Golovin, S.V. Systerov, Electromechanical surface hardening of tubing steels. Metal Science and Heat Treatment. 59(3–4) (2017). 173–175.

[6] L.V.Fedorova, S.K. Fedorov, Y.S. Ivanova, A.M. Lompas,Technology of Parts Wear Resistance Improvement by Using Electromechanical Surface Hardening. Universities New Bulletin. Mashinostroenie Publishing. (2017), №9 (690).85–92.

[7] L.V. Fedorova A.V. Morozov, V.A. Freeling. Increasing the efficiency of electromechanical hardening of holes of smooth cylindrical movable joints experiencing a one-sided radial load. Repair, restoration, modernization. (2012) №8, 49–52.

[8] A.V. Morozov, G.D. Fedotov, Development of the classification of electromechanical processing of holes by a moving high-temperature strip source. Strengthening technologies and coatings. (2015), №3, 44–50.

[9] A. Morozov, G. Fedotov, D. Musharapov. The results of bench tests for the durability of straight-line movable spline joints after hardening by electromechanical hardening. Vestnik of Ulyanovsk state agricultural academy. (2019), № 2 (46). 19–23.

[10] A. Morozov, G. Fedotov, K. Kundrotas. The influence of volumetric electromechanical mandreling on the lead yield from the matrix material on the bronze bearing bushing surface. Conference Series: Materials Science and Engineering Collection of materials of the International Conference on Modern Trends in Manufacturing Technologies and Equipment: Mechanical Engineering and Materials Science. (2020). C. 033058.

[11] A. Morozov, G. Fedotov, L. Fedorova, D. Musharapov, L Khabieva. The providing durability of the movable square-sided spline joints by electromechanical treatment of the working surfaces. Matec web of conferences The proceedings International Conference on Modern Trends in Manufacturing Technologies and Equipment: Mechanical Engineering and Materials Science. (2019). C. 00117.

[12] G. Carslow, D. Jaeger. Thermal conductivity of solids. Moscow, Science, 1964.

[13] L.A. Kozdoba. Methods for solving nonlinear heat problems. Moscow, Science, 1975.

Application of optimization methods to reduce the mass of body parts of minibuses made of layered composite materials

A Kostenko[1] and V Zuzov[1,2]

[1]Bauman Moscow State Technical University, 5 Second Baumanskaya Street, Moscow, 105005, Russian Federation

[2]E-mail: zuzvalery@rambler.ru

Annotation. This article proposes an approach to reducing the mass of parts from layered composite materials, including parametric and topological optimization of parts using the finite element method. The proposed approach made it possible to shape the body for a minibus with the least amount of time, which confirms its effectiveness.

Introduction

The finite element method (FEM) in general, and topological optimization, in particular, are well studied and are often used in the design of automobiles and other equipment [1]–[10]. However, this statement relates more to solid materials, but the calculations of layered composites, despite the increase in their popularity, are performed much less frequently. This is due to the insufficient study of optimization techniques for these types of materials [11]–[20]. At the same time, using the right approach, it is possible to significantly change the properties of these materials even without changing the geometry of the parts, reducing their thickness, and, hence, the mass [21]–[30]. This is especially true when designing load-bearing structures of automobiles made of fiberglass, such as the Shuttle minibus.

The purpose of this article is to evaluate the effectiveness of the combined approach, including parametric and topological optimization of load-bearing shell elements of body parts made of layered composites (in terms of thickness, quantity and nature of laying layers), from the standpoint of reducing their mass and reducing time costs.

Mathematical model

For the initial assessment, we will choose the object of the study, which is part of the Shuttle minibus floor structure. Its dimensions are 100x300 mm, its thickness is 3.2 mm, and its weight is 164 g. The floor experiences bending-torsional loads acting in different planes. For the plate, we select the design scheme shown in Figure 1: the support is carried out on two pivotally-fixed supports, and four concentrated forces act on the plate from above. The task is to reduce the weight of the plate by 70% with minimal compliance of the structure (since rigidity for the floor is crucial) and acceptable stresses.

Let's consider two options:

IOP Conf. Series: Materials Science and Engineering **963** (2020) 012018 doi:10.1088/1757-899X/963/1/012018

– The plate is made of a homogeneous orthotropic material (table 1). For it, we perform the traditional topological optimization;

– The plate is made of a layered material, 8 layers of which are laid in accordance with the scheme shown in table 2. For this plate we perform a three-stage optimization.

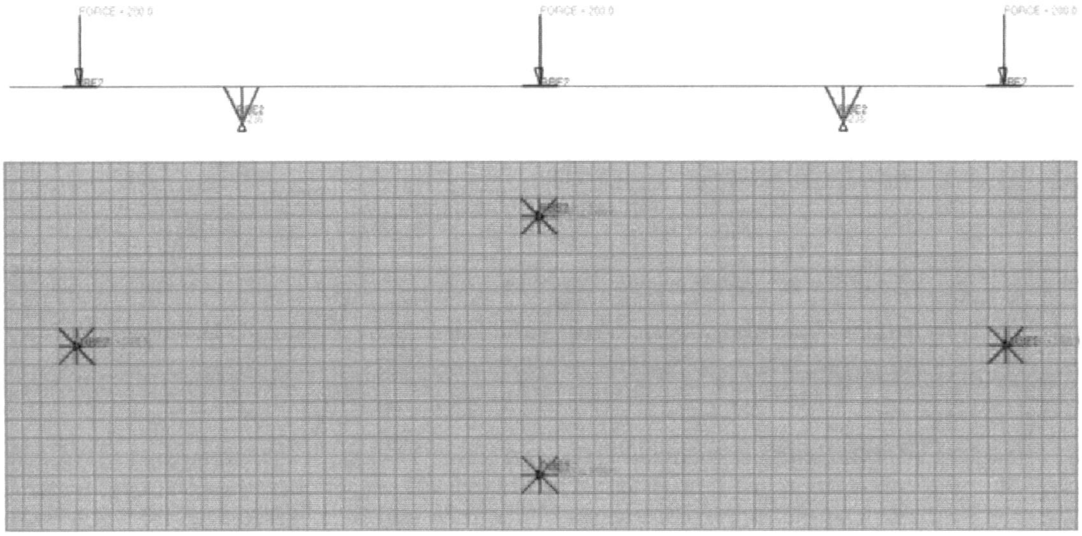

Figure 1. Calculation scheme

Table 1. Characteristics of the material used

Materialtype	Designation	E1, MPa	E2, MPa	μ	G, MPa	ρ, t/mm^3
orthotropic	MAT8	110000	8500	0,3	4000	1,7·10-9
σ_{ut} along the fibers, MPa	σ_{uc}along the fibers, MPa	σ_{ut} across the fibers, MPa		σ_{uc} across the fibers, MPa		$\sigma_{ush,}$ MPa
1500	900	58		170		80

Table 2. Layout of composite material

№ layer	1	2	3	4	5	6	7	8
Stacking	0°	45°	–45°	90°	0°	45°	–45°	90°
Layerthickness,mm	0,4	0,4	0,4	0,4	0,4	0,4	0,4	0,4

Comparison of source objects

First, we perform the calculation of the original plate. For an object of homogeneous material, we obtain displacements of 1.22 mm, and stresses of 93.9 MPa. For a layered material, with equal mass, we have a displacement of 0.98 mm and a stress of 61.1 MPa. The calculation results are presented in Figure 2. Thus, already at this stage it is clear that the use of layered material can provide tangible advantages in the development of structures: the rotation of several layers without changing the shape of the product significantly reduces stress and displacement in the structure.

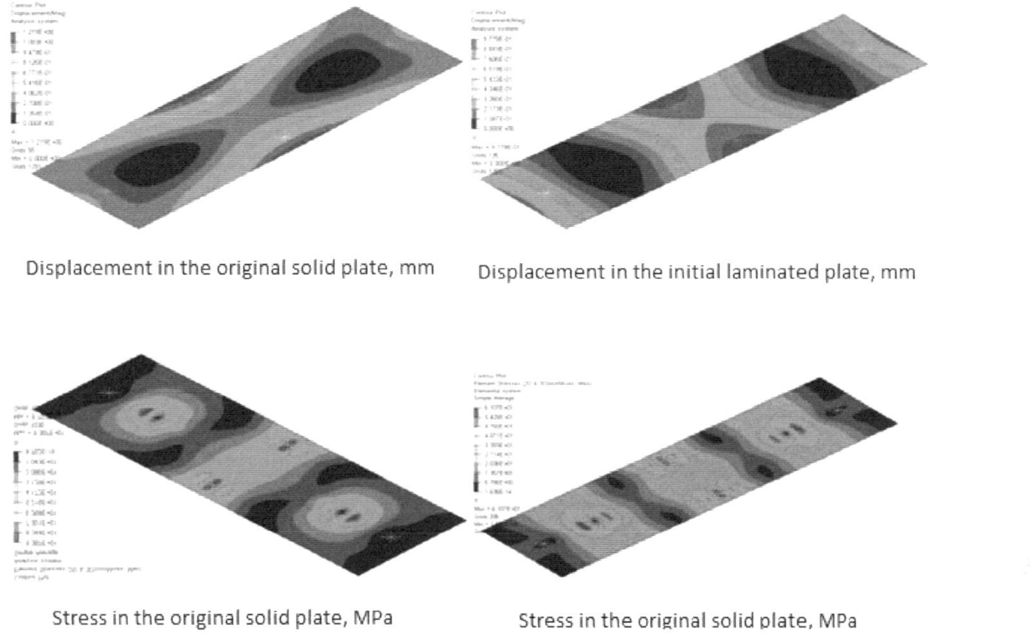

Figure 2. Results of calculations of the initial plates

Continuous plate topological optimization

The final mass of the object is set at 30% of the original. The objective function is to minimize compliance. To converge, the solution took 22 iterations. As a result, it was possible to reduce the weight to 49 grams, while the displacements amounted to 3.01 mm, and the stresses increased to 181.7 MPa. The results are presented in Figure 3.

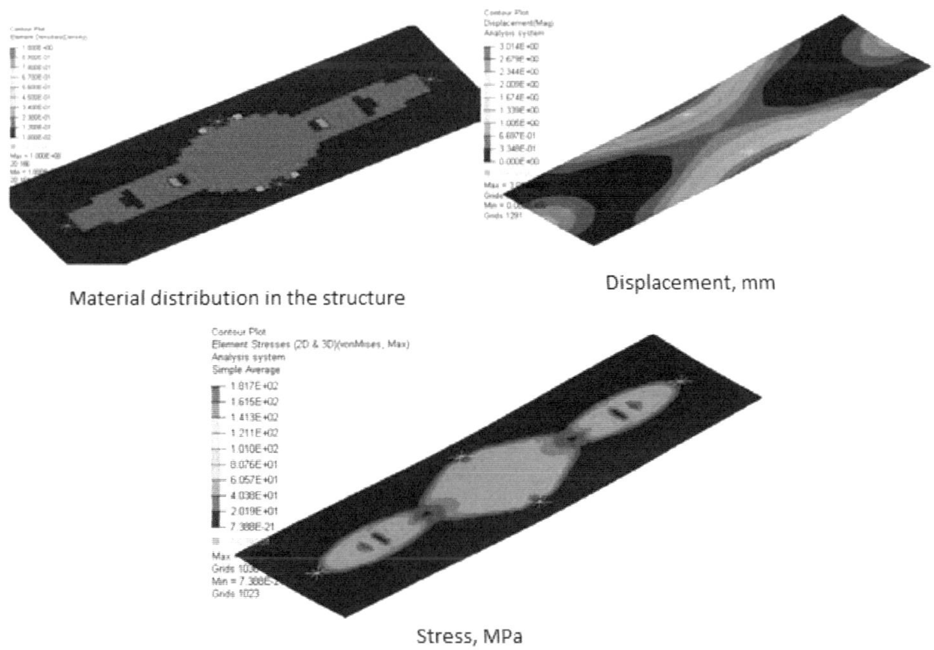

Figure 3. Results of topological optimization of a continuous plate

Layered plate optimization

We will perform optimization of the laminated plate in three successive procedures. Free-Size optimization allows algorithms to freely change the number of layers and their thickness, decreasing to values tending to 0 in the case when the layer does not carry a load. The final mass of the plate is set at 30% of the original. The objective function is to minimize compliance. The solution converged in 20 iterations. The total mass of 49 g, the maximum displacement of 2.53 mm, the maximum equivalent stress of 282.9 MPa. The solver broke the composite into 32 layers. The structure of the obtained composite is shown in Figure 4 and in table 3.

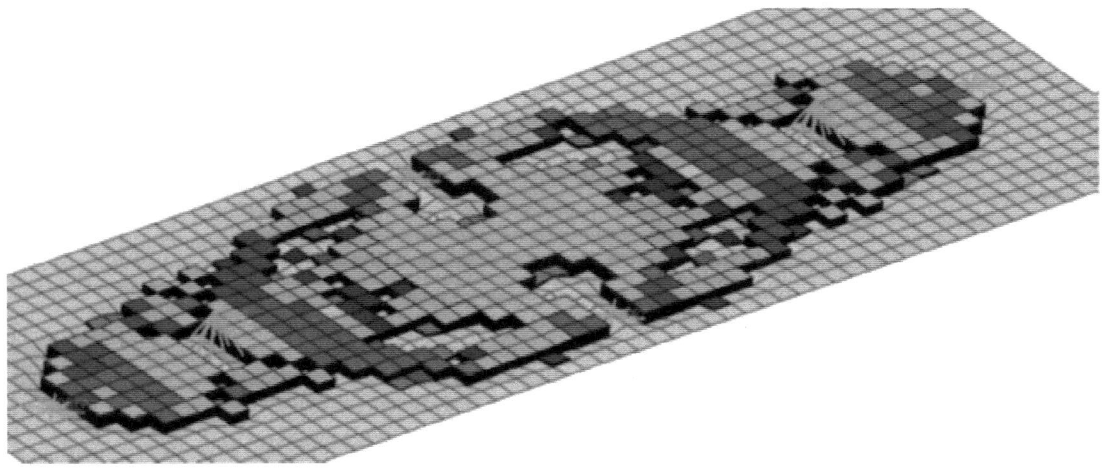

Figure 4. Composite structure (Free-Size optimization)

Table 3. Layout after Free-Size optimization

№ layer	1	2	3	4	5	6	7	8
Stacking	0°	45°	–45°	90°	0°	45°	–45°	90°
Layerthick ness,mm	0,00518	0,00646	0,0640	0,00610	0,00500	0,00656	0,00551	0,00531
№ layer	9	10	11	12	13	14	15	16
Stacking	0°	45°	–45°	90°	0°	45°	–45°	90°
Layerthick ness,mm	0,13351	0,12730	0,13594	0,13652	0,14388	0,09563	0,08252	0,05818
№ layer	17	18	19	20	21	22	23	24
Stacking	0°	45°	–45°	90°	0°	45°	–45°	90°
Layerthick ness,mm	0,25161	0,25271	0,24520	0,24271	0,24324	0,27939	0,28975	0,29090
№ layer	25	26	27	28	29	30	31	32
Stacking	0°	45°	–45°	90°	0°	45°	–45°	90°
Layerthick ness,mm	0,00971	0,01352	0,01246	0,01467	0,00788	0,01842	0,02223	0,04561

The second stage is Size optimization. The initial data are the results obtained in the previous step. The point of this stage is to achieve the same technologically feasible layer thicknesses. The maximum mass of the plate is set at 100% of the original. The objective function is to minimize compliance. The target thickness of one layer is 0.2 mm. The solution converged in 5 iterations. The total mass of

49 g, the maximum displacement of 2.77 mm, the maximum equivalent stress of 160.2 MPa. The solver has reduced the number of layers to 15, and the total plate thickness is now 3 mm. The thicknesses and orientations of the layers are presented in table 4, and the distribution of the material is shown in figure 5.

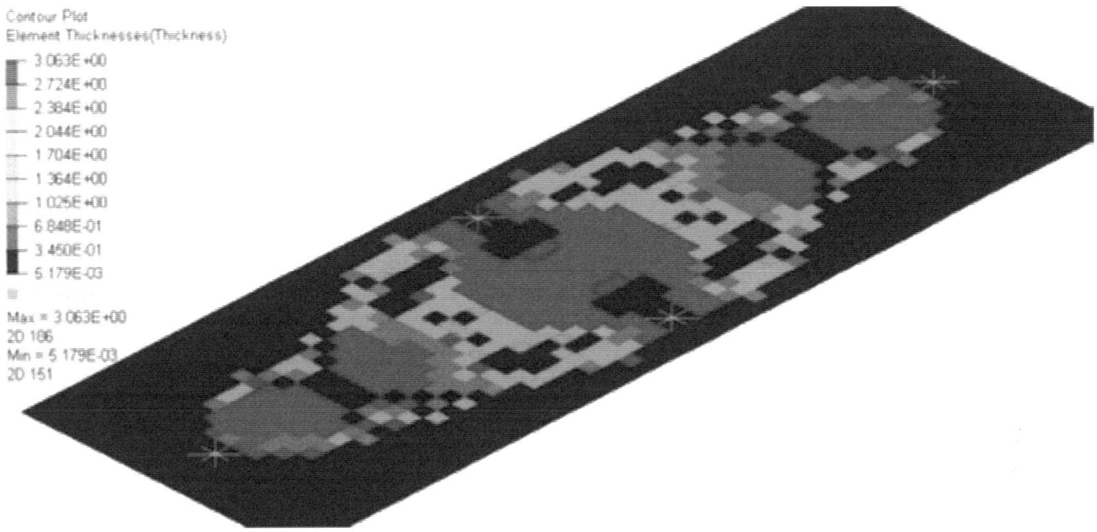

Figure 5. Distribution of material in the structure (Size optimization)

Table 4. Layout after Size Optimization

№ layer	1	2	3	4	5	6	7	8
Stacking	0°	45°	–45°	90°	0°	45°	–45°	90°
Layerthickness,mm	0,2	0,2	0,2	0,2	0,2	0,2	0,2	0,2
№ layer	9	10	11	12	13	14	15	
Stacking	0°	45°	–45°	90°	0°	45°	–45°	
Layerthickness,mm	0,2	0,2	0,2	0,2	0,2	0,2	0,2	

The third stage is Shuffle optimization. Here, the solver varies only the mutual arrangement of the layers, achieving the optimal location. The objective function is to minimize compliance. The solution converged in 2 iterations. The total mass of 49 g, the maximum displacement of 2.69 mm, the maximum equivalent stress of 150.2 MPa. The calculation results are shown in Figure 6, and the final thickness and orientation of the layers are presented in table 5.

Table 5. Layout after Shuffle Optimization

№ layer	1	2	3	4	5	6	7	8
Stacking	45°	–45°	0°	90°	45°	–45°	0°	90°
Layerthickness,mm	0,2	0,2	0,2	0,2	0,2	0,2	0,2	0,2
№ layer	9	10	11	12	13	14	15	
Stacking	45°	–45°	0°	90°	45°	–45°	0°	
Layerthickness,mm	0,2	0,2	0,2	0,2	0,2	0,2	0,2	

IOP Conf. Series: Materials Science and Engineering **963** (2020) 012018 doi:10.1088/1757-899X/963/1/012018

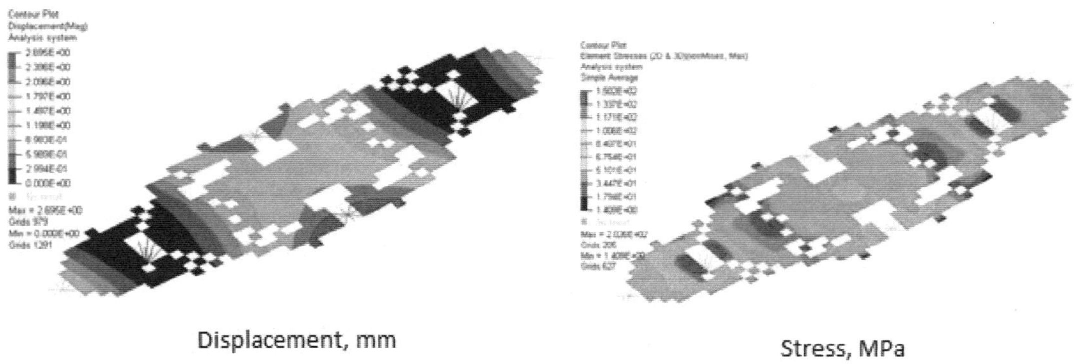

Figure 6. Shuffle optimization results

For clarity, we present table 6 with the calculation results.

Table 6. The results of the calculations

Model	Mass, g	Displacement, mm	Stress, MPa
Original, orthotropic	163	1,22	93,9
Original, layered	163	0,98	61,1
Aftertopologicaloptimization	49	3,01	181,7
AfterFree-SizeOptimization	49	2,53	282,9
AfterSizeOptimization	49	2,77	160,2
AfterShuffleoptimization	49	2,69	150,2

Applying Combined Optimization to a Van Body

As you can see, the optimization of the layered material will achieve better outcomes than the calculations with orthotropic material. At the same time, optimization of composite materials is much more laborious both in terms of creating models and the time spent on calculations. For best results with minimal time, it is proposed to combine these two methods. Let us consider the proposed algorithm more clearly by the example of preliminary optimization of the Shuttle minibus body made of laminated fiberglass, shown in Figure 7.

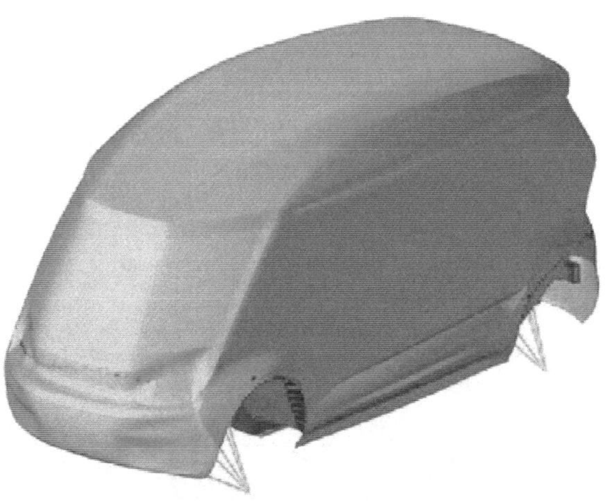

Figure 7. Model for optimizing the body of the Shuttle minibus

First, a finite element model is created, including a frame, suspension, batteries, as well as the estimated mass distribution. The body is modeled on the basis of a design project, door openings are determined. Then, orthotropic material is set, and the traditional topological optimization of the body is performed on the basis of a dedicated package of design cases, for example, parking, breakdown of the front wheel, torsion, skidding, as well as static simulations of tipping, side and frontal impacts. Based on the results obtained, the future power framework is built and included in the calculation model (Figure 8).

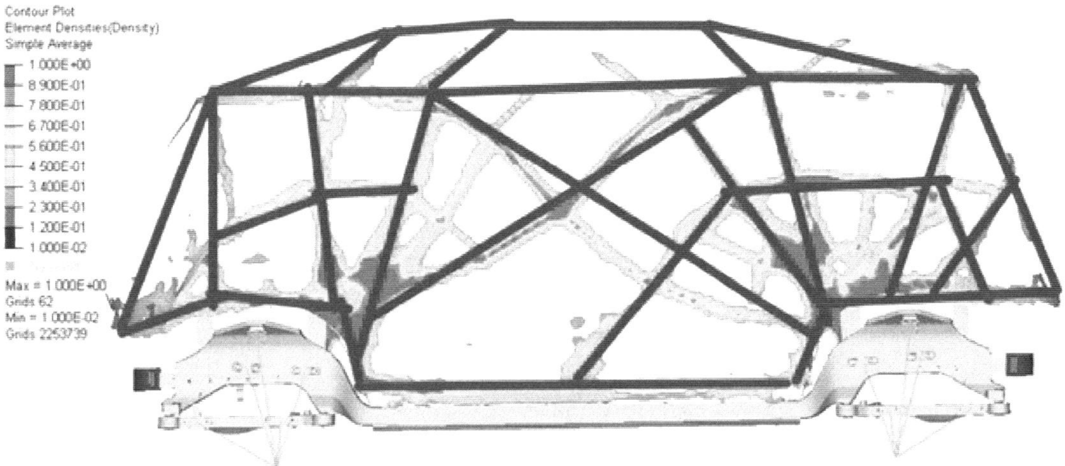

Figure 8. Creating a power frame

The homogeneous model of the body material is replaced by a layered one, which is attached to the designed frame. Using the previous load conditions, the optimization of a layered body is performed: Free-Size, Size and Shuffle. As a result, an optimized design is obtained, on the basis of which the body is built, intended for further deeper research.

Conclusions

The use of a combined approach based on a combination of parametric and topological optimizations as applied to structures made of composite materials made it possible to obtain better parameters for a layered material compared to a homogeneous one. In the considered example, it was possible to reduce the plate mass to 49 g, and at the same time achieve maximal displacements of 3.01 mm and stresses of 181.7 MPa for a homogeneous material, and 2.69 mm and 150.2 MPa for a layered material, respectively.

The proposed combined approach, combining topological optimization with the optimization of layered composites, made it possible to form the appearance of a body with a supporting frame for a minibus with the least time costs, which confirms its effectiveness.

References

[1] Hashemi E, Pirani M, Khajepour A, Kasaiezadeh A 2016. *Vehicle System Dynamics* **54** 1736–61
[2] Wei Y, Liu Y, Li X, Oertel C 2016. *Vehicle System Dynamics* **54** 463–73
[3] Taheri M, Ahmadian M 2016. *Vehicle System Dynamics* **54** 653–66
[4] Rodriguez J, Freeman P T, Wagner J, Pidgeon P, Alexander K 2016. *Int. J. of Automotive Technol.* **17** 71–81
[5] Xia X, Xiong L, Sun K, Yu Z P 2016. *Int. J. of Automotive Technol.* **17** 991–1002
[6] Gorelov V A, Komissarov A I 2016. *Procedia Engineering* **150** 1322–28.
[7] Gorelov V A, Komissarov A I, Miroshnichenko A V 2015. *Procedia Engineering* **129** 300–7
[8] Keller A V, Gorelov V A, Anchukov V V 2015. *Procedia Engineering* **129** 280–7.

[9] Kotiev G O, Padalkin B V, Kartashov A B, Diakov A S 2017. *ARPN J. of Eng. and Appl. Sci.* **12** 1064–71.
[10] Volskaya N S, Zhileykin M M and Zakharov A Y 2018. *IOP Conf. Series: Materials Science and Engineering***315** 012028
[11] Vdovin D, Chichekin I 2016. *Procedia Engineering***150** 1276–79
[12] Ejsmont J, Taryma S, Ronowski G, Swieczko-Zurek B 2016. *Int. J. of Automotive Technol.* **17** 237
[13] Klubnichkin E E, Klubnichkin V E, Kotiev G O 2018. *IOP Conf. Series: Materials Science and Engineering***386** 012025
[14] Zhileykin M M,Kotiev G O, Nagatsev M V 2018. *IOP Conf. Series: Materials Science and Engineering***315**012031
[15] Sarach E B, Kotiev G O, Beketov S A 2018. *MATEC Web of Conf.***224** 04009
[16] Skotnikov G I, Jileykin M M and Komissarov A.I 2018. *IOP Conf. Series: Materials Science and Engineering***315** 012027
[17] Diakov A S, Kotiev G O 2018. *MATEC Web of Conf.***224** 02096
[18] Kotiev G O, Butarovich D O, Kositsyn B B 2018. *IOP Conf. Series: Materials Science and Engineering***315** 012014
[19] Evseev K B, Kartashov A B, Dashtiev I Z and Pozdeev A V 2018. *MATEC Web of Conf.* **224**02039
[20] Klubnichkin V E,Diakov A S, Klubnichkin E E, Zakharov A Y, Vakhidov U Sh, Suchenina A S and Basmanov I V 2019.J. of Phys.: Conf. Series **1177** 012048
[21] Altair. University. Tutorial: Composite optimization of a skateboard with Optistruct, 2017. — 37 c
[22] Banichuk N.B. Optimization of structural elements from composite materials // NB Banichuk, V.V. Kobelev, R.B. Rickards — M.: Mechanical Engineering, 1988. — 224 s.
[23] Chernyaev A.V. Optimization of elements of aircraft structures from composite materials in a discrete setting // A.V. Chernyaev, SSU, 2009 — 151 s.
[24] Goncharov R.B. Topological optimization of a car bumper design under impact from the standpoint of passive safety / RB. Goncharov, V.N. Zuzov // Proceedings of the Moscow State Technical University MAMI. — 2018. — No. 2 (36). — S. 2–9.
[25] Vdovin D.S. Method of topological optimization in the design tasks of safe cabins of agricultural and construction equipment / D.S. Vdovin // Proceedings of the Moscow State Technical University MAMI. — 2018. — No. 4 (38). — S. 21–29.
[26] Andres Tovar Renaud Crashworthiness design using topology optimization / Tovar Andres, M. Patel Neal, Byung-Soo Kang, E. John // Journal of mechanical design. — 2009. — No. 131. — P. 1–12. —– DOI: 10.1115 / 1.3116256.
[27] Shaban B.A. Features of modeling frame elements of car bodies and cabs in the study of passive safety. Science and education / Shaban B.A., V.N. Zuzov // MSTU. N.E. Bauman, 2012, no. 11. URL: http://technomag.edu.ru/en/doc/ 486675.html, doi: 10.7463 / 1112.0486675
[28] Goncharov R.B. Problems of searching for optimal solutions to ensure passive safety of truck cabs with minimum weight / RB. Goncharov, V.N. Zuzov // Proceedings of the Moscow State Technical University MAMI, 2019, No. 1, p. 2–14.
[29] Frantsuzov A.A. Application of the method of topological optimization in design problems of lifting equipment / A.A. French, Y.I. Shapovalov, D.S. Vdovin // Engineering. Engineering and mechanical engineering. — 2017. — No. 2 (42). — S. 99–108. DOI: 10.21685 / 2072-3059-2017-2-9.
[30] Kishov E.A. Topological optimization of power structures by convex linearization / E.A. Kishov, V.A. Komarov // Bulletin of Samara University. Aerospace engineering, technology and engineering. — 2018. — T. 17. — No. 1. — S. 137–149. DOI: 10.18287 / 2541-7533-2018-17-137-149.

IOP Conf. Series: Materials Science and Engineering **963** (2020) 012019 doi:10.1088/1757-899X/963/1/012019

To the concept of improving the quality of technology for the arctic zone of the russian federation

L Kushnarev[1,2] and S Kushnarev[1]

[1]Bauman Moscow State Technical University, 5 Second Baumanskaya Street, Moscow, 105005, Russian Federation

[2]E-mail: kushnarevl@mail.ru

Abstract.
The restoration and development of the technological potential of machine-building enterprises to the modern technical level with the existing approaches requires not only significant investment, but also a robust strategy of actions to achieve the targets, the main of which is a comprehensive increase in the quality of manufacturing and increase in the production volume of machine-building products. Only a quick and steady increase in the quality of machinery and equipment, at least to the level of the best world analogues, will ensure the steady and intensive development of the entire machine-technological complex of the country. This is especially true for improving the quality of equipment designed for use in harsh arctic conditions.
Materials and methods. The current level of domestic equipment quality differs significantly from foreign counterparts. The results of organizational and technological studies indicate the need to improve existing forms and methods of organizing production and maintenance by machine manufacturers throughout the entire life cycle.
Monographic research, analysis and synthesis, abstract and logical, technical and economic assessments are used as scientific research methods.
Research results. The goal is solving the problem of improving the quality of domestic technology. The proposed organizational and technological solutions provide an accelerated output of mechanical engineering in terms of product quality and production volumes in accordance with the requirements of technological modernization of the country's economy.
Discussion and conclusion. To achieve the required level of quality of technological machines and equipment, authors propose to improve the quality of manufacturing and reliability indicators of machines and equipment. The solution to this problem is the reorganization of the existing production and technological system and the introduction of proprietary engineering and technical service (maintenance) of equipment throughout the entire life cycle.

Introduction

The basis of the socio-economic development of the Arctic zone of the Russian Federation is the effective functioning of military, industrial and social facilities that ensure the defense capability of the northern borders, exploration and production of natural resources, and vital functions of personnel and the population. Harsh Arctic conditions impose special requirements not only on people living and working in these conditions, but also on the means of material and technical support used, forms and methods of organizing labor and production [1–4].

The peculiarities of the Arctic zone development are the presence of significant sparsely populated territories, a low level of technical equipment of production and the social sphere with highly reliable machines and equipment, objects of repair and maintenance base, a shortage of highly qualified technical personnel, limited opportunities for the development of specialization and cooperation in organizing engineering and technical support.

The purpose of organizational and technological research is to comprehensively improve the quality of machinery and equipment to the level of the best world analogues based on the development and implementation of a proprietary production system and the provision of engineering and technological support for engineering products at all stages of the life cycle. The implementation of the following principles is equally applicable to the organization and development of engineering and technical support for military, industrial and social structures during operation of machinery and equipment, provided in accordance with their areas of activity and tasks [3–5].

1. It is advisable first to focus efforts and resources on intensifying the development of already explored and developed areas. This allows to use them purposefully and as efficiently as possible using the concentration of all types of resources.

2. The development of the production and operational base for engineering and technical support for the operation of machinery and equipment should be based on the maximum concentration of production of goods, work and services, a rational level of specialization and autonomous functioning of military, industrial and social facilities.

3. It is advisable to determine the production, technological and economic parameters of objects of engineering and technical support for all areas of activity, their structure, buildings, technical equipment, and staffing with qualified personnel, taking into account real production conditions and sizes based on the development of pilot and standard organizational and technological projects.

4. The structure and parameters of engineering support facilities are determined by the need for modern technology, machinery and equipment for technical service, which depends on the level of their quality and, first of all, the quality of manufacture and reliability parameters. The cost of their operation, downtime for technical reasons, the efficiency of production processes and enterprises in general directly depend on the quality of manufacturing and reliability parameters of the machines and equipment produced.

Therefore, in the production of technology (including military), machinery and equipment operating in severe climatic conditions of the Arctic, a significantly higher level of quality is necessary. And if by the parameters of the technical level, by the technical and tactical characteristics, the vast majority of domestically produced machinery and equipment meets or exceeds the level of the best world analogues, the quality indicators and reliability parameters are lower, which significantly reduce the efficiency of their intended use. This requires solving certain organizational and technological problems of improving the quality of engineering products.

Methods and techniques

The fundamental requirement for conducting real organizational and technological research is to ensure a high degree of accuracy and reliability of the results at all stages of work. Therefore, modern methods were used to justify and optimize the parameters, collect, systematize and process the initial and statistical information about the production processes and use of machinery and equipment for its intended purpose, ensuring their performance and restoring the resource; research and evaluation of requirements for preventive maintenance and sudden failures; research and evaluation of the wear processes of critical work surfaces of the joints of machine parts and equipment, their components and assemblies; monitoring and analysis of the processes of creating and using a machine, assessing the technical condition and efficiency of machines and equipment in real use in the production of goods, works and services [4–10].

Research results

The solution to the problems of improving quality is the reorganization of the existing system of manufacturers of machinery and equipment into proprietary production and technological systems (firms) that can provide engineering and technical support for their products from the creation of the machine to its disposal. All quality improvement processes are managed at all stages of the life cycle of machinery and equipment. Due to this, a proprietary production and technological system (firm) is able to ensure the achieved world (and higher) level of quality of domestic equipment, control it and manage it.

The problems of improving the quality (competitiveness) of Russian technology, machinery and equipment are systemic and require a comprehensive solution of production, technological, organizational, economic and legislative problems and issues. Their solution will provide the required level of quality not only of technology for the Arctic, but also of the entire national economic complex of the country. Therefore, it is advisable to single out a set of problems of improving quality into a special state program.

The main requirements for the quality of machinery and equipment for the Arctic are as follows. The design of machines and equipment should provide a high level of ergonomic requirements and, first of all, safety and comfort of personnel during their use and maintenance; have a high level of reliability and, above all, failure-free performance; the frequency of planned preventive measures and the mean time between failures should not be lower than that of the best world analogues. In the manufacture of the main critical parts and joints, units and assemblies of machines and equipment, it is advisable to use promising innovative materials, high-precision machining technologies and hardening of working surfaces to increase their resource. During operation, new highly effective lubricants and other working fluids designed specifically for harsh Arctic conditions should be used. Together, these measures can provide an increase in the resource of machinery and equipment by 1.5–2 times.

Taking into account the peculiarities of the operation of machinery and equipment in the Arctic, the equipment should be delivered with a specific set of tools for monitoring the technical condition, performing maintenance operations and eliminating the consequences of simple failures by operators and drivers directly at site or at the facilities of the repair and maintenance base [11–14].

As the main method of organizing technical maintenance of machines and equipment, it is advisable to carry out maintenance or replace faulty parts and assemblies using the self-service method (using their own resources) with the involvement of the forces and means of maintenance and repair workshops. When performing technologically complex repair work, the aggregate repair method provides the greatest efficiency.

6. Designing of objects for the repair and maintenance base of technological machines and equipment is carried out taking into account the actual operating conditions of machines and equipment, their parameters and design, needs for maintenance and repair, and the level of specialization of work. When creating objects, it is advisable to use block-modular structures, from which buildings and structures of the repair and maintenance base should be made easily and practically without special technical means. Materials of block-modular structures must ensure high environmental safety and low thermal conductivity.

7. The facilities of the repair and maintenance base should be equipped with modern universal diagnostic and control and repair and technological equipment, machines, devices and tools in accordance with their real needs.

8. For the efficient use of machinery and equipment, maintenance and repair tools and objects of the repair and maintenance base, it is advisable to organize engineering and technical services staffed by qualified personnel, whose main task should be to ensure the operability of the entire complex of machines and equipment of an enterprise or group of enterprises. As criteria for assessing their activities, we establish the level of serviceability of the serviced machinery and equipment [15–17].

9. In order to reduce the downtime of machinery and equipment for technical reasons at the facilities of the repair and maintenance base, it is advisable to have an irreducible (constantly replenished) supply of units, assemblies and parts, operational and repair technical materials.

Replenishment, control over the expenditure and safety of spare parts, operational and repair and technical materials are assigned to the engineering and technical service.

10. For a rational and highly efficient organization of production and economic activities of the engineering and technical services of enterprises and other mechanized structures, it is advisable to carry out organizational and technological design of services and their types depending on the areas of machinery and equipment used and its quantity at the enterprise.

11. When switching to the proprietary method of organizing the technical servicing of machinery and equipment, engineering and technical services may act as technical representatives of equipment manufacturers directly at site.

12. The effectiveness of the functioning of military, industrial, social structures, as well as structures providing engineering and technical support (service) depends on the completeness and timeliness of deliveries of high-quality and reliable equipment, fuel and lubricants and other operational materials and liquids, spare parts and repair technical materials, availability and creation of volumes of insurance reserves of these resources. This requires solving a large number of complex logistic problems by types and volumes of supplied resources, directions of consumers' activities and their territorial distribution, the presence and condition of roads and stable communication, the availability and condition of logistics centers, their technical equipment and many other factors.

Deliveries of machinery and equipment, material and technical resources for operation and ensuring operability are carried out by various supply trading and intermediary organizations, unsystematically as a rule, according to one-time applications. Machines, assemblies, units, operational and repair materials supplied by them usually have deviations from the requirements of GOST and TU for manufacturing. As a result, the cost of the supplied material and technical resources significantly increases, the quality of these products decreases, since the direct connection between the manufacturer and the consumer of machines and equipment is lost and there is no warranty technical service, the costs of machine downtime for technical reasons are huge [17–19].

13. The implementation of this concept requires the development of an innovative integrated system of training, retraining and advanced training of specialists: highly qualified specialists, engineering and technical workers and workers in the mass professions. The main innovative directions of development of the integrated training system should be:

comprehensive improvement of the quality of domestic machinery and equipment for the Arctic;

reorganization of existing manufacturing plants into proprietary production and technological systems (firms) capable of ensuring the world level of quality of domestic equipment;

development and organization of proprietary engineering and technical services for manufactured machinery and equipment;

development and implementation of promising and resource-saving technologies and technical means of machine use (production and technical operation);

development and implementation of organizational and economic mechanisms to stimulate the quality improvement of manufactured machines, the use of resource-saving technologies and technical means, the introduction of effective forms and methods of organizing the production and maintenance of machinery and equipment.

The practical implementation of these requirements will reduce the need for equipment and its downtime for technical reasons, reduce the cost of maintaining machinery and equipment, minimize the need for a repair and maintenance base, investment in its creation and highly qualified repair and maintenance personnel.

References

[1] Adler Yu.P. A new direction in statistical quality control is Taguchi methods. Moscow, Znanie; 1988. [In Russian]
[2] Ford G. Moya zhizn' [My life]. Moscow: AST. 2014.

[3] Ishikawa K. Japanese methods of quality management. Short translation from English. Moscow, Economics. 1988.[In Russian]
[4] Kiselev M.I., Pronyakin V.I., Tulekbaeva A.K. Technical diagnostics functioning machines and mechanisms // IOP Conference Series: Materials Science and Engineering. 2018. V. 312(1). 012012.
[5] Gubaydulina R.H., Gruby S.V., Davlatov G.D. Analysis of the Lifecycle of Mechanical Engineering Products // IOP Conference Series: Materials Science and Engineering. 2016. V. 142(1). 12060. DOI: 10.1088/1757-899X/142/1/012060
[6] Protopopov A., Jakovich C. Compromise resource-efficiency curve for a centrifugal pump. *IOP Conf. Ser.: Mater. Sci. Eng.*, 2019, vol. 492, no. 1, art. 012034.
DOI: 10.1088/1757-899X/492/1/012034
URL: https://iopscience.iop.org/article/10.1088/1757-899X/492/1/012034
[7] Fedorov S., Albagachiev A., Isaenkova M., Yakovleva A., Zaripov V., MinushkinR.The effect of combined processing schedules on the value and nature of residual stresses in the surface layer of cylindrical friction pairs Всборнике: IOP Conference Series: Materials Science and Engineering electronic edition. 2019. C. 012024.
[8] N.N. Zubkov, A.I. Ovchinnikov, S.G. Vasil'ev, Tool–workpiece interaction in deformational cutting. Russian Engineering Research , 36(3) (2016) 209-212.
[9] N.N. Zubkov, V. Poptsov, S. Vasiliev., A. Batako, Steel Case Hardening Using Deformational Cutting. Journal of Manufacturing Science and Engineering, 140(6) (2018) 061013.
[10] N. Zubkov, V. Poptsov, S. Vasiliev, Surface Hardening by Turning without Chip Formation. Jordan Journal of Mechanical and Industrial Engineering, 11(1) (2017) 13–19.
[11] Harrington J.H. Quality Management in American Corporations. Translation from English. Moscow, Economics. 1990. [In Russian]
[12] Hofmann D. Measuring and computing quality assurance systems. Translation from German. Moscow, Energoatomizdat. 1991. [In Russian]
[13] Taylor F.W. The Principles of Scientific Management, Harper & Brothers, N.Y.. 1911.
[14] Kushnarev L.I. [etc.]. The quality and reliability of domestic equipment is the basis of its competitiveness // Tractors and agricultural machinery. T. 11. P. 35–37. [In Russian]
[15] Kushnarev S.L., Chepurina E.L. Organization of proprietary technical service for agricultural machinery // Tractors and agricultural machinery. T. 10. P. 3–4. [In Russian]
[16] Kushnarev L.I. Proprietary technical service of machinery and equipment. Problems. Search. Solutions (monograph). Saarbrücken: Palmarium Academic Publishing. 2014.
[17] Chepurina, E. L.,Kushnarev L. I., A. V. Chepurin, S. L. Kushnarev, the Quality and reliability of domestic equipment is the basis of its competitiveness //Niva of the Urals. — № 11 (133). — December 2015. — Pp. 52–54
[18] Chepurina, E. L., Kushnarev L. I., S. L. Kushnarev. Problems and directions of development of engineering and technical support of rural producers //Repair, restoration, modernization № 1. — 2016. — P. 3–9.
[19] Kushnarev L. I. to the solution of a problem of increase of competitiveness of domestic equipment. Repair, restoration, modernization. — № 5. — 2017. — P. 3–8

IOP Conf. Series: Materials Science and Engineering **963** (2020) 012020 doi:10.1088/1757-899X/963/1/012020

Practical usage of study results of the porous materials weldability from corrosion-resistant steels when creating stamp-welded products with desired properties

A Tretyakov

Bauman Moscow State Technical University, 2-nd Baumanskaya, 5, 105005, Moscow, Russian Federation

E-mail: tretyakov@bmstu.ru

Abstract. It is shown that a significant factor limiting the use of porous powder materials for the manufacture of die-welded products is low tensile plasticity and insufficient weldability. Porous meshing materials (PMM) made of steel 12Cr18Ni10Ti (X10CrNiTi18-10) have the necessary technological and operational properties for the manufacture of permeable die-welded elements. The conditions for the formation of a weld in PMM fusion welding are affected by the porosity and the ratio of the thermal conductivity coefficients along the thickness of the welded edges and in the direction perpendicular to the movement of the welding source. Technologies for argon-arc welding of a heat exchanger-evaporator sleeve, electron-beam welding of filter elements with a housing, which are implemented in the manufacture of preproduction models, have been developed.

Introduction

The development of modern engineering requires the creation of new power plants and devices with high efficiency coefficient and functioning reliability. The transition to more intense operating parameters of units operation made it necessary to solve a number of problems, one of which is the creation of technological processes for the manufacture of structures that provide protection from high temperatures, oxidizing environments, the ingress of mechanical particles, as well as their rational cooling.

An increase in the temperatures and speeds of energy carriers led to an increase in heat fluxes to the walls, the most effective way of cooling which is to supply gas to the boundary layer through a porous shell [1,2]. In addition, the use of devices with permeable elements makes it possible to reliably maintain the required temperature on the wall surface, reduce friction forces, and reduce or eliminate their mechanical erosion.

The desire to increase the reliability of modern pneumatic and hydraulic systems leads to increased requirements for the purity of the liquids and gases used in them. In the process of creating the design of filters, an important step is a reasonable choice of permeable elements [3,4], which should provide the necessary fineness of cleaning, have high permeability and the necessary mechanical properties in the entire range of operating temperatures and loads. At the same time, particles washed out of the permeable material during operation should not get into the cleaned media. In addition, filters must meet the requirements of corrosion resistance. In this case, the porous preforms used for the manufacture of permeable elements of various configurations must have the necessary technological properties [5].

The effective functioning of structures largely depends on a reasonable choice of porous preforms, the type and dimensions of which are determined from the conditions of optimal compliance of their properties and the requirements for them. Porous materials include fabrics made of artificial and natural fibers [6], mesh made of metal wires, blanks obtained from metal powders [3,4,7] and fibers, perforated sheets [8] and some others.

The analysis of the structures of the units and their operating conditions has allowed us to establish that the most promising are porous materials based on metal, which have the required permeability, pore size, and high mechanical and technological properties.

The use of filter elements as blanks was found in woven mesh of corrosion-resistant steels. The fineness of cleaning liquids and gases with a single layer of a grid reaches 20...40 μm [3]. Some improvement in filtering performance can be achieved as a result of multi-layer mesh briquettes.

Studies of the stampability and weldability of woven meshes showed their low technological properties due to significant anisotropy and the movement of wires relative to each other during plastic deformation. In addition, the disadvantages of the meshes include the possibility of increasing the bore sections of the cells during sheet stamping, fusion welding and product operation.

The creation of permeable materials based on metal fibers (PFM) was of great interest, primarily due to the possibility of achieving high porosity, which can reach 0,98. Investigations of the effect of compaction parameters on the structure and properties of PFM by A.G. Kostornov and his co-workers [9]–[11] showed that these materials have high hydraulic and mechanical properties.

The data presented indicate the feasibility of using PFM blanks for the manufacture of capillary structures of heat pipes and intake devices that supply liquids against gravity or in zero gravity, as well as permeable elements with a fineness of filtration 2...10 μm. In addition, these materials with a porosity of 0,76...0,80 have a 2...6 times higher damping ability than aluminum with an equal mass-to-area ratio. Due to the high anti-vibration and sound-absorbing properties, fiber materials have found application in the manufacture of structural elements used in the operation of precision instruments, electronic equipment, machine tools and turbines. However, the lack of data on the formability and weldability of fiber materials does not currently allow us to recommend them for the manufacture of stamped products with desired properties.

Experimental studies of the hydraulic, filtering, mechanical and technological properties of porous powder materials (PPM) made of corrosion-resistant steels have shown that the main advantage of this type of permeable workpieces is the ability to retain mechanical particles up to 2 μm in size when filtering liquids and gases [3,7]. A significant factor limiting the use of PPM for the manufacture of permeable products is low formability since the elongation of these materials at break does not exceed 1,5%.

The possibility of manufacturing permeable products of large sizes and complex shapes from PPM is inextricably linked with the problem of obtaining welded structures by connecting porous elements. At the same time, the joints of the product elements must have a minimum seam width, maximum relative strength in the absence of cracks and other through defects in the seam or in the heat affected zone, and also minimal oxidation of the internal and external surfaces. The results of the study of the weldability of PPM under argon-arc welding (ARAW) showed that when the porosity of the base material is $\Pi = 0,4...0,5$ ($P = 0,4...0,5$), due to the presence of closed pores in them, the main defect is the through «fistula» in the weld formed during ejection metal in the process of melting. In this case, the relative strength of the butt joints does not exceed 0,7...0,9. In addition, the use of ARAW is limited, since wide seams are formed, which significantly reduce the filtration surface.

The experiments on electron beam welding (EBW) of PPM blanks using standard equipment showed the possibility of obtaining equal strength joints. However, as a result of the uneven structure characteristic of highly porous PPM, burn-throughs occur in butt welded joints in areas with large pore sizes. To obtain high-quality welds in the manufacturing process of products from PPM grade PCr18Ni15 (powder made from high alloy steel Cr18Ni15), it was proposed in [12] to use the remaining linings from compact steel 12Cr18Ni10Ti, the melting temperature of which is lower than that of the welded workpieces.

The analysis of designs, manufacturing techniques and operating conditions of elements for cleaning from mechanical contaminants of liquids and gases [3,4,12], heat protection systems of rocket engines [3,13], shells of turbine blades of combined-cycle plants [14]–[16] and gas turbine engines [1,17,18], capillary intake devices [9], etc. showed that the porous mesh materials (PMM) made of 12Cr18Ni10Ti steel possess optimal properties. PMM is made by hot rolling in vacuum envelopes of a briquette of woven nets. These materials provide specified operational characteristics and have high mechanical [19, 20, 21] and technological properties [22], which makes it possible to produce die-welded structures with desired properties and the required configuration.

The existing results of theoretical and experimental studies in the field of pressure welding and hot rolling of multilayer sheets [22]–[25] were used to create sheet porous mesh materials.

Research Methodology. The difference between PMM and porous powder materials is a pronounced anisotropy of properties. The ratio of the coefficients of the effective thermal conductivity of the edges in thickness ($\lambda_\perp$) and in the direction perpendicular to the movement of the welding source (λ_W) has a significant effect on the formation of a weld under conditions of fusion welding. In addition, when welding highly porous workpieces, which are characterized by low thermal conductivity in thickness, the formation of defects of various types is possible depending on the size.

Thus, for workpieces with a high value λ_W as a result of intensive heat removal in the sheet plane, the penetration depth decreases and the temperature gradient increases along the width of the weld pool, which, in turn, can lead to a gap in the weld during crystallization (Fig. 1). And when welding the same workpieces in the direction, which is characterized by a low value, there is no fusion of individual wires in the edges of the connected elements with the weld metal. Such defects arise as a result of the action of surface tension forces at the boundary of the molten metal of the melted wires and the weld pool.

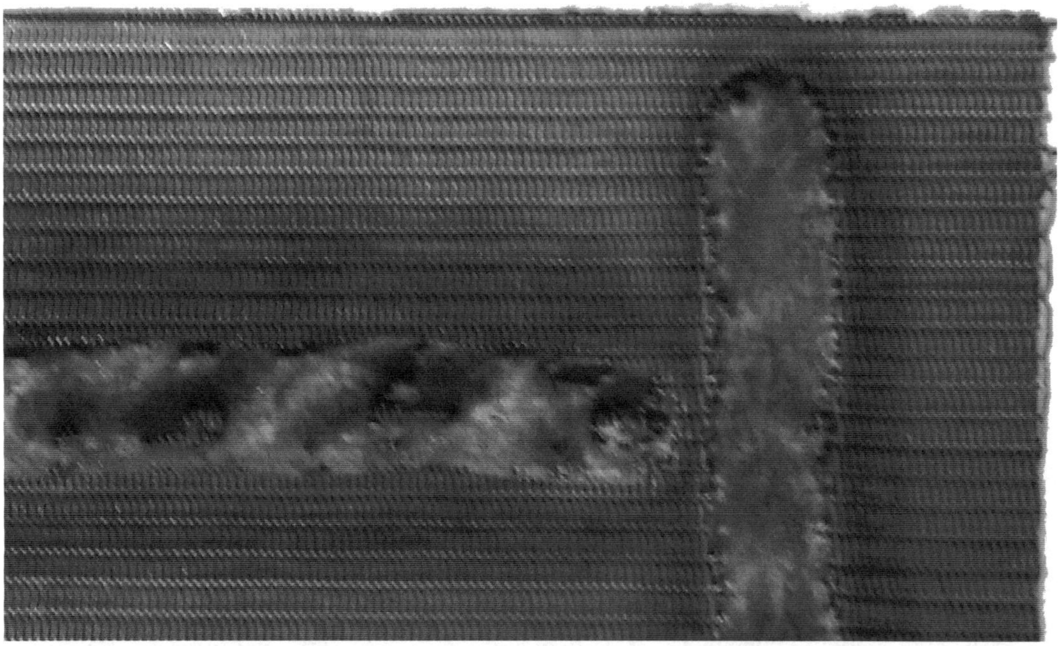

Figure 1. The nature of the penetration of PMM based on P60 filter meshes with a parallel arrangement at П = 0,4 (P = 0,4) in conditions of automatic argon-arc welding.

In addition, the experiments showed that for workpieces with high thermal conductivity anisotropy in the sheet plane, the width of the impermeable zone (weld) does not depend on the direction of welding, but increases with decreasing thermal conductivity in the direction of the thickness of the welded edges [22].

A study of the weldability of PMM from steel 12Cr18Ni10Ti showed that with a decrease in porosity and an increase in the ratio, the quality of welded joints increases. For the manufacture of highly porous welded structures, it is advisable to use PMM with a lower value of thermal conductivity anisotropy in the sheet plane. To this end, it is advisable to use PMM based on meshes with square cells located at an angle of 45° [22].

The study of the mechanical properties of tensile butt joints made by fusion welding made it possible to establish that with an increase in the porosity of the base material, their relative strength decreases. Improving the quality of joints of highly porous workpieces can be achieved by using special edge preparation, providing an increase in their thermal conductivity and a decrease in porosity.

One of the effective ways to improve the quality of welded joints of workpieces from multilayer PMM with a porosity of more than 20% is to seal the welded edges, which can be realized by cold deformation between the rotating rollers of the seam welding machine. The effect of edge compaction on the relative strength of butt joints was studied on PMM samples 2...5 mm thick with a porosity of 0,3...0,5. Sealing the edges provides a decrease in porosity to 0,20...0,25. The test results showed that cold deformation of the edges made it possible to increase the strength of butt welded joints made by automatic ARAW by 15...20% (Fig. 2).

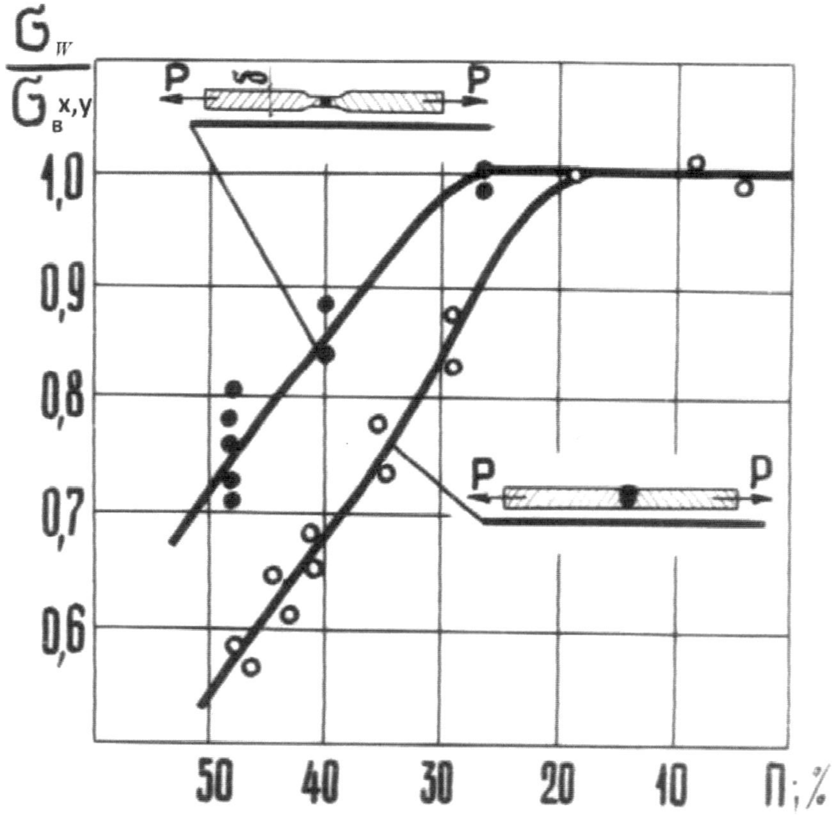

Figure 2. The effect of porosity of the base material and rolling of the edges on the relative strength of welded joints.

The variety of parameters that determine the properties of welded joints of PMM blanks greatly complicate the ability to assess the impact of each of them on the quality of products. The choice of the optimal welding technology is based on a comparison of several options and can be most effectively carried out by analyzing mathematical models.

The rational values of the ARAW and EBW modes of butt joints are determined on the basis of communication equations obtained using the theory of similarity and the method of regression analysis. Using the obtained equations and criterion dependencies for calculating the linear energy and the width of the weld when welding compact steel 12Cr18Ni10Ti, we obtained the equations for the approximate calculation of the energy required for the formation of butt joints in EBW, as well as the width of the weld in ARAW:

$$q/\sqrt{V_W} = 20 \cdot h_P \cdot T_M \cdot (1-P) \cdot \sqrt{\lambda_\perp \cdot C_C \cdot \rho_C}, \quad (1)$$

$$S_S = 0{,}17 \cdot q/T_M \cdot (1-P) \cdot \sqrt{\lambda_\perp \cdot C_C \cdot \rho_C \cdot V_W \cdot h_P}, \quad (2)$$

where q — he effective power of the welding source; T_M — the melting temperature of the material; V_W — welding speed; ρ_C — density of compact material; C_C — heat capacity of compact material 12Cr18Ni10Ti; h_P — the thickness of the blanks from PMM; S_S — the width of the seam.

Research results and discussion. The performed ARAW experiments on porous 12Cr18Ni10Ti steel workpieces made it possible to establish that high-quality welded joints can be obtained at the lowest possible current at which the weld root is penetrated. Automatic welding of butt joints with a speed of more than 30 m / h requires a process with a higher current strength, which for workpieces with porosity of more than 30% leads to the formation of undercuts in the fusion zone, and at V_W <10 m / h the weld width sharply increases. Therefore, the optimal welding speed should be in the range from 10 to 30 m / h, and with an increase in the porosity of the workpieces, the speed decreases. To reduce the probability of undercut formation in the fusion zone, it is advisable to weld products with a porosity of more than 0,2 with filler material.

In order to reduce the width of the seam, the possibility of using pulsed argon-arc welding was investigated. The experiments showed that the pulsed mode is suitable for welding workpieces with porosity of less than 0,2, and at higher values burns and undercuts are formed. The use of filler material partially eliminates these defects, but leads to a significant increase in the width of the seam.

An analysis of the obtained dependences and the results of experimental studies allowed us to develop technological processes for manual and automatic argon-arc welding with a non-consumable direct current direct current electrode of direct polarity of structural elements with porosities from 0,05 to 0,5. The study of the dimensions of the weld joints showed that expression (2) gives overestimated values. However, it can be used for approximate calculations of the width of the tightness of the zone of welded products at the design stage.

A study of the mechanical properties of butt welded joints under uniaxial and biaxial tension made it possible to establish that with an increase in the porosity of the base material, a decrease in relative strength occurs, the value of which at a porosity of more than 20% can be described by the following empirical dependence:

$$\sigma_W/\sigma_{BX,y} = 1{,}3 - 1{,}5 \cdot \Pi,$$

where $\sigma_{BX,y}$ — tensile strength of PMM under uniaxial tension in the direction of the wire mesh; σ_W — the strength of the butt welded joint with the same porosity in the direction of the wire mesh.

Electron beam welding is the optimal type of welding of porous workpieces when creating filter elements for fine cleaning of liquids and gases from mechanical impurities. To develop welding technology and study the strength of welded joints, studies were carried out in which the beam current strength, welding speed, and workpiece porosity were varied. The intervals of variation of the parameters given in table.1, was determined based on the analysis of expression (1). The value of the relative strength of welded joints $\sigma_W / \sigma_{BX,y}$ was taken as the optimization parameter.

Table1. Main levels and intervals of variation

Parameters	Levels			Intervals of variation	Dimension
	lower–1	zero 0	upper +1		
Beam current	45	55	65	10	μA
Welding speed	20	35	50	15	m / h
Porosity	0,15	0,25	0,35	0,1	–

In the process of welding, the diameter of the focal spot, the magnitude of the focal length, and the accelerating voltage were not changed and were maintained respectively equal to $d_S = 0,5$ mm, $l_0 = 120$ mm, $V_V = 30$ kV. The value $\sigma_{BX,y}$ for PMM with a porosity of 0,35 and 0,15 was 190 and 300 MPa, respectively. As a calculation model, a polynomial of the first degree was adopted, the calculation of the coefficients of which allowed us to obtain the following regression equation:

$$\sigma_W / \sigma_{BX,y} = 0,804 + 0,075 \cdot \Pi - 0,023 \cdot \Pi \cdot V_W + 0,0057 \cdot V_W . \quad (3)$$

An analysis of equation (3) showed that when the porosity of the base material is less than 0,25%, an increase in the relative strength can be achieved by increasing the welding speed, and when the porosity is more than 0,25, a decrease in the welding speed leads to an increase $\sigma_W / \sigma_{BX,y}$. In this case, the beam current equal to 0,45...0,65 mA, slightly affects the strength of the connection.

Metallographic studies revealed the presence of undercuts in the places where the wires are not fused with weld metal, the number of which increased with increasing porosity. The width of the welds obtained on samples with a thickness of 2,8...3,8 mm when varying the parameters of the welding mode changed insignificantly and amounted to 2,5...3,0 mm.

The study of the influence of the thermal welding cycle on the change in the hydraulic and structural characteristics of porous billets showed that the decrease in the permeability of welded joints is proportional to the weld area, while the pore sizes in the fusion zone do not change.

In the manufacture of welded structures, one must take into account the requirement of manufacturability, which is ensured by a rational choice of the workpiece material, the shape of the elements to be welded, the type of connection, the type of welding and measures to improve the quality of products. To obtain welded structures with high performance, one should strive to use PMM with lower porosity and minimal anisotropy of properties in the sheet plane.

In the case when it is necessary to obtain products from highly porous materials, an increase in the quality of welded joints can be achieved by increasing the thermal conductivity in the direction of the thickness of the edges to be welded, and if possible, using diffusion welding.

The results of the studies were used in the design and manufacture of die-welded bushings of heat exchangers-evaporators and filter elements with a given refinement and permeability.

The development of aviation and space technology requires the creation of thermal control systems for aircraft. Particularly noteworthy is the use of devices such as a heat exchanger-evaporator using a porous material in which the heat supplied to the outer surface is absorbed as a result of the evaporation of the refrigerant.

Of considerable interest for the manufacture of a permeable element (sleeve) of the heat exchanger is PM based on metal meshes. The manufacturing technology of heat exchangers-evaporators using PMM is developed in accordance with the requirements of the technical specifications, which consisted of the following:

– the design of the sleeve of the heat exchanger (Fig. 3) is a pipe made of PMM with a size of 30x2 mm and a length of 180 mm;

–the sleeve should be made of steel 12Cr18Ni10Ti;

–the power of the heat exchanger must be at least 160 watts.

To ensure these requirements, studies were conducted using sheet PMM made by welding by rolling filter meshes of plain weave P60 (GOST 3187-76) [3]. The porosity of the blanks was 0,20 and 0,45. To obtain a heat exchanger sleeve, a technology has been developed for the manufacture of welded pipes, the molding of which was performed on a rolling mill 10-60 of the design of VNIIMETMASH. Automatic ARAW of the formed continuously moving billet was carried out on a mill in a special chamber.

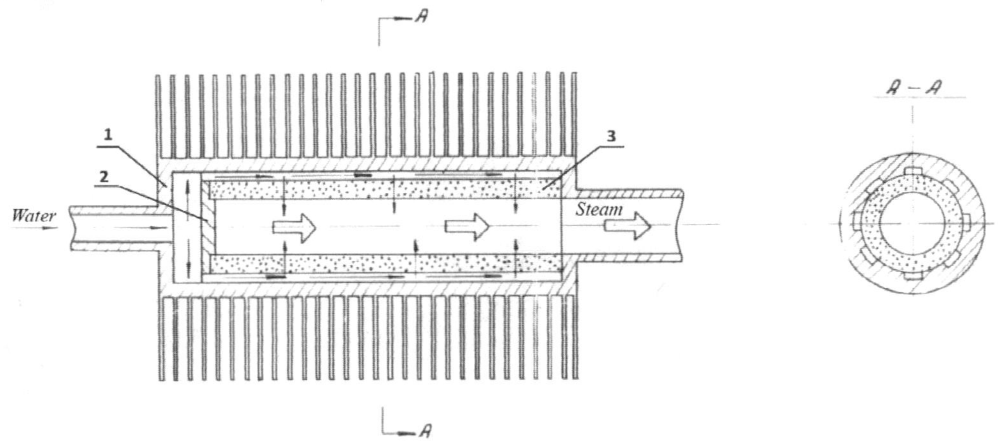

Figure 3. The scheme of the heat exchanger-evaporator: 1 — housing; 2 — sleeve; 3 — flange.

The calibration used after welding made it possible to produce welded pipes of the required sizes (Fig. 4). The choice of the welding mode of pipes was carried out according to the results of studies of the technology of argon-arc welding of butt joints.

Figure 4. Welded pipe from PMM.

The quality of the welded pipe joint was evaluated by external inspection and metallographic studies, which showed that the formation of pipes from a material with the studied porosity provides the desired geometry without breaking the welded joints of the wires, i.e. without stratification of preparations. The manufactured pipes were connected to the flanges of compact steel 12Cr18Ni10Ti manual ARAW. The sleeve thus obtained was pressed into the heat exchanger body. A general view of the heat exchanger-evaporator is shown in Fig. 5.

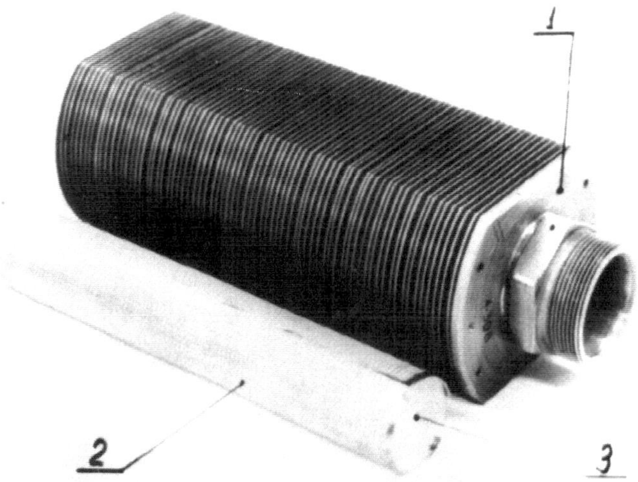

Figure 5. General view of the heat exchanger-evaporator:
1 — housing; 2 — pipe from PMM; 3 — flange.

The test of the obtained units was performed in laboratory conditions according to the following scheme. The tested heat exchanger was installed in the duct, in which the fans provided the required air flow. A pressure P = 0,133 Pa was created in the internal cavity of the heat exchanger. As a refrigerant, distilled water was used, which was supplied through channels in the heat exchanger body, where it evaporated. The capacity of the heat exchanger-evaporator was estimated by the maximum amount of completely evaporated water without the formation of ice grains.

The test results are given in table. 2, it was established that the use of PM based on metal meshes with porosity of 0,20 and 0,45 in heat exchangers allows to obtain the required power. In this case, an increase in heat exchanger power can be achieved by increasing the porosity of the sleeve material and increasing the flow rate of the refrigerant.

Table2. Evaporator Heat Exchanger Test Results

Sleeve Porosity	Temperature air in front heat exchanger, °C	Air consumption, l / s	Depth of vacuum, Pa	Power of the heat exchanger-evaporator,W
0,20	40	22,5	0,133	180
0,45	40	22,5	0,133	200
0,45	40	40,0	0,133	266

The studies of weldability of PMM and the developed EBW technology of porous preforms were used in the design and manufacture of die-welded filter elements with a fineness of 10...12 microns, a sketch of which is shown in Fig. 6.

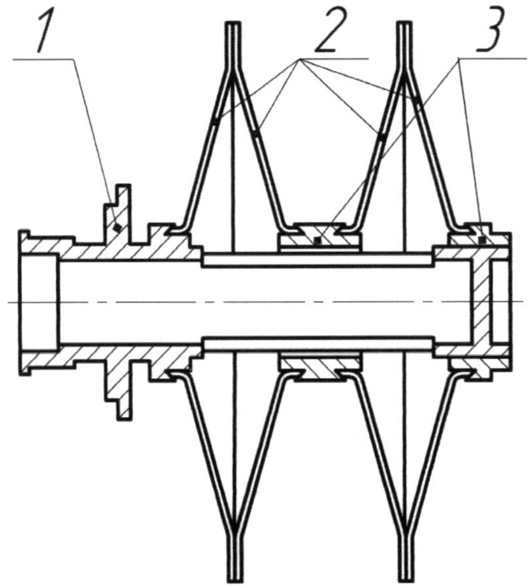

Figure 6. A sketch of the filter, providing fineness of cleaning 10...12 microns: 1 — axis; 2 — filter elements; 3 — sleeves.

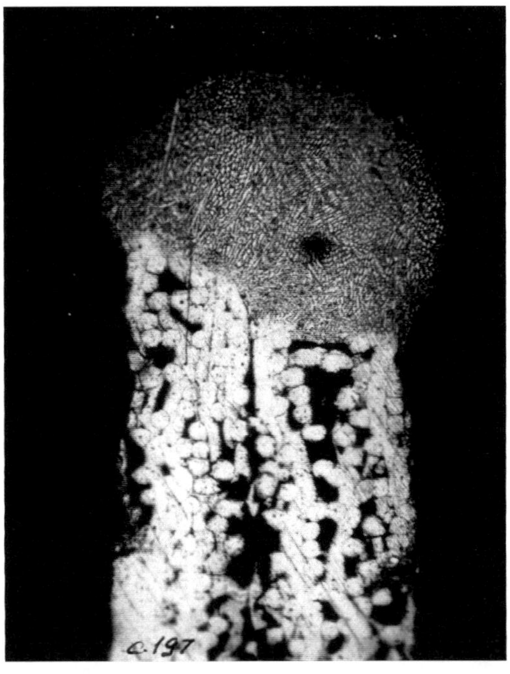

Figure 7. The macrostructure of the end connection of the filter elementsx40

The technological process for the manufacture of filters includes EBW of porous elements with an axis, sleeves and between each other. Filter elements were made by sheet stamping PMM with a thickness of 0,5 mm and П=0,45 (P = 0,45), obtained by welding by rolling nets with square cells, the size of which was a = 0,04 mm (GOST3826-82), located at an angle of 45°. Porous preforms after ultrasonic cleaning in freon were installed in special devices that ensure the assembly of welded elements and their rotation.

Welding was carried out at a current of 5...6 mA at an accelerating voltage of 50 kV with a focal spot diameter of 1 mm. The welding speed was 6,3 m / h when connecting the filter elements with the axle and sleeves, and 22 m / h between them. Figure 7 shows the macrostructure of the welded end joint of the porous elements. The quality control of welded joints of the filter under hydrostatic testing showed that the probability of obtaining a defect-free welded joint is 0,96.

Conclusion

1. The weldability of porous materials is determined by the chemical composition, shape, size of the structural elements and porosity of the welded edges. With increasing porosity and particle size of the powder, the mechanical properties of the PPM decrease, and the ranges of fusion welding, characterized by cracking, expand.
2. It has been established that during PMM fusion welding, in addition to the porosity, the conditions for the formation of welded joints are significantly affected by the ratio of the coefficients of effective thermal conductivity in thickness $\lambda_{\perp}$ and in the direction perpendicular to the movement of the welding source λ_W. As the ratio $\lambda_{\perp} / \lambda_W$ increases and the porosity of the edges decreases, the quality of the welded joints increases.
3. The EBW technology of filter elements of die-welded filters with the required properties and ARAW of the permeable sleeve of the heat exchanger-evaporator made of PMM based on corrosion-resistant steel 12Cr18Ni10Ti were developed. The results of bench tests showed that the developed technological processes make it possible to obtain products that fully meet the requirements of the technical specifications. The introduction of filters in the units of hydro and pneumatic systems increased the operational reliability of power plants.

References

[1] ZeigarnikYu.A., Polyakov A.F., Stratiev V.K., Tretyakov A.F., ShekhterYu.L. Tests of porous mesh materials as shells for turbine blades, high-temperature gas turbines. Moscow, OIVT RAN, 2010, No. 2–502, 64 p.
[2] Bunker R.S. Gas turbine cooling. Moving from macro to micro cooling. Proceeding of the ASME TURBO EXPO, 2013. 3p.
[3] Porous permeable materials: Ref. ed. / Ed. Belova S.V. M .: Metallurgy, 1987. 335 p.
[4] Sparks T., Chase G. Filters and Filtration. Handbook. Elsevier, 2013. 444 p.
[5] Kolesnikov A.G., Safonov A.N. Modern developments of new materials and technological processes for engineering. TyazheloeMashinostroenie, 1998, no.2, pp. 17–21.
[6] Volkov S.S. Ultrasound welding of synthetic fabric for technical purposes. WeldingInternational, 2009, vol. 23, no. 10, pp.193–197.
[7] Belov S.V.,Shmelev L.S., Spiridonov V.S., Ksenofontov A.V., Prikhodko N.G. Structural hydraulic, and filter characteristics of type FNS porous sheets. Soviet Powder Metallurgy and Metal Ceramics, 1986, 25(3), pp. 256–241.
[8] Zubkov N.N., Polykov A.F., ShekhterYu.L. The Hydraulic characteristics of porous materials for a system of transpiration cooling. High Temperature, 2010, 48(2), pp. 231–237.
[9] Kostornov A.G. Capillary transport of low–viscous liquids in porous metallic materials under action of gravitation force. PoroshkovayaMetallurgiya, 2003, no. 9–10,pp.13–26.
[10] Kostornov A.G., Krichenko O.V.,Brodnikovskii N.P., Guslienko Y.A., Klimenko V.N. High-porous material made from alloy steel fibers: production, structure, and mechanical properties. Powder Metallurgy and Metal Ceramics, 2008, 47 (5–6), pp. 295–298.
[11] Kostornov A.G., Shapoval A.A., Lalor M.J., Mgalobishvili O., Legros J.C. A study of heat transfer in heat pipe evaporators with metal fiber capillary structures. Journal of Enhanced Heat Transfer, 2012, 19(1), pp. 43–51.
[12] Ovchinnikov V.V., Antonov A.A., Gureev M.A., Magnitov V.S. Special features of welding filtration of the cartridge type produced from porous metallic materials. Welding International, 2002, no.10, pp. 828–831.

[13] Pelevin F.V. Heat transfer in meshed metallic materials with interchanel transpiration and two-dimensional intermesh flow of heat transfer fluid. High Temperature, 2018, 46 (2), pp. 208–216.

[14] VaraksinA.Yu., Polyakov A.F., Reviznikov D.L.,Stratyev V.K., Tretyakov A.F. Investigation offlow in the vicinity of porous materials on a cylindrical body subjected to flow. High Temperature, 2002, 40(6), pp. 843–849.

[15] Polyakov A.F., Stratyev V.K., Tretyakov A.F., ShekhterYu.L. Heat transfer in envelopes made of porous network materials. Thermal Engineering, 2009, 56(3), pp. 227–334.

[16] Polyakov A.F., Stratyev V.K., Tratyakov A.F., ShekhterYu.L. Generalization ofexperimental data on heattransfer in permeable shells made of porous reticular materials. ThermalEngineering, 2010, 57(6), pp. 516–521.

[17] Wolf J., Moskowits S. Development of the transpiration air-cooled turbine for high — temperature dirty gas stream. Journal of Engineering for Power, 1983, 105(10), pp. 821–825.

[18] Bunker R.S. The role of materials and manufacturing technologies as enablers in gas turbine cooling for high performance engines. Ceramic Transactions, 2009, 205, pp. 3–2

[19] Tretyakov A.F., Specificstrength of porous materials based on metal wire cloths. Strenght of Materials, 1994, 26(2), pp.116–119.

[20] Liu P., He G., Wu L.H. Fabrication of sintered steel wire mesh and its compressive properties. Material Science and Engineering, 2008, vol. 489, no.1–2, pp. 21–28.

[21] Xu G., Liu Y., Luo X., Ma J., Li H. Experimental investigation of traspiration cooling for sintered wovenwire mesh structures. International Journal of Heat and Mass Transfer, 2015, vol.91, pp. 898–907.

[22] Tretyakov A.F. Technological heredity in the manufacturing process of products from porous mesh materials with desired properties. Communication 2. Regularities of the influence of plastic deformation and consolidation of wire mesh on the technological and thermophysical properties of porous mesh materials. Production of rolled products, 2013, No. 6, P. 29–34.

[23] Kolesnikov A.G., Plokhikh A.I., Komisarchik Y.S., Mikhaltsevich I.V. A study of special features of formation ofsubmicro-and nanosize structure in miltilayer materials by the method of hotrolling. Metal Science and Heat Treatment, 2010, 52 (5–6),pp. 273–278.

[24] Schmiedt A., Luecker L., Kolesnikov A.G, Plokhikh A., Walther F. Production and microstructurebased fatiguemultilayer materials produced by hot pack rolling. Materialpruefung/Materials Testing, 2007, 59(2), pp. 123–129.

[25] Vlasova D.N., Plakhikh A.I., Semenov M.V., Polyanskii B.M., Safonov M.D. Effect of diffusion processes during pack rolling on multilayer material stability. Metallurgist, 2018, 62 (5–6), pp. 432–439.

IOP Conf. Series: Materials Science and Engineering **963** (2020) 012021 doi:10.1088/1757-899X/963/1/012021

Ensuring the performance of equipment during the warranty period of operation

S Kushnarev

Bauman Moscow State Technical University, 5 Second Baumanskaya Street, Moscow, 105005, Russian Federation

E-mail: kushnarevl@mail.ru

Annotation. The problem of increasing the reliability of domestic technological machines and equipment remains relevant and requires a speedy solution in the current conditions of acute competition in the market of equipment and technologies. The low level of reliability of equipment reduces the efficiency of using machines and the quality of their technological processes, increases operating costs. The solution of this problem, as the results of research show, is impossible without participation in ensuring the operability of technological machines and equipment of manufacturers of modern domestic equipment. Providing engineering and technical support for manufactured machines, taking part in technical service, equipment manufacturers have the opportunity to study the reasons for the low reliability of units, units, connections and take measures to exclude their appearance. By setting the wear rate of the working surfaces of the critical interfaces in real-world operating conditions of the machine, it becomes possible to control its reliability and efficiency

The purpose of the research

The purpose of the research is to increase the efficiency of using modern high-performance equipment by attracting the forces and resources of equipment manufacturers and technical service companies to ensure a high level of failure-free machines and study the causes of failures. The low level of reliability of domestic equipment requires increased expenditures of all types of production and technical resources in the operation of even new machines and equipment, leading to unproductive losses of working time and production. Consumers, when purchasing equipment, pay for quality manufactured machines, hoping that they will serve them properly not only during the warranty period of operation.

Introduction

One of the main reasons for the low quality of manufacturing and reliability indicators of modern machines and equipment of domestic production is the absence and failure to ensure in practice the direct participation of manufacturers in restoring their performance in the event of sudden failures not due to the fault of the operating party. The consumers of machinery and equipment today almost don't even pay attention to the presence in the contract of supply of equipment items on the guarantees of the manufacturers, trouble-free operation of the machine, about the promised terms of elimination of consequences of failures and losses associated with downtime of the machine due to technical reasons, etc., because they had never been carried out [1–3]. It just so happened from time immemorial (or rather since the 60-70-ies of the last century), the main executor of works on technical maintenance of machines and equipment in warranty and post-warranty periods of operation were and are themselves consumers of technology to the Department of the chief mechanic(OGM) and (or) engineering and technical services (its). Although the structure of the economy have created the most powerful in the world repair and service industries in the face of enterprise system type "Goskomteleradio", almost all of mashinno-technological production sectors to service and repair machines and equipment companies used a method of self-service.

The system of planned economy initially did not provide for the involvement of manufacturers of equipment to eliminate the consequences of failures and ensure the performance of machines and equipment, even during the warranty period of operation. A significant reduction in productivity and operational efficiency of modern high-performance and expensive machines and equipment indicated the urgent need and feasibility of implementing a proprietary method of technical service to improve the quality of manufacturing and reliability parameters of modern equipment. One of the most negative consequences of the low quality of domestic machinery is the low level of competitiveness of Russian-made machinery and equipment [4–7]. It is almost impossible to solve the problem of improving the quality of manufacturing and reliability indicators without involving manufacturers of equipment, machinery and equipment without proper organization of the complaint work and ensuring the required level of performance of machinery and equipment during the warranty period of operation [7–10].

Research result

Currently, in practice, the created machines and equipment are designed to perform certain technological processes of production of products, works and services without due consideration and requirements of the actual production conditions of enterprises of various natural and climatic zones of the country. This is indicated and emphasized by the data provided in the reports on the results of tests and observations of new samples of machinery and equipment: pilot testing, tests at MIS and specialized laboratories, real operating conditions, etc. As a result, it turns out that the machine is not ready or intended for use in real (complex) production conditions. Manufacturing plants and other structures involved in the creation of machines and equipment are not interested in improving the reliability of manufactured equipment, reducing the number of failures, downtime of machines and equipment for technical reasons, and losses of products and their quality in real conditions of enterprises, firms and holdings [11–13].

To eliminate such negative phenomena, it is necessary urgently, if not to introduce a full-scale branded technical service of machines and equipment of domestic production, then, at the first stage, at least to involve manufacturers in eliminating the consequences of sudden failures that occur in new machines during the warranty period of operation. The introduction of branded technical service provides for the development and use of mechanisms that stimulate the improvement of manufacturing quality and the level of reliability of machines and equipment.

In order to ensure a high level of reliability and operability of machines and equipment during the initial warranty period, it is necessary to approach the (seemingly simple) technological processes of technical service with high requirements and obligations. Since the basis of uptime for the entire service life of the machine are the timeliness, completeness and quality while planned preventive measures, it is expedient to provide for their execution under the control of the manufacturer, at least in the initial period of operation (Fig. 1).

Organizational and technological design of technical service processes should begin with the justification and selection of a rational option for the distribution of functions, processes and work among all participants in ensuring effective machine use — manufacturers, consumers of equipment and technical service enterprises.

Obviously, in the current conditions, the first step should be to create a full-fledged service for pre-sale and warranty maintenance and repair of equipment at the manufacturers of machinery and equipment. Production and technological parameters of the services of the chief mechanic or engineering services of enterprises — consumers of equipment should be determined based on the actual flow of requirements for maintenance AND Troubleshooting that occur during the warranty period of operation. The number of complaints currently received from consumers does not reflect the actual needs and cannot be used to justify these services [13, 14].

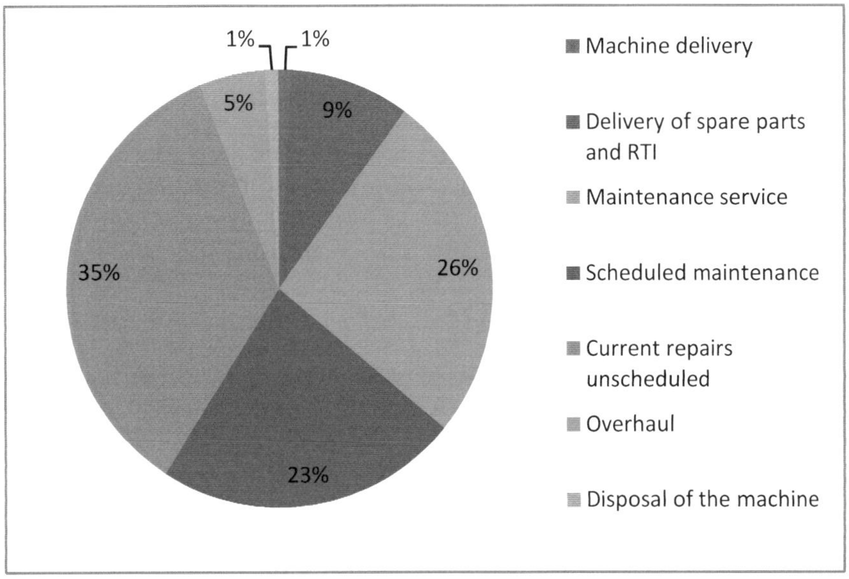

Fig. 1. Structure of labor costs in the operation of machinery (tractors)

The interests of the manufacturer, in this case, can be represented directly by its employees, who go to the location of the failed machine, establish as part of a qualified Commission, the causes of failures, opportunities and ways to eliminate the consequences of failures, decide on the earliest restoration of the machine or equipment and ensure its implementation, prepare a report on the work done with proposals and measures to exclude the occurrence of such failures in the future at all stages of the life cycle of the machine and equipment.

The most direct and active participation in the organization of work to eliminate the consequences of failures should be taken by territorial, district and regional services of Gostekhnadzor. In addition, Gostekhnadzor services should be assigned the functions of quality control over the operation of new equipment, both production and technical. On the one hand — the services of "Gostekhnadzor» they act as a controlling state body that ensures the correct, safe and efficient operation of machinery and equipment in real production conditions of enterprises and firms. On the other hand, participating in the work to eliminate the consequences of failures, they monitor the manufacturer's compliance with their warranty obligations. Thus, they protect the rights of consumers of equipment and participate in the process of improving the quality of manufacturing and reliability parameters of machinery manufactured by machine builders.

Proposed solution

As a direct main executors of works on technical service can be used by own strength and means of the enterprises of consumers of technology — chief mechanic service or engineering service to manufacturers of goods, works and services, or be involved in the next maintenance enterprise, which is the official technical representatives of the manufacturer (Fig. 2). From the point of view of efficiency and cost — effectiveness, the first option is the most appropriate-by the forces and means of its own OGM or its. The possibility of using this option for organizing maintenance work requires the necessary repair and technical base, its technological equipment and personnel directly on the farm. Otherwise, it is advisable to perform technical service work in the company's technical center of the manufacturer of technical services [15, 16].

IOP Conf. Series: Materials Science and Engineering **963** (2020) 012021 doi:10.1088/1757-899X/963/1/012021

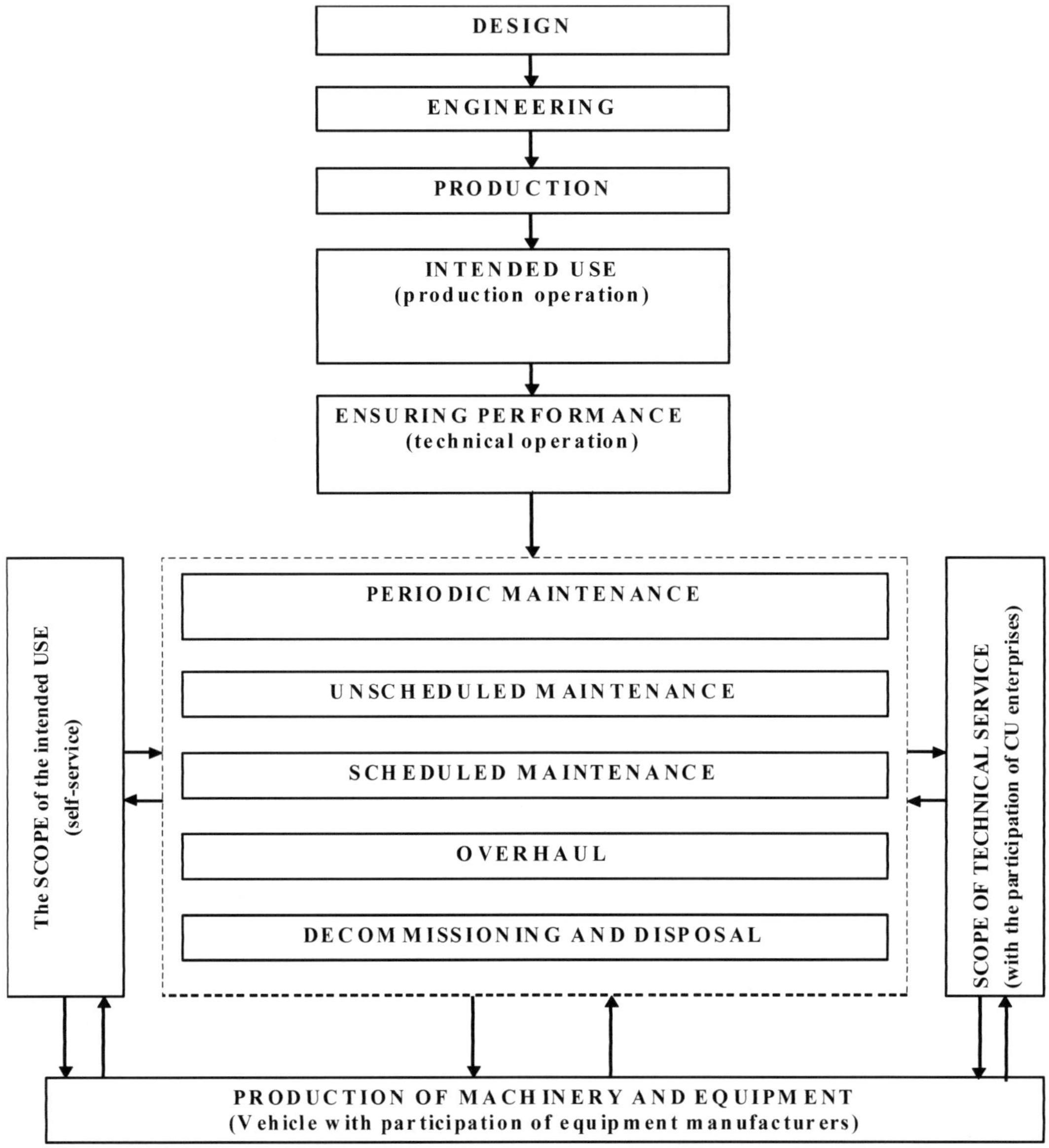

Fig. 2. Organization of maintenance and study of the causes of failures during the warranty period of equipment operation

After establishing the causes of sudden failure and its consequences, issued a report (complaint), signed by representatives of the manufacturer, the consumer operating the machine (equipment), maintenance enterprise and "STS". The main sections of the report are: characteristics of failure, reasons for failure, performers and ways to eliminate the consequences of failure, organization of work, recommendations to production (designers, designers, technologists) and (or) consumers to exclude such failures in the future. This report is immediately sent to the manufacturer (manufacturer) or to the main (head) company technical center, from where it is sent to all structures of the company, as well as to related companies and suppliers of components. The updated list of structures for mailing materials is determined by the manufacturer of equipment.

Management of the manufacturing company, together with the departments of advertising work, technological, technical control, planning and organization of technical service during the pre-sale and warranty period of operation of new equipment, determines possible production reasons that caused the failure and develops measures to eliminate them. If this requires improving the design of products, their modes of operation, lubrication of the working surfaces of parts, etc., then designers and designers are involved. Efforts are made by all structures that ensure the quality of manufacturing and operational reliability of manufactured equipment. If necessary, changes and additions to the technological process of manufacturing parts and Assembly units are made as soon as possible. If necessary, structural materials are replaced, dimensional tolerances are tightened, working surfaces of critical parts are strengthened, etc. The working documentation may include additional requirements to strengthen control over the operation of certain components and units during the operation of the machine, and the spare parts kit may be supplemented with details to quickly eliminate the consequences of failures [14–20].

The need and feasibility of rapid and timely elimination of the consequences of failure are dictated by modern conditions not only market competition, but also a better consumer protection technology. In contracts for the supply of agricultural, construction and other equipment, manufacturers ' obligations to ensure the quality of products and guarantees to eliminate the consequences of failures no later than a certain period of time gradually appear.

Innovative development of mechanical engineering, as the results of organizational and technological research show, is impossible without full ensuring the priority of consumer rights, primarily for high-quality equipment, and not its constant maintenance and repair. This requires the development and implementation of proprietary technical services and organizational and economic mechanisms to encourage improvement of the quality of engineering products. To improve the economic liability of equipment manufacturers have developed and are preparing to introduce the methodology for determining costs and detriment the consumer on occurrence of the failure and downtime of machines and equipment not his fault. And this, with the existing level of quality and reliability of modern domestic equipment, threatens machine-building enterprises, if not with bankruptcy, then with huge economic difficulties.

Conclusion

Thus, market competition, ensuring the priority of the rights of consumers of equipment, tightening economic responsibility for the production and supply of low-quality products require every possible improvement in the quality of domestic machinery and equipment. The only correct way to solve the problems of improving the quality of domestic equipment is to introduce a proprietary method of technical service. Only in corporate of technical service possible to organize at the proper level of technical service of machinery and equipment of domestic production in warranty period, what is the first step towards the creation of competitive engineering.

References

[1] Aleshin N.P. et al. Using nondestructive testing methods for in-production quality control of additive manufactured parts // Russian Journal of Nondestructive Testing. 2016. V. 52(9). P. 532–537. DOI: 10.1134/S1061830916090023

[2] Vasiliev A.S. Controlled Forming of Machine Components Operating Characteristics // Procedia Engineering. 2016. V. 150. P. 975–979.

[3] Nikolaev A.B. et al. Tools of production and logistics support life cycle of high-tech products // Life Science Journal 11 (10 SPEC. ISSUE). 2014. P. 238–242.

[4] Fedyukin V.K. Upravleniyekachestvomprotsessov [Process Quality Management]. St. Petersburg: Peter. 2005.

[5] Agostin V. Y. About the process approach in quality management // Standards and quality. 2010. Vol. 11. P. 36–38.

[6] Chernoivanov V. I., Severny A. E., Kushnarev L. I., and others. Problems of technical service in the agro-industrial complex of Russia (monograph). Moscow: GOSNITI, 2000. — 309p.

[7] Planned preventive MAINTENANCE and repair of machinery and equipment. Moscow: Mashinostroenie, 2001. — 369 p.
[8] Complex system of technical maintenance and repair of machines in agriculture. Part 1. — Moscow: GOSNITI, 1985. — 143 p.
[9] Technical operation of agricultural machines (with normative materials). — Moscow: GOSNITI, 1993.
[10] Cherepanov S. S. Maintenance and repair of machines in agriculture: (Fundamentals of scientific organization) / S. S. Cherepanov. — M.: Kolos, 1978. — 288 p.
[11] Strategic management of the company's organizational and economic stability: Logistics-oriented business design / A.D. Kanchaveli, A. A. Kolobov, I. N. Omelchenko, etc. Edited by A. A. Kolobov and I. N. Omelchenko. Moscow: Bauman Moscow state technical University, 2001, 600 p.
[12] Kushnarev L. I. Chepurina E. L., Kushnarev S. L. Problems and directions of development of engineering and technical support of rural commodity producers //Repair, restoration, modernization № 1. — 2016. — P. 3–9.
[13] Lipkovich E. I. Technical equipment of farms, organization of mechanized works and technical service of machines / E. I. Lipkovich, L. I. Kushnarev, L. M. Sergeeva // Engineering and technical support of the agro-industrial complex. — 1996. — N1. — P. 8–12.
[14] System of technical maintenance and repair of agricultural machines based on the results of diagnostics. — M.: Informagrotech, 1995. — 64 p.
[15] Chepurina E. L., Kushnarev L. I. Role and place of agricultural machinery manufacturers in corporate technical service //Machinery and equipment for the village. — 2013. — № 7. — P. 38–40.
[16] Kushnarev L. I., Didmanidze O. N. State and directions of innovative development of the engineering and technical service of the agro-industrial complex //International technical and economic journal, no. 1, 2014, Pp. 31–40.
[17] Kushnarev L. I. Chepurina E. L., Chepurin A.V., Kushnarev S. L. Quality and reliability of domestic equipment-the basis of its competitiveness //Tractors and agricultural machines. — 2015. — № 11. Pp. 35–37.
[18] Kushnarev L. I. [et al.]. Modernization of the technical service system of the agro-industrial complex. Scientific publication. Moscow: MESKH. — 2017.
[19] Chepurina E. L. Sevastyanova D. L. Improving the reliability of machinery and equipment in animal husbandry// Scientific and informational support of innovative development of the agro-industrial complex. Materials of the IX International scientific and practical conference "Informagro-2017". Moscow: "Rosinformagrotech". — 2017.
[20] Chepurina E. L. State of the organization of technical service of livestock machinery and equipment./International technical andeconomicjournal, no. 4, 2013. Spectrum LLC. —Pp. 61–67.

Increasing the creative potential of production in the context of prevailing trends

G Ganina[1], A Mukhin[1], Yu Ostrovsky[1] and A Yakovleva[1,2]

[1]Bauman Moscow State Technical University, 5 Second Baumanskaya Street, Moscow, 105005, Russian Federation

[2]E-mail: Yakovleva525@mail.ru

Abstact. The article examines the prevailing trends in modern production and their role in forming ideas about further development. It is shown that involvement in the process of understanding the trends of the regularity of labor activity transformation leads to an understanding of the place of self-development in the production strategy. A comprehensive analysis of trends, combined with the use of new approaches to the description of labor activity, helped to reveal and justify a new view of the strategy for improving production. This made it possible to describe at the conceptual level a new type of production systems, namely self-developing production systems. The presence of a model of real transformation of labor activity in the description of such production makes it possible to create a system with an internal self-development device. Such self-developing production systems do not need external influences and are invariant in a variety of social environments.

Introduction

Currently, among economists, production managers and specialists of industrial enterprises, issues of strategic development are of increasing interest, taking into account changes in technical equipment, employment problems and other social issues. Most often, issues related to the emergence of global trends in the life system, in particular, in production activities, are discussed.

It is noteworthy that the analysis of trends in the discussed problem is complicated by the fact that factors include both the parameters of technology and the economy of direct production activity, and factors of social and social sound. In this regard, the analysis of trends in production activities is most often carried out at the level of emotional judgments, which does not approach an objective assessment of the impact of trends on production efficiency.

In these circumstances, it is advisable to resort to a well-known technique when the analysis of a global problem is carried out step by step:

– splitting the global system into parts (cells);
–analysis of the problem at the cell level;
– spreading the results of cell analysis to the entire system,

This article attempts to deduce a solution to the problem of production development, taking into account the analysis of various trends and patterns used in the goals adopted by the community at present and in the foreseeable future.Since trends reflect only the direction of controlled or spontaneous development, the most interesting problem is the impact of the identified trend on the solution of specific tasks.Due to the fact that in the future, the creative potential of the production system will affect the efficiency of production, this article is devoted to this problem.

Analysis of trends in production activities

In order to delve deeply into the essence of trends in production activities, it is advisable to identify the causes that determine the nature of trends, and, if possible, classify these causes by certain characteristics. Since trends do not appear to occur spontaneously, but are the result of someone else's activity, we can assume that it is possible to manage trends if you know the reasons for their occurrence.

Attempts to systematize trends have been made in all spheres of human life; all research inevitably faces a huge scale of tasks and the only acceptable way is to divide the problem into components for which it seems easier to find a solution [1].

In relation to production activities, it is advisable to start analyzing trends by dividing the signs of trends into two types:

–trends that are caused by the external environment (exotendencies);

–trends, the causes of which need to look inside the company (intendencia).

A characteristic exotendence observed in production at present is an increase in the load from consumer demand, which manifests itself in different ways depending on the interaction with production.

Among a large number of attendance possible to distinguish those that are visibly manifest in practice. First of all, we are talking about such exotendencies as a permanent increase in the variety of manufactured and, consequently, consumed products. Another current trend is to accelerate the change of manufactured and consumed products.These trends are related, but they should be considered separately. These trends have an impact on the production, where you can also select the current time of attendance.

Among the most important attendance include two, in our opinion, important:

– accelerating with each passing year, automation of production processes;

–the outsourcing as a new phenomenon in the established production systems.

In themselves, Exo-and endotendencies are of great interest, but the problem remains how they interact with each other, bringing a tangible effect or, conversely, unconditional harm.

Methodically, it is convenient to represent the interaction mechanism using a matrix (see table 1).

Table 1. Trend matching matrix

			Intendencia	
			Outsourcing	Automation
			x_1	x_2
Intendencia	Variety of products	y_1	Degree ofcompliance y_1x_1	Degree of compliance y_1x_2
	The rate of change of demand	y_2	Degree ofcompliance y_2x_1	Degree of compliance y_2x_2

It should be noted that the analysis of the correspondence of various trends in the life of society is carried out by many specialists from different points of view; most often this is done at the level of the economy of various subjects [2-10], even when it concerns purely industrial activities [3,9,11]. However, we are interested in the problem of analyzing the compliance of trends at the level of organizational and technological approach to the activities of enterprises.

The degree of compliance of the yixi type is easiest to find, based on the previously described method of evaluating product-product compliance [2]. It is easy to show that according to this method, the values of characteristics given in the table will allow us to conclude that these trends correspond to each other.

However, recently, at the intersection of these trends, another cause for deep concern has emerged: the trend of a steady increase in the number of employees released from routine work, but not engaged in any further activity.

A new problem arises, which we propose to solve on the basis of a previously identified pattern [3].

The task of effective use of released employees for creative activities

The article [3] describes a model of a production system that can claim to be self-developing in the future. This role can be provided primarily by the model of transformation of labor activity (1), based on two types of thinking in decision-making: intuitive and logical.

$$T_i \xrightarrow{H} P_i \xrightarrow{A} T_j \quad (1)$$

Intuitive thinking and, consequently, decision-making is present in creative work (Ti), and formal-logical in routine work (Ri).

The model presented by (1) is fairly transparent and does not require additional explanation. However, it contains one nuance that has to be considered in practice.

The fact is that the automation of routine work does allow you to release workers, which there are a huge number of examples, but it does not guarantee that the released immediately engage in creative activities (TJ). This is still the case in most industrialized countries — the problem of "release" is becoming almost the most important one in the context of rapidly advancing automation.

Let's try to show a solution to this problem based on the following assumptions:

–production development is subject to the "transformational model" [3];

– the driving force of production transformation is creative work, making the production system " self-developing»;

– all people have creative abilities; the division into routine or creative workers depends on the existing industrial relations (forms of ownership).

Taking these assumptions into account, we will develop the model (1), bringing it to a form that includes a mechanism for the possible transformation of P into T into a real one. Taking into account the created mechanism, we obtained a model of real transformation of labor activity based on intuitive and formal-logical decision-making in production (Fig. 1).

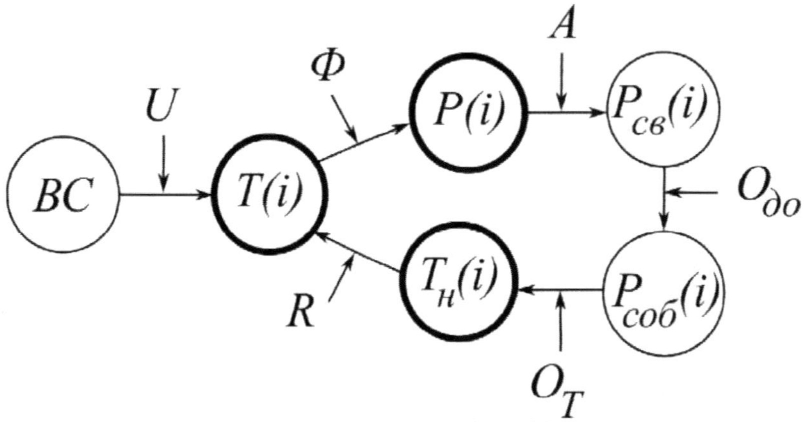

Fig. 1. Model of real transformation of labor activity

Here: * BC-external environment containing the conditions (meaning) of production activity (for example, production of necessary products);

* T (i) — a creative individual; a person who generates actions for the production of products based on intuitive thinking;

* P(i) — an individual who performs routine actions for the production of a product in accordance with a formalized prescription f (technological process);

* Pсв (i) — an individual released as a result of automation (A) from routine work and released from wage labor as part of an enterprise organized on the basis of public or private ownership;

* Pcob(i) — owner of an individual creation (creation) based on the acquired experience;

* Tн (i) — a creative individual who creates a new creation, and is the result of the transformation of Pcob(i).

An important role in the model presented in figure 1 is played by factors-the driving forces of individual transformations:

• Φ-formalization of actions created by T (i) using passive or active experiments;

* A-automation of routine actions, which leads to the release of the individual, both from routine work and from the dictates of property;

* Odo— organization of voluntary associations in order for individuals of RSV (i) to be able to carry out activities to create their own creations;

* O_T— organization of labor in voluntary associations, which contributes to the transformation of the individual owner of Pcob (i) into a new creative individual Tн (i) , as the basis for replenishing the creative potential of the production system;

• R-expanding the diversity of satisfied human needs by replenishing creative potential;

* U — an infinite variety of human needs arising from the state of the external environment of the sun and the capabilities of the creative individual T(i) at the current time.

Comparing models (1) and Fig. 1, we come to the conclusion that for a real transformation of labor activity, it is necessary to find a suitable mechanism for the transformation of RSV(i) → Pcob (i) and Pcob (i) → Tн(i).

Research in the field of ergodynamics has shown [7] that such a mechanism in the most acceptable form can be represented as an energy dynamic cooperative (EDC) [4]. After reviewing the work [4], an attentive reader will easily find that the proposed scheme of the EDC includes the organization of labor in voluntarily United cooperatives.Without reducing the role of conclusions on the formation of Odo and FROM" inside " EDC, we note the most important-the model of real transformation of labor activity, presented in figure 1, reflects a fundamentally new type of production system — self-developing.

Transformation of labor activity as the basis of a self-developing production system

The transformation of the theoretical model of the transformation of labor activity (1) into a real model (Fig. 1) is made in order to "design" the main core of the self-developing production system (SRS). The recent discussion of the place and role of self-developing systems[5], [6] is mainly aimed at presenting the image and driving forces of the SRS.

At present, the discussion of this problem has brought at least clarity about the conceptual apparatus of the SRS.

Conclusion

It turns out that the analysis of existing trends in production is necessary not only to develop objective measures to improve efficiency. A comprehensive analysis of trends, combined with the use of new approaches to the description of labor activity, helped to reveal and justify a new view of the strategy for improving production.

This made it possible to describe at the conceptual level a new type of production systems, namely self-developing production systems. The presence of a model of real transformation of labor activity in the description of such production makes it possible to create a system with an internal self-development device. Such self-developing production systems do not need external influences and are invariant in a variety of social environments.

References

[1] Latysheva, V. V. Fundamentals of sociology and political science: textbook for secondary vocational education / V. V. Latysheva. - 2nd ed., ISPR. and additional — Moscow: Yurayt publishing house, 2019. — 304 p. — (Professional education). — Text: electronic / / EBS yurayt [site]. — URL: https://urait.ru/bcode/437410 (date accessed: 01.10.2019)

[2] Ganina G. E., Ostrovsky Yu. a., Mukhin A.V. Accounting for production efficiency in the system of technical and economic indicators of an industrial enterprise / / Controlling. — 2014. — № 52. — Pp. 26–35.

[3] Mukhin A.V. Identification of production. Modeling and optimization of production systems. Textbook. M., Bauman Moscow state technical University. Access mode: http://wwwcdl.bmstu.ru/ibm2/index1.htm (accessed 16.09.2019).

[4] Mukhin A.V., Ganina G. E., Ostrovsky Yu. a., Yakovleva A. P. a New approach to the formation of cooperatives that ensure the continuous development of innovative production. // Bulletin of the South Russian state technical University (Novocherkassk Polytechnic Institute). Series: Socio-economic Sciences. — 2019. — № 2. — P. 15–21.

[5] Stepin B. C. self-developing systems and post-non-classical rationality / / Questions of philosophy, 2003, no. 8, Pp. 5–17.

[6] Akatov N. B. Initiation of organizational self-development // Bulletin of Perm University. Series: Economics. — 2012. — №2(13). — Pp. 118–123

[7] many-Sided role of material production in society / / a. Mukhin, G. Ganina, S. Mozgin, Yu. Ostrovsky and A. Yakovleva, IOP Conference series: Material Science and Engineering 2019. — Volume 589, Issue 1. — Art. no 012024 DOI: 10.1088 / 1757 899X/589/1/012024

[8] A. E. Brom, O. V. Belova, A. Sissinio. Energy equipment lifecycle costs: Fmeca and Lifecycle Costing models as "decision-making" tools to reduce costs over the entire life of the equipment. Procedure Engineering, Vol. 152, 2016, pp. 173–176, doi:10.1016/j. proeng.2016.07.687.

[9] Irina Omelchenko, Pavel Drogovoz, Eugene Gorlacheva, MrSabadeco and Olga Yusufov. The efficiency of production and distribution systems of a new generation based on cognitive factors of production is modeled.VGD Conf. Series: materials Science and engineering 630 (2019) 012020, IOP Publishing, doi: 10.1088 / 1757-899X/630/1/012020-Yes.

[10] Kushnarev L. I. improving the competitiveness of the machine-building industry / / IOP Publishing, IOP Conf. Series: materials Science and engineering a 560 (2019) 012074 doi: 10.1088 / 1757-899X/560/1/012074 012074

[11] Dmitry Novikov, Alexander Veretennikov, Alexey Borodin and Alexander Snedkov / / Factory of the future-a new solution and a new quality in scientific instrumentation. IOP Publishing: IOP Conf. Series: materials Science and engineering a 497 (2019) 012133 doi: 10.1088 / 1757-899X/497/1/012133

Experience in calculating the annular discharge of a centrifugal pump based on the hydrodynamic modeling method

O Martynyuk[1] and A Petrov[1,2]

[1]Bauman Moscow State Technical University, 5 Second Baumanskaya Street, Moscow, 105005, Russian Federation

[2]E-mail: alexeypetrov@bmstu.ru

Abstact. The problems of small-sized centrifugal pumps for thermal stabilization systems and the need to optimize them in order to increase efficiency, wear resistance and service life are considered. The types of outlets used in such pumps and their advantages are defined. Using hydrodynamic modeling in the STAR CCM+ software package, the values of efficiency and radial force on the impeller rotor at different feed rates for centrifugal pumps with annular and spiral discharge are obtained. The data obtained were analyzed and processed, as well as compared with the results of similar studies in previous years. The advantages and disadvantages of each outlet are determined and the conclusion is made.

Introduction

Pumps of thermal stabilization systems play an important role in the operation of the system. They are often used in the military, aviation and space industries. Therefore, it is necessary to ensure their high service life and reliability. Therefore, it is necessary to optimize their flow part in order to reduce wear and increase efficiency [1–4]. To do this, the outlet must provide the least radial force, since it negatively affects the service life of the pump due to increased load on the bearings. It should also be taken into account that the radial force increases significantly when the pump operation mode deviates from the optimal one, which is possible when the viscosity increases or decreases, or the equipment becomes clogged. The most common in centrifugal pumps are spiral bends that provide uniform distribution of pressures and speeds in the design mode of operation [5, 6]. In small, household, multi-stage or pumping abrasive-containing liquid pumps, ring outlets are sometimes used. They are characterized by an increased radial force compared to spiral outlets in the optimal mode of operation of the centrifugal pump, lower efficiency and larger dimensions compared to pumps with a spiral tap. This article provides a comparative analysis of the spiral and annular outlet by comparing the efficiency and radial force on the rotor of a centrifugal wheel.

Methods

Based on the existing model small-sized centrifugal pump, hydrodynamic modeling of two outlets in the STAR CCM+ software package was performed [7–10]. This research method allows us to solve various problems of continuum mechanics. its possibilities are presented in [11–16]. The bends are built in the SOLIDWORKS program, and the cross-section shape is determined by graphical integration, based on the condition of optimal throughput for this pump, and is shown in figure 1: on the left for the spiral outlet, on the right for the ring one.

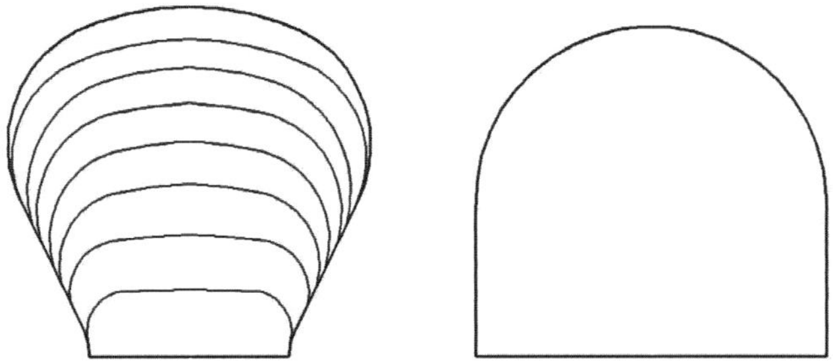

Fig.1. The spiral and ring outlets

Mathematical model:

- Continuity equation (mass conservation equation):

$$\frac{\partial \rho}{\partial t} + div(\rho \cdot \vec{u}) = 0,$$

where ρ — density;

t — time;

$\vec{u}$ — velocity vector

- Reynolds equations (equations of conservation of the amount of motion averaged over time):

$$\frac{\partial}{\partial t}(\rho \cdot u_j) + \frac{\partial}{\partial x_j}(\rho \cdot u_i \cdot u_j) = -\frac{\partial p}{\partial x_i} + \frac{\partial}{\partial x_i}\left[T_{ij} - \rho \cdot \overline{u'_i u'_j}\right],$$

where $T_{ij} = 2\mu s_{ij} - \frac{2}{3}\frac{\partial u_j}{\partial x_j}$ — the tensor of viscous stresses;

μ — dynamic coefficient of liquid viscosity, Pa·s;

$s_{ij} = \frac{1}{2}\left(\frac{\partial u_i}{\partial x_j} + \frac{\partial u_j}{\partial x_i}\right)$ — strain rate tensor;

$\overline{\rho u'_i u'_j}$ — Reynolds stresses;

u'—pulsation component of the speed;

p — the averaged value of the pressure.

- The transport equation of turbulent kinetic energy:

$$\frac{\partial}{\partial t}(\rho k) + \frac{\partial}{\partial x_j}(\rho u_j k) = P_k - \rho\beta^* k\omega + \frac{\partial}{\partial x_j}\left[(\mu + \sigma_k \mu_t)\frac{\partial k}{\partial x_j}\right],$$

where k — kinetic energy of turbulence;

P_k — energy generation of turbulence by the velocity shear;

σ_k, β^* — the coefficients of the circuit;

μ_t — coefficient of turbulent viscosity.

- Transfer equation for the relative rate of dissipation of this energy:

$$\frac{\partial}{\partial t}(\rho\omega) + \frac{\partial}{\partial x_j}(\rho u_j \omega) = a_k^{\omega} p_k - \rho\beta\omega^2 + \frac{\partial}{\partial x_j}\left[(\mu + \sigma_\omega \mu_t)\frac{\partial \omega}{\partial x_j}\right] + 2\cdot(1 - F_i)\sigma_{\omega 2} \cdot \frac{1}{\omega}\frac{\partial k}{\partial x_i}\frac{\partial \omega}{\partial x_i},$$

Where:

ω — velocity of kinetic energy dissipation of turbulence;

a, β, σ_ω — the coefficients of the circuit;

F_i — weighting function.

Parameters of the mathematical model:

1. Models for building the calculation grid.
   1.1. Generator polyhedral cells.
   1.2. Generator surface mesh.
   1.3. Generator of the prismatic layer.
   1.4. Extruder (constant speed in the normal direction).
2. Reference values of the calculated grid.
   2.1. Base size 1.5 mm
   2.2. The thickness of the prismatic layer is 0.75 mm.
   2.3. The number of prismatic layers is 5.
   2.4. Stretching of the prismatic layer 1,3.
   Extrusion:
   2.5. The amplitude is 90 mm.
   2.6. The number of layers is 100.
   2.7. Stretching 5.
3. Physical models.
   3.1. Three-dimensional.
   3.2. Material: water (single-phase liquid).
   3.3. Constant density.
   3.4. The Reynolds-Averaged Navier-Stokes equation.
   3.5. k-omega turbulence models.
   3.6. Mode of calculation: unsteady implicit.
4. Boundary conditions.
   4.1. The rotational speed of 10,800 rpm.
   4.2. Inlet pressure is 0 PA.
   4.3. The speed of the output 3.148 m/s.

The constructed grid in figures 2 and 3.the Grid for a pump with a spiral outlet consists of 138144 cells, and for a pump with an annular tap of 225783 cells.

Fig. 2. Pump grid with spiral outlet

IOP Conf. Series: Materials Science and Engineering **963** (2020) 012023 doi:10.1088/1757-899X/963/1/012023

Fig. 3. Pump grid with ring outlet

Results

The values of efficiency and radial force are found. The results are shown in table 1.

Table 1. Results of hydrodynamic modeling

Parameter		Pump with spiral outlet	Pump with ring outlet
Q=0.25·Qn	Efficiency, %	51	61
	Radial force, H	12	0,96
Q=0.5·Qn	Efficiency, %	74	66,5
	Radial force, H	9	1,55
Q=0.75·Qn	Efficiency, %	81	71
	Radial force, H	5,5	1,5
Q=Qn	Efficiency, %	84	73
	Radial force, H	0,9	1,4
Q=1.25·Qn	Efficiency, %	74	73
	Radial force, H	5	1,86
Q=1.5·Qn	Efficiency, %	50	71
	Radial force, H	10	2,3

As an example, figure 4 shows the pressure distribution.

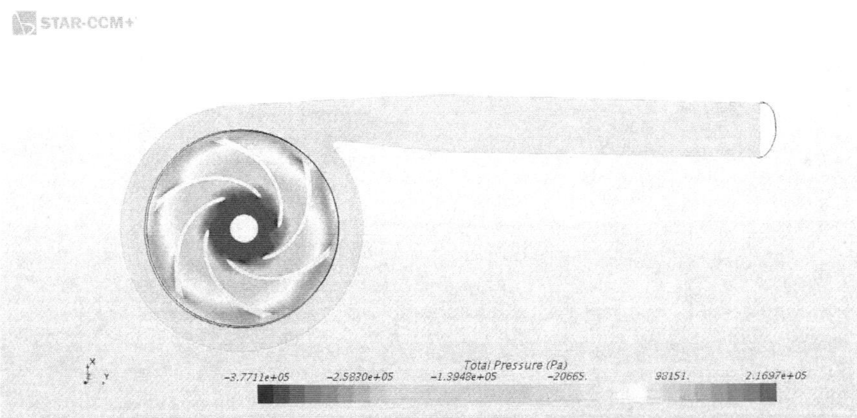

Fig. 4. Pressure distribution in a pump with an ring outlet at Q=0.25×Qn.

Based on the results obtained, the graphs shown in figures 5 and 6 are obtained.

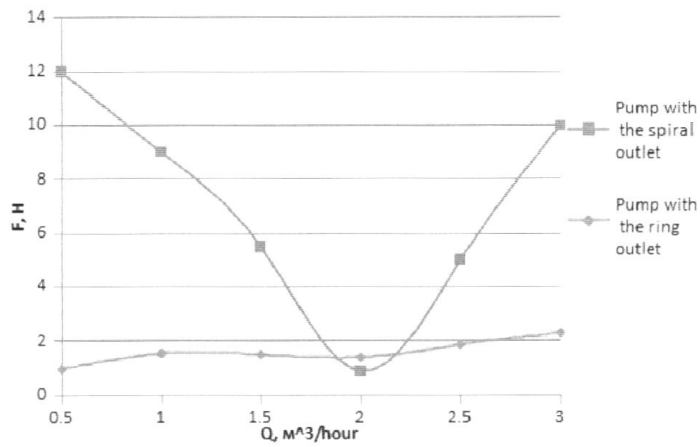

Fig. 5. The dependence of the force from the flow

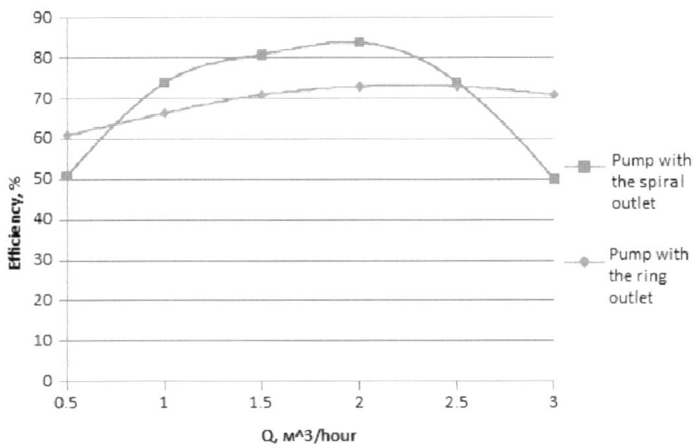

Fig. 6. The dependence of the efficiency from the flow

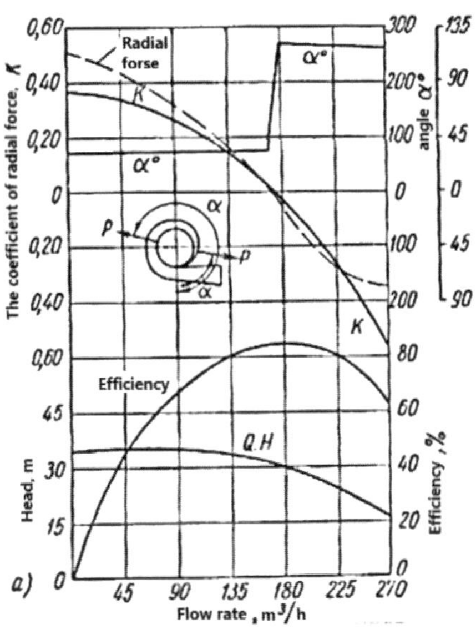

Fig. 7. Characteristics of the pump with a spiral outlet

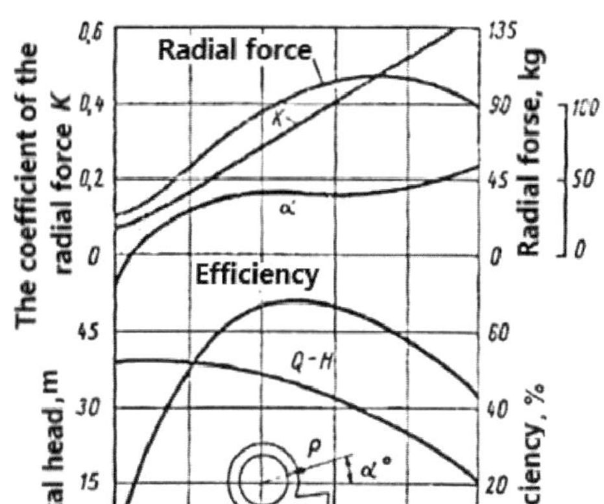

Fig. 8. Characteristics of the pump with a ring outlet

For comparison the graphs for pump volute and annular drainage [16] in figures 7 and 8, pump wheel diameter D2 is 267 mm, wheel width at outlet b2=32 mm, speed n=1760 rpm.

Discussions

The results of the study showed that the annular branch in non-optimal operating modes provides the smallest deviation of the radial force from the value at the optimum point, while in the spiral branch it increases several times. The pump with a spiral tap showed a higher efficiency (by 10%), but it falls sharply in non-calculating modes, in contrast to the efficiency of the pump with a ring tap. It is necessary to conduct a further study of the taps of small-sized pumps

References

[1] Lomakin V. O. Petrov A. I. Stepanyuk A. I. Optimization of geometric parameters of the oil main pump discharge of the NM type / / Science and education: scientific publication of Bauman Moscow state technical University 2012. no. 3
[2] Beizman a.m. Monasikhin V. A. Lomakin V. O. Improving the energy efficiency of a pump working with a replaceable rotor by adding a cylindrical insert to simulate the extension of the tap tongue. Expediency of application / / Hydraulics 2018. № 6
[3] Lomakin V. O. Artemov A.V. Petrov A. I. Determining the influence of the main geometric parameters of the pump discharge NM 10000-210 on its characteristics / / Science and education: scientific publication of Bauman Moscow state technical University 2012. № 8
[4] Lomakin V. O. Cheremushkin V. A. Influence of the shape of the impeller blades on the pressure of the centrifugal pump // Engineering Bulletin 2016. no. 1
[5] Lomakin A. A. centrifugal and axial pumps. Moscow. Leningrad. Mashinostroyeniye Publ. — 1966. 364 P.)
[6] Zimnitskiy V. A. vane pumps: guide.Leningrad. Mashinostroyeniye Publ. — 1986. 334 P.)
[7] V Lomakin et al. 2019 VGD Conf. Serial.: Mother. Sci. Eng. 492 012012
[8] N. Yegorkina and Petrov 2019 VGD Conf. Serial.: Mother. Sci. Eng. 492 012015
[9] toDobrokhodov and Petrov 2019 VGD Conf. Serial.: Mother. Sci. Eng. 492 012016
[10] nIsaev 2019 IOP Conf. Serial.: Mother. Sci. Eng. 492 012026
[11] Lomakin V. O. Petrov A. I. Verification of calculation results in the package of hydrodynamic modeling of ster CCM + flow part of the centrifugal pump Ah 50-32-200 / / Izvestiya higher educational institutions. Mechanical engineering 2012. —Spets.the release of the Spec.issue "Works of students and young scientists of Bauman Moscow state technical University". — P. 6–9
[12] tValiev and Petrov 2019 VGD Conf. Serial.: Mother. Sci. Eng. 492 012038
[13] V. Cheremushkin and V. Lomakin 2019 IOP Conf. Serial.: Mother. Sci. Eng. 492 012039
[14] m. Saprykina and V. Lomakin 2019 IOP Conf. Serial.: Mother. Sci. Eng. 589 012017
[15] Petrov A. I. Fraz A. S. Gavryushina O. S. Hydrodynamic modeling of flow in the flow part of a centrifugal pump running on a high-viscosity liquid / / Hydraulics 2018. — № 5
[16] Stepanov A. I. centrifugal and axial pumps. Moscow.Mashgiz Publishing House, 1960, 462 P.

Hardware-in-the-loop modeling applied to FESTO hydraulic learning system

B Kulakov[1] and D Kulakov[1,2]

[1]Bauman Moscow State Technical University, 5 Second Baumanskaya Street, Moscow, 105005, Russian Federation

[2]E-mail: kulakov@bmstu.ru

Abstract. The issue of Hardware-in-the-loop modeling in Matlab/Simulink using a physical model of an electrohydraulic servo drive implemented on FESTO learning system is considered. As an example, the synthesis of state components feedback coefficients for the electrohydraulic servo drive control is discussed. The paper provides the block-diagrams for connecting FESTO physical model to Matlab software and algorithms that allow using Matlab for automatic optimization of the controller parameters directly during the operation of the electrohydraulic servo drive physical model. Thus, already at the stage of optimizing the parameters of the controller, its operation as part of a real hydraulic system is studied.

Introduction

Numerical optimization of the controller parameters usually uses a mathematical model of the control object [1–7]. In [2], it was shown that it is possible to determine the electrohydraulic servo drive (EHSD) mathematical model parameters by analyzing the experimentally determined frequency characteristics of its physical model implemented on FESTO hydraulic learning system with the Matlab/Simulink software environment connected to it. There was also considered the LQR synthesis of the EHSD in Matlab/Simulink and the possibility of studying its work on the EHSD physical model. As a result, it becomes possible to conduct all stages of the synthesis of hydraulic system controllers on one research system: determining the parameters of the mathematical model, synthesizing the controllers, and studying their operation on physical models (Fig. 1).

In general, the behavior of the EHSD changes depending on its mode of operation [8, 9]. Therefore, when solving a certain class of problems, it is advisable to study a controller operation as a part of a real hydraulic system at the stage of its synthesis [10, 11]. That is, to use Hardware-in-the-loop (HIL) modeling in the controller synthesis process [12–18]. Also, HIL modeling with elements of automatic conduct of the experiment series can be used for structural and parametric identification of mathematical models of the hydraulic systems under study and their control systems [11].

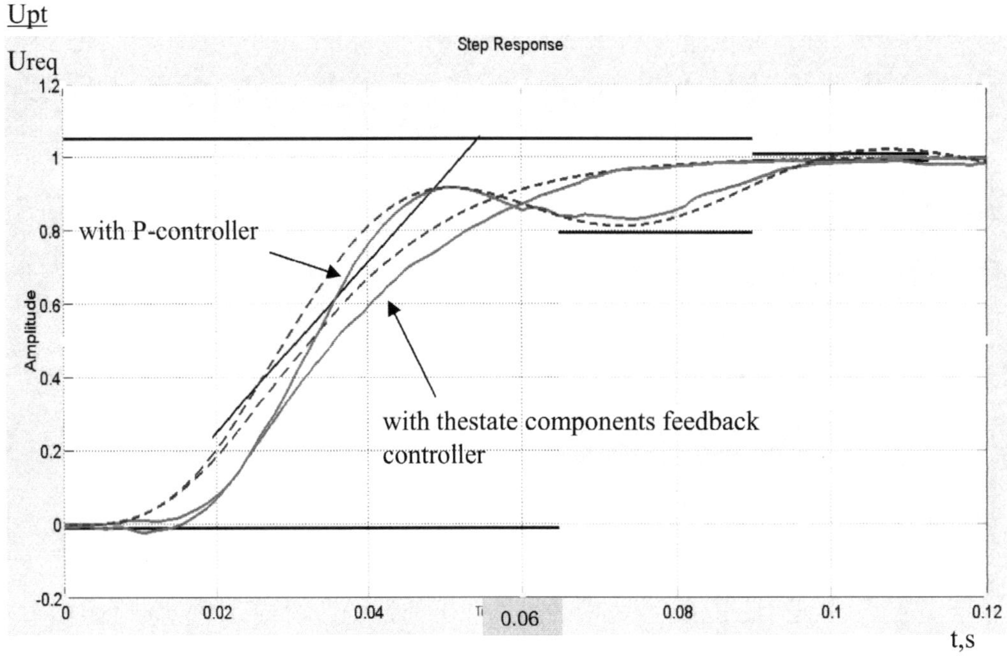

Fig. 1. Theoretical (dashed) and experimental (solid) EHSD transients on FESTO research system at different settings of the state components feedback coefficients.

Synthesis of the FESTO EHSD controllers in the HIL modeling mode

To optimize the controller parameters directly on the EHSD physical model, a hardware-software complex of HIL modeling based on FESTO hydraulic learning system was developed. The controller parameters are determined in Matlab software package using optimization methods. In optimization process automatic mode, the operation of the EHSD physical model with various controller parameter values is repeatedly investigated. The quality criteria are determined by analyzing changes in the state components feedback of the physical model that working out the specified input signals.

The block-diagram of the hardware-software complex is shown in Fig. 2.

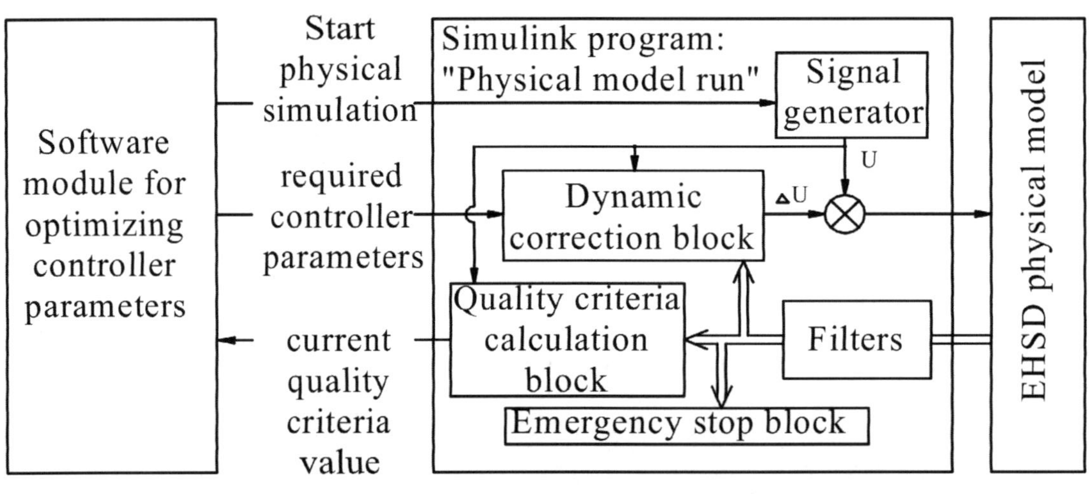

Fig. 2. The block-diagram of the hardware-software complex.

The optimization module, in this example, determines the state components feedback coefficients that are optimal according to the integral quality criterion ITAE. $I=\int_0^{\infty}\left|U_{req}-U_{pt}\right|\cdot t\cdot dt$. During the optimization process, the Simulink program «Physical Model Run» (PMR) is repeatedly run (Fig. 2). It provides sequential launch of the real EHSD physical model with changeable parameters of its controller.

Simulink PMR program contains:

– the block for setting the controller parameters;

– the module that generates desirable position of the hydraulic cylinder rod (for example, a square wave);

– the module that calculates the quality criterion needed to optimize the synthesized controller parameters.

– the block for emergency shutdown of the physical model in case of the EHSD state will be out of the allowed area.

The PMR program also provides bringing the EHSD physical model to the initial state for each iteration of determining the controller parameters.

The composition of the hardware-software complex may vary depending on the type of problems solved using the research complex of HIL modeling.

The physical model of the EHSD based on FESTO learning system with remotely adjustable feedback coefficients

The scheme of FESTO hydraulic system inclusion in the HIL is shown in Fig. 3.

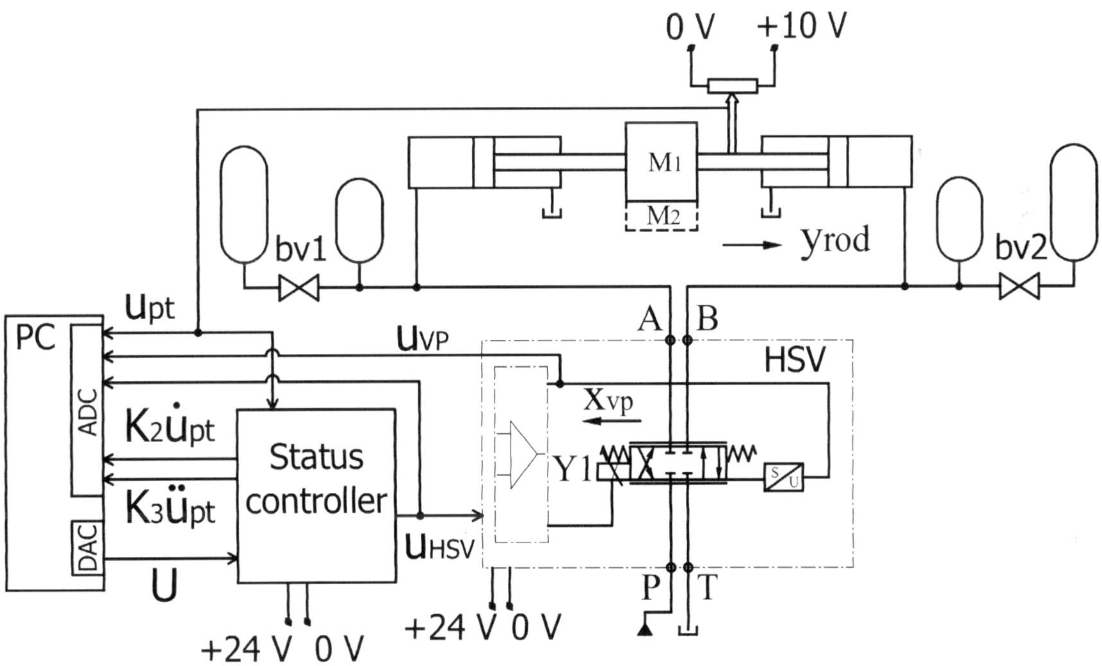

Fig. 3. Connection diagram of FESTO learning system to personal computer (PC) while studying the parameters of the EHSD controller with the parameters of the control object, which can be changed during its operation.

The notation conventions for Fig. 3: HSV — the hydraulic servo valve; bv1, bv2 — the ball valves; U_{HSV} — the HSV control signal (in effect a control piston setpoint position value); U_{VP} — the measured position of the HSV control piston; $M_1 = 27.5\,\text{kg}$, $M_2 = 10\,\text{kg}$, — the masses of the inertial load (main and additional, respectively). Status controller — the unit, shortly described below, y_{rod} — the position of the hydraulic cylinder rod; U_{pt} — the rod position feedback signal.

FESTO hydraulic learning system includes an electronic signal amplification and conversion unit (Status controller) (Fig. 4) (Part Number 162253). It implements the main feedback and feedbacks for the first and second derivatives of the main feedback signal. The direct circuit and feedback coefficients for the first and second derivatives are set by status controller potentiometers.

Three main feachures of the status controller were used while working on FESTO hydraulic research system:

1. control signal power amplifier;
2. analog state components feedback controller with adjustable coefficients;
3. the first and second derivatives of the main feedback signal.

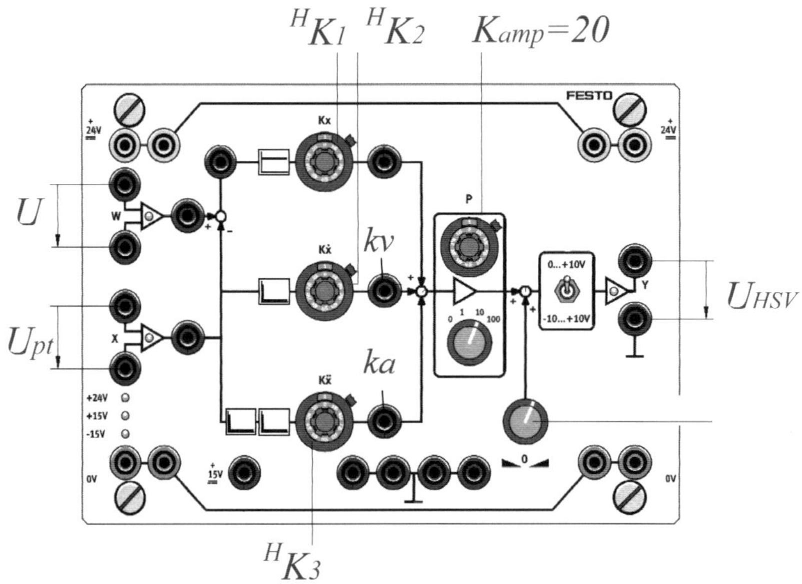

Fig. 4. Status controller front panel.

If the physical model is used directly in the process of synthesizing a controller running on the computer, then there is a need to change the feedback coefficients of the linear controller during the operation of the EHSD from the program on the computer. This can be achieved by implementing a part of the controller on the control computer [1]. The software part of the controller will make a dynamic correction of the control signal by adding an additive to it (Fig. 5). The additive ΔU provides a change in the the state components feedback coefficients of the EHSD from those pre-set $^H\overline{K}$ on the status controller unit (Fig. 4) to the required ones $\overline{K}$.

$$\Delta U = f_{\Delta U}\left(U, U_{pt}, \dot{U}_{pt}, \ddot{U}_{pt}, {}^H\overline{K}, \overline{K}\right). (1)$$

Let us define the structure and parameters of the linear controller with dynamic correction. Let us assume that the status conroller unit has coefficients $\left[{}^HK_1 \quad {}^HK_2 \quad {}^HK_3\right]$.They will remain unchanged. During the operation of the physical model it is necessary to set the state componentsfeedback coefficients equal to $\left[K_1 \quad K_2 \quad K_3\right]$.

Let us write down the expressions for the signal sent to the HSV as functions defined by the feedback coefficients.

$$U_{HSV} \overset{\Delta}{=} K_{amp} \cdot K_I \cdot U - K_{amp} \cdot \left(K_I + K_2 \cdot s + K_3 \cdot s^2\right) \cdot U_{pt} \quad (2)$$

$$U_{HSV} = f\left(U + \Delta U, U_{pt}\right) =$$

$$= K_{amp} \cdot {}^{H}K_I \cdot \left[U + \Delta U\right] - K_{amp} \cdot \left({}^{H}K_I + {}^{H}K_2 \cdot s + {}^{H}K_3 \cdot s^2\right) \cdot U_{pt} . \quad (3)$$

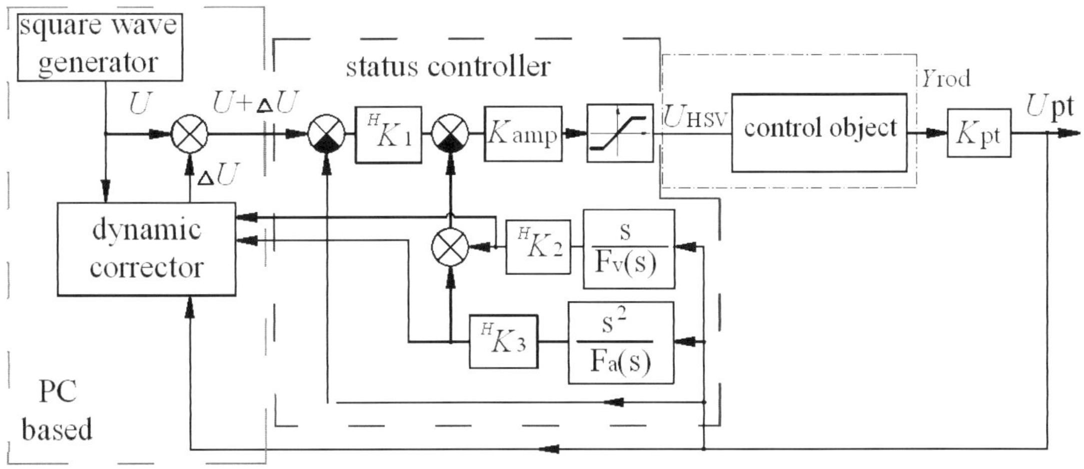

Fig. 5. The structure of the configuration of the state components feedback coefficients by using dynamic correction of the control signal.

Given the U_{HSV} equality in both equations, we can write an expression defining the additive $[\Delta U]$:

$$\left[\Delta U\right] = \frac{\left(K_I - {}^{H}K_I\right) \cdot U + \left[\left({}^{H}K_I - K_I\right) + \left({}^{H}K_2 - K_2\right) \cdot s + \left({}^{H}K_3 - K_3\right) \cdot s^2\right] \cdot U_{pt}}{{}^{H}K_I} \quad (4)$$

If we use the output signals of the status controller unit that compute the first and second derivatives of the feedback signal (Fig. 5), the resulting expression that defines the correction additive to the input signal at the input of the status controller unit takes the form:

$$\left[\Delta U\right] = \left[\frac{{}^{H}K_I - K_I}{{}^{H}K_I} \quad \frac{{}^{H}K_2 - K_2}{{}^{H}K_I} \quad \frac{{}^{H}K_3 - K_3}{{}^{H}K_I}\right] \cdot \begin{bmatrix} U_{pt} - U \\ \dfrac{kv}{{}^{H}K_2} \\ \dfrac{ka}{{}^{H}K_3} \end{bmatrix} . \quad (5)$$

U_{pt}, kv, ka signals are taken from the corresponding outputs of the status controller unit (Fig. 4).

Figure 6 shows transients of the EHSD with two implementations of the same controller with coefficients $\left[K_1 \quad K_2 \quad K_3\right]$=[1.64 0.00325 0.00016]. In one case, these coefficients were set on the status controller. The status controller input received a step signal U (unchanged for time t> 0). In the second case, the coefficients ${}^{H}\overline{K}$=[1.58 0.0154 0.000353] were pre-set on the status controller, and the final coefficients of the controller were provided by an additive ΔU calculated using the formula (5) (the $U + \Delta U$ Signal is shown in Fig. 6). As you can see from the graphs, the transition processes of the EHSD completely coincide.

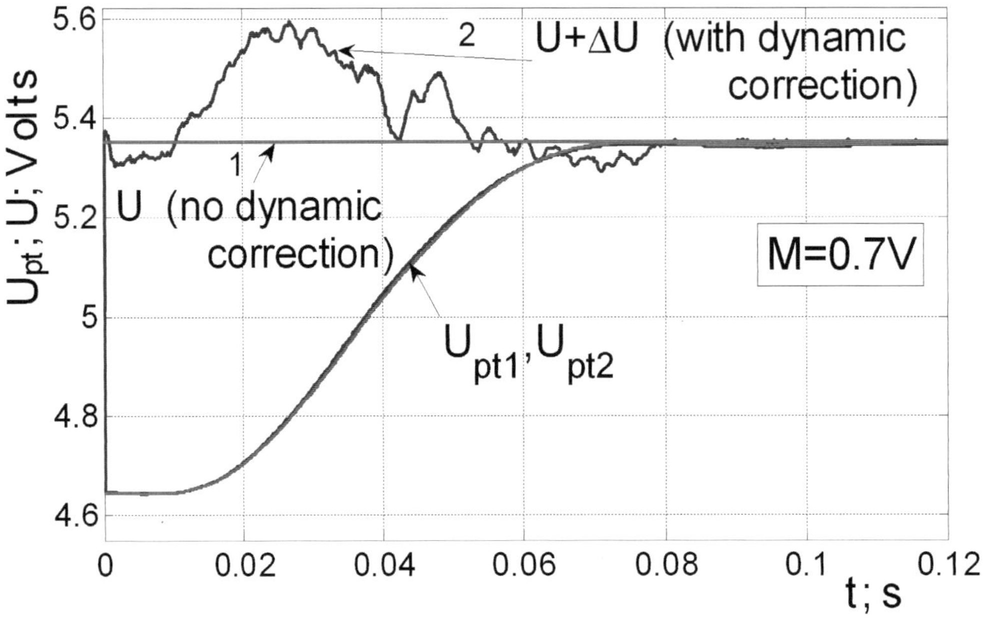

Fig. 6. The EHSD transients with two implementations of state components feedback coefficients.

Conclusions

Using the example of Matlab environment and FESTO hydraulic learning system, the possibility of creating a complex of HIL modeling have been considered. That allows to combine the mathematical description of the actuating mechanism and the physical model of the control object into a single complex for the purpose of conducting research in automatic and automated mode. The developed approaches and algorithms have shown that it is possible to synthesize controllers using automatic optimization of the controller parameters directly in the process of operation of the EHSD physical model.

This significantly expands the capabilities of both the software package and FESTO hydraulic learning system that implements the physical model of the EHSD when conducting training and research.

References

[1] B Kulakov and D Kulakov Research of the relay-linear control of an electro-hydraulic servo drive on FESTO learning system. 2020 IOP Conf. Ser.: Mater. Sci. Eng. 779 012027

[2] D Kulakov and B Kulakov Study of the dynamics of an electro-hydraulic servo drive physical model on FESTO learning system. 2019 *IOP Conf. Ser.: Mater. Sci. Eng.* **492** 012029

[3] D.N. Popov, N.G. Sosnovskiy, M.V. Siukhin Computer simulation of hydraulic systems with typical nonlinear characteristics // Science & Education of the Bauman MSTU, 2017, no. 07, pp. 63–74. DOI: 10.7463/0717.0001136

[4] Adeoye, A.O.M., Aderoba, A.A., Oladapo, B.I. Simulated Design of a Flow Control Valve for Stroke Speed Adjustment of Hydraulic Power of Robotic Lifting Device (2017) Procedia Engineering, 173, pp. 1499–1506. DOI: 10.1016/j.proeng.2016.12.230

[5] Kozlov, O.S., Skvortsov, L.M. MVTU software package in scientific research and applied developments (2016) Mathematical Models and Computer Simulations, 8 (4), pp. 358–368. DOI: 10.1134/S2070048216040116

[6] Goodwin G.C., Graebe S.F., Salgado M.E. Control system design. Upper Saddle River; L.: Prentice-Hall, 2001. 908 p.

[7] Pupkov K.A., Egupov N.D. et al. *Metodyklassicheskoyisovremennoyteoriiavtomaticheskogoupravleniya: uchebnik v 5 tomah. Tom 5. Metodisovremennoyteoriiavtomaticheskogoupravleniya*[Methods of Classic and Modern Control Theory: textbook for universities includes 5 volumes. Vol. 5. Methods of Modern Control Theory]. Moscow, Bauman MSTU Publ., 2004. 784 p. (in Russian)

[8] Wang, S., Xu, Q., Lin, R., Yang, M., Zheng, W., Wang, Z. Feedback linearization control for electro-hydraulic servo system based on nonlinear disturbance observer (2017) Chinese Control Conference, CCC, № 8028135, pp. 4940–4945. DOI: 10.23919/ChiCC.2017.8028135

[9] Shumilov, I.S. The steering hydraulic drives with centralized or independent power supplies for the main aircrafts (2015) Proceedings of 2015 International Conference on Fluid Power and Mechatronics, FPM 2015, pp. 596–602. DOI: 10.1109/FPM.2015.7337186

[10] Pupkov K.A. *Modelirovaniyeiispitaniye system avtomaticheskogouprableniya* [Automatic control systems modeling and testing] Moscow, RUDN Publ., 2014. 114 p. (inRussian)

[11] Kulakov, D.B., Semenov, S.E., Kulakov, B.B., Shcherbachev, P.V., Tarasov, O.I. Hydraulic Bipedal Robots Locomotion Mathematical Modeling (2015) Procedia Engineering, 106, pp. 62–70. DOI: 10.1016/j.proeng.2015.06.009

[12] Dooner, M., Wang, J., Mouzakitis, A. Dynamic modelling and experimental validation of an automotive windshield wiper system for hardware in the loop simulation (2015) Systems Science and Control Engineering, 3 (1), pp. 230–239. DOI: 10.1080/21642583.2015.1012309

[13] Schulte, H., Georg, S. Hardware-in-the-Loop test-bed for benchmarking of fault tolerant control schemes for wind turbines (2015) IFAC-PapersOnLine, 28 (21), pp. 1375–1382. DOI: 10.1016/j.ifacol.2015.09.717

[14] Tejado, I., Serrano, J., Pérez, E., Torres, D., Vinagre, B.M. Low-cost Hardware-in-the-loop Testbed of a Mobile Robot to Support Learning in Automatic Control and Robotics (2016) IFAC-PapersOnLine, 49 (6), pp. 242–247. DOI: 10.1016/j.ifacol.2016.07.184

[15] Hancer, M., Bitirgen, R., Bayezit, I. Designing 3-DOF Hardware-In-The-Loop Test Platform Controlling Multirotor Vehicles (2018) IFAC-PapersOnLine, 51 (4), pp. 119–124. DOI: 10.1016/j.ifacol.2018.06.058

[16] Traphöner, P., Olma, S., Kohlstedt, A., Fast, N., Jäker, K.-P., Trächtler, A. Hardware-in-the-loop simulation for a multiaxial suspension test rig with a nonlinear spatial vehicle dynamics model (2019) IFAC-PapersOnLine, 52 (15), pp. 109–114. DOI: 10.1016/j.ifacol.2019.11.659

[17] Centelles, D., Soriano, A., Marin, R., Sanz, P.J. Wireless HROV Control with Compressed Visual Feedback Using Acoustic and RF Links (2020) Journal of Intelligent and Robotic Systems: Theory and Applications. DOI: 10.1007/s10846-020-01157-5

[18] Guo, A., Zhou, Z., Zhu, X., Zhao, X., Ding, Y. Automatic control and model verification for a small aileron-less hand-launched solar-powered unmanned aerial vehicle (2020) Electronics (Switzerland), 9 (2), № 364. DOI: 10.3390/electronics9020364

Gradient identification method of the model parameters for electrohydraulic servo drive of FESTO learning system

B Kulakov[1], D Kulakov[1,2] and Long Qi[1]

[1]Bauman Moscow State Technical University, 5 Second Baumanskaya Street, Moscow, 105005, Russian Federation

[2]E-mail: kulakov@bmstu.ru

Abstract. Gradient identification method of the model parameters for electrohydraulic servo drive is considered. The method is adapted to be implemented by an experimental complex consists from FESTO hydraulic learning system and Matlab/Simulink. The implementation of the considered identification algorithm in the Matlab/Simulink package is shown.

Introduction

While the electrohydraulic servo drive (EHSD) is working, its parameters may vary. It can be caused by changing mass of the load, oil bulk modulus [1], friction model parameters [2], electrohydraulic amplifier characteristics. One way to obtain the EHSD required control quality is increasing robustness property of controlled hydraulic drive with controller to be synthesized, another way — using adaptive controller, i.e. controller that changes its parameters for adaptation to changing actuator parameters. The works [3–4] were dedicated to a question of using FESTO hydraulic learning system for research static and dynamic EHSD characteristics on the base its physical model. The questions of controller synthesis with required quality criterion was considered in [5–11]. The required control quality can be provided due to the high robustness of the synthesized controller. However, when a real drive is operating, its parameters can change significantly, and in this case, it becomes advisable to identify the changing parameters and adjust the controller parameters directly during the drive operation, i.e. use an adaptive controller. Adaptive control issues, including EHSD, are discussed in [5,6,12–18]. The paper [5] describes a method of synchronous detection that allows identifying parameters by applying a high-frequency search signal to an object under study and determining the sign of change in the configurable parameter by comparing the phase of the search signal and the same-frequency component of the quality criterion signal. The work [13] describes the use of the above method with the disclosure of the analytical dependence that determines the gradient of change in the required quality criterion. The quality criterion used was a mismatch between the states of the model of object and the object whose parameters were identified. In this paper, we consider the application of a gradient identification algorithm to the EHSD using work [13], with the prospective goal of using it as part of indirect adaptive control. Just as in [3,4], the EHSD based on FESTO research system is considered.

Thanks to FESTO electronic status controller unit, it is possible to measure the state vector of the object under study in this work, and it is possible to avoid the need for a full-fledged simulation of the control object with a state vector independent of the object.

Mathematical formulation of the EHSD as a control object

Figure 1 shows open loop mathematical model block-diagram of the EHSD that is assembled on FESTO hydraulic research system. The mathematical model have been used to determine the state components feedback coefficients [3].

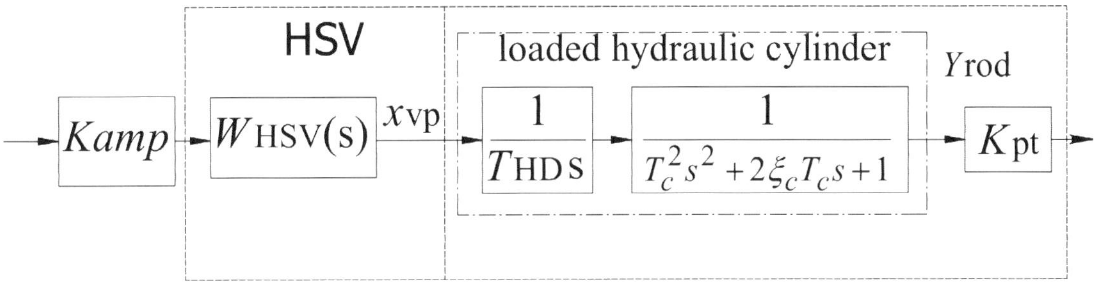

Fig. 1. The EHSD open loop mathematical model block-diagram.

K_{amp} —electronic amplifier gain;

$W_{HSV}(s)$ — hydraulic servo valve transfer function;

ξ_c — the hydraulic cylinder damping coefficient;

T_{HD} — the hydraulic drive time constant;

T_c — the time constant of the loaded hydraulic cylinder;

K_{pt} — the potentiometric feedback sensor gain.

The loaded cylinder parameters are defined by the expressions [8]:

$$T_c = \sqrt{\frac{mV}{2BF_c^2}};\ \xi_c = \frac{T_{dc}}{2T_c};\ T_{dc} = \frac{K_{fr}V}{2B \cdot F_c^2} + \frac{K_{Qp}m}{F_c^2}. (1)$$

To conduct research on the use of a gradient algorithm for identifying parameters of the EHSD mathematical model, we present the mathematical model of the hydraulic drive with state components feedback to the form Fig. 2.

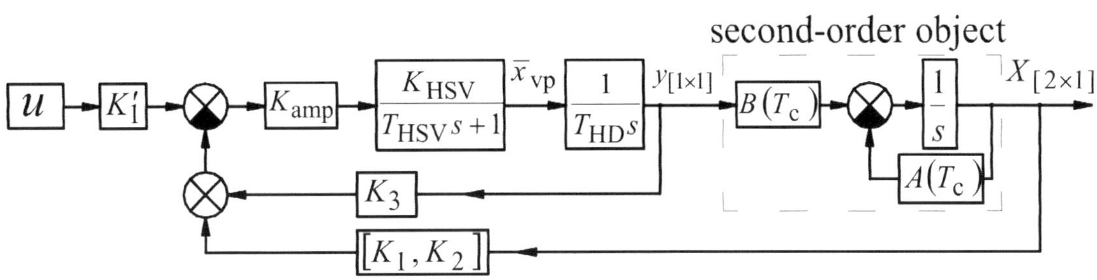

Fig. 2. The EHSD mathematical model block-diagram.

State space representation of the second order object — object under study (Fig.1) has the form

$$\dot{\mathbf{X}} = A\mathbf{X} + Bu, (2)$$

where

$$A=\begin{pmatrix} 0 & 1 \\ -\dfrac{1}{T_c^2} & -\dfrac{2\xi_c}{T_c} \end{pmatrix},\ B=\begin{pmatrix} 0 \\ \dfrac{K_{pt}}{T_c^2} \end{pmatrix}. (3)$$

Gradient descent algorithm for the EHSD parameters identification

The essence of this method is to configure the desired parameters of the model in the process of system operation, by changing them in the negative direction of gradient of an objective function.

$$\dot{\Theta}=-\Gamma\cdot\nabla_{\Theta m}q(E)\ (4)$$

Here:

$\Theta_{м}$ — the defined parameters vector of the object mathematical model;

Γ — matrix of weight coefficients $\Gamma=\Gamma^T>0$;

$$q(\mathbf{E})=\mathbf{E}^T\mathbf{E}\ (5)$$

– the objective function;

$$E\overset{\Delta}{=}X_o-X_m\ (6)$$

– the discrepancy between the state vector of the object under study and the vector X_m of the mathematical model;

X_o — the state vector of the object under study;

X_m — the one-step forecast vector of the object under study state based on its mathematical model.

We assume that the difference in the structure of the control object and its model, together with the initial mismatch of the desired parameters and their evaluation, leads to a single extremum of the objective function in the identification process [13].

For numerical calculation of the objective function gradient, we will use high-frequency changes in the configurable parameters with a small amplitude (search signals), which are superimposed on the "slowly" changing (adjusting) parameters of the mathematical model.

In this paper, we consider the application of an identification algorithm to determine the parameter T_c of the linearized mathematical model of the hydraulic cylinder, since it changes most strongly when the load mass and volume of the hydraulic cylinder cavities change (1).

To define the objective function changes, the following expressions were entered:

$$X_{m1\,i}=X_{o\,i-1}+F_m\left(X_{oi-1},y_{i-1},\Theta_{mi-1}\right)\cdot dt;\ (7)$$

$$X_{m2\,i}=X_{o\,i-1}+F_m\left(X_{oi-1},y_{i-1},\Theta_{mi-1}+\delta\Theta_{mi-1}\right)\cdot dt;$$

where:

$F_m(\ldots)$ — function of the mathematical model right parts;

y — the signal according to Fig. 2;

Θ_m and $\delta\Theta_m$ — the definable parameter and its small change.

The determination of the state vector components of the model and the object under study using the formulas (7) are shown in Fig. 3.

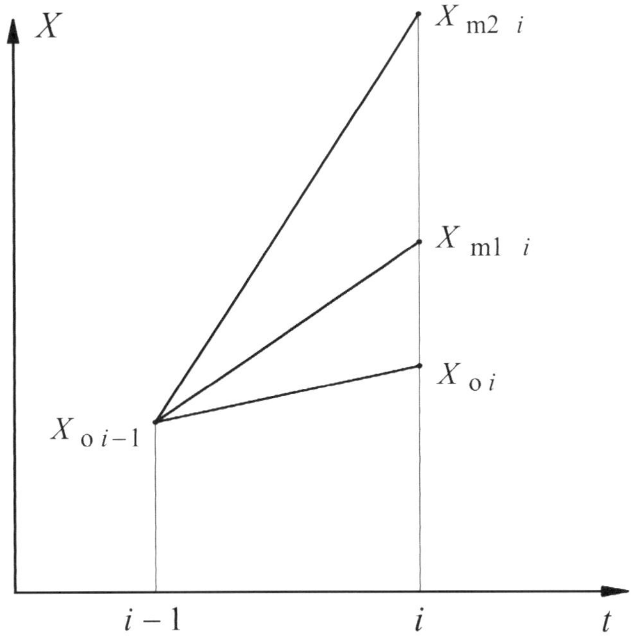

Fig. 3. Illustration of the components definition $X_{oi}, X_{oi-1}\ X_{m1i},\ X_{m2i}$.

NotationinFig.3:

X_{oi}, X_{oi-1} — the current and previous state of the object, respectively;

X_{m1i} — the current state of the model as a forecast relative to the previous state of the object X_{oi-1} by a known function of the right parts of the model with the previous value of the parameter being determined Θ_{mi-1};

X_{m2i} — differs from X_{m1i} in that $\Theta_{mi-1} + \delta\Theta_{mi-1}$ is taken in place of Θ_{mi-1}.

For a linear mathematical model written in the state space, equations (7) take the form:

$$X_{m1\,i} = X_{o\,i-1} + \left[A_m\left(\Theta_{mi-1}\right) X_{oi-1} + B_m\left(\Theta_{mi-1}\right) y_{i-1} \right] \cdot dt \quad (8)$$

$$X_{m2\,i} = X_{o\,i-1} + \left[A_m\left(\Theta_{mi-1} + \delta\Theta_{mi-1}\right) X_{oi-1} + B_m\left(\Theta_{mi-1} + \delta\Theta_{mi-1}\right) y_{i-1} \right] \cdot dt$$

The discrepancies between the object state vector under study and the state vectors of its mathematical model obtained in accordance with (7) are defined as:

$$E_2 = X_{oi} - X_{m2i},$$

$$E_1 = X_{oi} - X_{m1i}. \quad (9)$$

The gradient of the objective function will be determined in accordance with the expression:

$$\frac{\partial q(E)}{\partial \Theta_m} \cong \frac{E_2^T E_2 - E_1^T E_1}{\delta\Theta_m} \quad (10)$$

Thus, substituting expressions (8)–(10) in (4), you can get the equation of the defined parameter change:

$$\dot{\Theta}_{mi} = f\left(X_{oi,} X_{oi-1}, y_{i-1}, \Theta_{mi}, \delta\Theta_{mi} \right) \quad (11)$$

Investigation of the identification algorithm for the EHSD mathematical model

To study the gradient algorithm for identifying EHSD parameters described above, a program was developed in Matlab/Simulink, the block diagram of which is shown in Fig. 4.

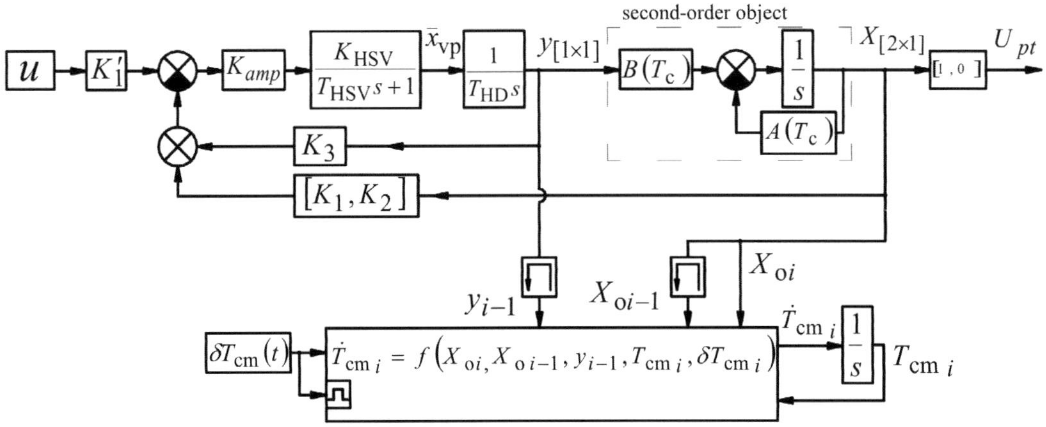

Fig. 4. The block diagram that shows the identification algorithm adapted to Matlab/Simulink.

Some numerical simulation results of the block-diagrams in Figure 4 are shown in Figure 5. Thedrivefollows-upsomeharmonicinput. During the simulation, a stepwise change in the time constant T_c is modeled (it can be implemented on FESTO hydraulic learning system by changing load mass and additional volumes connected to the cavities of the hydraulic cylinders). Figure 5 shows the parameter adjustment of the mathematical model during the EHSD simulates.

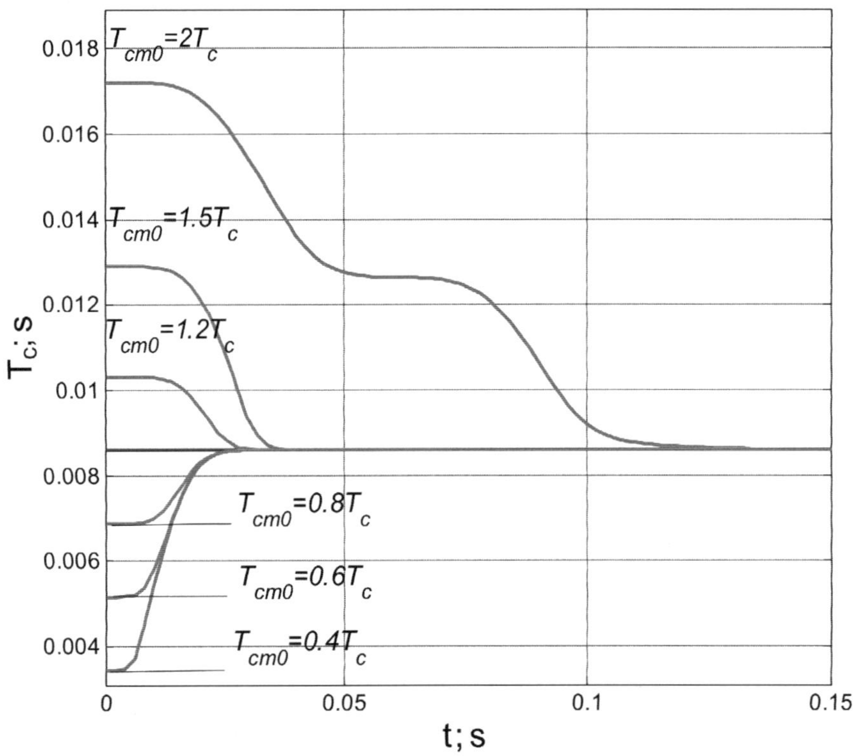

Fig. 5. Adjustment of the mathematical model parameter T_c for its various step changes.

Conclusions

In this paper, a modification of the gradient search method for the parameter of the control object under study based on [13] in relation to the EHSD of FESTO research system [3] is performed. The modified method is applied to the linearized mathematical model EHSD, but it is not limited in principle to the linearity of the mathematical model. The study shows good convergence of the method based on the results of numerical modeling in the Matlab/Simulink environment. It is shown that the considered method is suited for implementation on FESTO hydraulic research system, which is provided by the presence of an element base that allows experimentally determining the control object state components: valve position measurement, rod position and its velocity and acceleration determination. The considered algorithm expands the possibilities of using the Festo experimental complex for educational and research purposes.

References

[1] Wang, S., Xu, Q., Lin, R., Yang, M., Zheng, W., Wang, Z. Feedback linearization control for electro-hydraulic servo system based on nonlinear disturbance observer (2017) Chinese Control Conference, CCC, № 8028135, pp. 4940-4945. DOI: 10.23919/ChiCC.2017.8028135

[2] Lomakin V.O., Shumilov I.S., Shcherbachev P.V. *Matematicheskoyemodelirovaniyeelectrogiravlicheskogoslediaschegoprivodavisocotochnogoregulirovaniyavraschatelnogodvizheniya.* [Mathematical modeling of the electrohydraulic rotary motion servo drive]. BMSTU Journal of Mechanical Engineering. 2011. № 10. pp. 38–45. (in Russian)

[3] D Kulakov and B Kulakov Study of the dynamics of an electro-hydraulic servo drive physical model on FESTO learning system. 2019 *IOP Conf. Ser.: Mater. Sci. Eng.* 492012029

[4] B Kulakov and D Kulakov Research of the relay-linear control of an electro-hydraulic servo drive on FESTO learning system. 2020 IOP Conf. Ser.: Mater. Sci. Eng. 779 012027

[5] Krasovsky A.A. et al.*Spravochnikpoteoriiavtomaticheskogoupravleniya* [Handbook on the theory of automatic control]. Moscow: NaukaPubl., 1987. 712 p. (inRussian)

[6] Pupkov K.A., Egupov N.D. et al.*Metodyklassicheskoyisovremennoyteoriiavtomaticheskogoupravleniya: uchebnik v 5 tomah. Tom 5. Metodisovremennoyteoriiavtomaticheskogoupravleniya*[Methods of Classic and Modern Control Theory: textbook for universities includes 5 volumes. Vol. 5. Methods of Modern Control Theory]. Moscow, Bauman MSTU Publ., 2004. 784 p. (in Russian)

[7] Goodwin G.C., Graebe S.F., Salgado M.E. Control system design. Upper Saddle River; L.: Prentice-Hall, 2001. 908 p.

[8] Popov D.N. *Dinamikairegulirovaniegidro- ipnevmosistem* [Dynamics and regulation of hydro- and pneumatic systems]. Moscow, Mashinostroenie Publ., 1987. 464 p. (in Russian

[9] N Sosnovsky and D Ganieva Mathematical modeling and analysis of the electrohydraulic actuator with the throttle control. 2019 IOP Conf. Ser.: Mater. Sci. Eng. 589 012016

[10] Sosnovskiy, N.G., Popov, D.N., Siukhin, M.V. Sensitivity of dynamic characteristics of electrohydraulic servo drive (2018) Proceedings - 2018 Int. Conf. on Industrial Engineering, Applications and Manufacturing, ICIEAM 2018, № 8728669. DOI: 10.1109/ICIEAM.2018.8728669

[11] Kulakov D.B., Semenov S.E., Kulakov B.B., et al. Hydraulic Bipedal Robots Locomotion Mathematical Modeling Elsevier publishing, Procedia Engineering Volume 106, 2015, Pages 62–70 https://doi.org/10.1016/j.proeng.2015.06.009

[12] Bortsov, Yu.A.,Polyakhov, N.D., and Putov, V.V., *Elektromekhanicheskiesistemy s adaptivnymimodal'nymupravleniem* (Electromechanical Systems with Adaptive and Modal Control), Leningrad: Energoatomizdat, 1984. 216 p. (in Russian)

[13] Pupkov K.A., Egupov N.D. et al. *Metodyrobastnogo, neyro-nechetkogoiadaptivnogoupravleniya*[Methods of Robast, neuro-fuzzy and Adaptive Control Theory: textbook for universities]. Moscow, Bauman MSTU Publ., 2002. 744 p. (in Russian)
[14] Lemeshko, M.A., Stradanchenko, S.G., Iliev, A.G., Zanina, I.A. Study of adaptive hydraulic drive of rotary drilling machine (2019) IOP Conference Series: Materials Science and Engineering, 680 (1), статья № 012006DOI: 10.1088/1757-899X/680/1/012006
[15] Pershin, V., Adigamov, K., Khinikadze, T. Study on unified hydraulic drive module with self-Adaptation by power and kinematic parameters (2019) E3S Web of Conferences, 135, № 02023. DOI: 10.1051/e3sconf/201913502023
[16] P Chaburko and Z Kossova Wet part hybrid optimization method of hermetic pump. 2019 IOP Conf. Ser.: Mater. Sci. Eng. 492 012011
[17] Wang, X., Li, W., Yao, J., Wan, Z. Acceleration Harmonic Estimation for Hydraulic Servo Shaking Table Based on Multi-Innovation Stochastic Gradient Algorithm (2020) Mathematical Problems in Engineering, 2020, № 3063469. DOI: 10.1155/2020/3063469
[18] Nguyen TrongToan, Nguyen Nam Trung Auto-tuning controllers of a class of plants using gradient descent algorithm. International Journal of Engineering and Applied Sciences (IJEAS) ISSN: 2394-3661, Volume-6, Issue-6, June 2019

IOP Conf. Series: Materials Science and Engineering **963** (2020) 012026 doi:10.1088/1757-899X/963/1/012026

Variable speed control in centrifugal pumps for closed systems

A Petrov[1] and Rafael Leonardo Gomez Dias[1,2]

[1]Bauman Moscow State Technical University, 5 Second Baumanskaya Street, Moscow, 105005, Russian Federation

[2]E-mail:gomes.diasr@student.bmstu.ru

Abstract. Energy consumption in any type of installation, whether for urban purposes, commercial or industrial, is a very important question; therefore, reducing energy consumption is of great importance. It is very important to control the rotational speed of the pumping equipment in accordance with the required consumption or requirement. For closed circuits (closed systems) it is important to ensure the required flow rate while maintaining the proper pressure or working head in centrifugal pumps. This type of control is ideal for installations with changeable curves, where different work processes are required at the same pressure or different process pressures in centrifugal pumps.

1. Introduction

HVAC pump systems with a total pump system capacity of more than 10 (Hp), including the control valves designed to modulate or opening and closing in step depending on the load, must be designed for variable fluid flow and should be able to reduce pump flow rate by up to 50% or less from the design flow rate [1]. When choosing a centrifugal pump, it is necessary to take into account factors such as pipelines and a hydraulic system through which cold water is supplied to each of the loads that are being parts of the system. Considerations about calculating of heat loads are not considered in this article, authors focus only on the selection of pumping equipment and the change in rotational speed in accordance with the volumetric flow rate. The pump system must supply water to each of the supply and exhaust systems, providing pressure in each of the coils, regardless of how far the system loads can be located relative to the centrifugal pump. We will select a pump for distributing cold water in a building that is used as an office of a large company. There is a control unit on each floor that distributes conditioning air on each floor; the greatest heat load of the system falls on the first level, which is being used as the main auditorium of the company. In table 1 we can see a summary of heat loads for each level in the building, the flow is determined in accordance with AHRI [2].

Knowing the volumetric flow rate of each processing unit, we can determine the total flow rate of the pump. A study of the losses of the system should be carried out then. All cooled and condenser water pipelines should be designed so that the estimated flow rate in each part of the pipeline does not exceed the values indicated in the table. 6.5.4.6 [1] for the corresponding total annual hours of operation. The choice of pipelines sizes for systems operating in variable flow conditions (for example, modulating two-way control valves on spools) and containing variable speed pump motors can be made from the “variable flow / variable speed” columns. The others should be made from “other” columns [1]. When we study the pressure loss in a closed hydraulic circuit, we must analyze each of the conclusions and distribution of pipes; in this particular case, we have the most distant equipment in relation to the pump at the first level, in addition to being the largest equipment in the system. We will assume the following equipment losses [2] (loading-and-unloading units, chiller, air-separator) (Fig. 1.) as well as friction losses along the pipes and fittings.

Table 1. Heat loads on the floor

Zone	Heat load [Btu/hr]	Heat load [kW]	Flow rate [GPM]*	Flow rate [Lt/s]
7th Floor	240,000	70.34	48	3.64
6th Floor	360,000	105.51	72	5.46
5th Floor	360,000	105.51	72	5.46
4th Floor	480,000	140.68	96	7.27
3rd Floor	420,000	123.09	84	6.36
2nd Floor	480,000	140.68	96	7.27
1st Floor	600,000	175.85	120	9.09

*Nominal water flow rate is determined by water temperature at nominal power. The normalized flow rate per unit of evaporator capacity is shown. AHRIStd 550-590, 2015

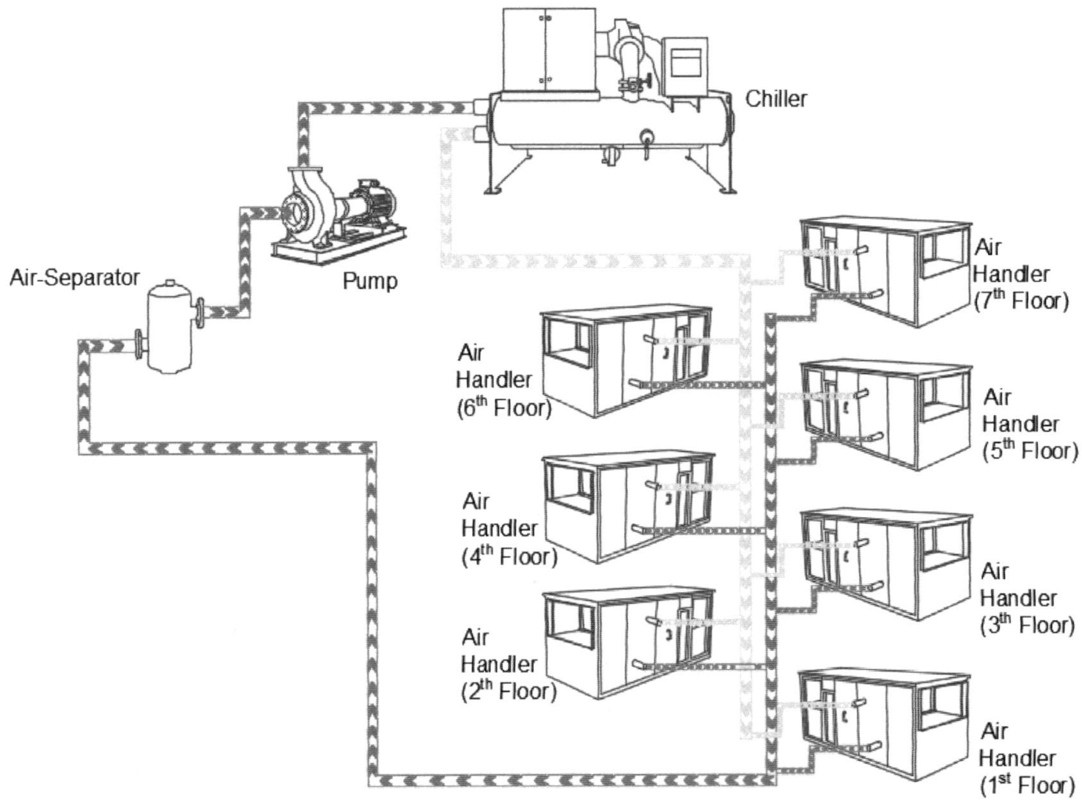

Fig. 1. HVAC Closed-loopsystem

The total loss is 57 [ftwg] or 170.03 [kPa]. Now, knowing the necessary head and volumetric flow rate of water, we can choose a convenient pump.

For a more detailed study of efficiency, velocities and pressure differentials, it is recommended to carry out the research using the methods of computational fluid dynamics (CFD) [3]. There are various types of software using CFD, such an analysis is beyond the scope of this article.

2. Selection of pumping equipment

The pump should be selected from the manufacturer in this industry, the pump curve and the operating point should be obtained [4]. This point corresponds with the operating of the system at a maximum load of 862 [kW] (Table 1). However, one of these systems is characterized by operation at partial loads, so the pump must change its rotational speed in order to be able to supply the required amount of water at partial loads. In Fig. 2, we can see the pump curve and the system curve. This operation moment happens when the pump is operating at 100% of its capacity.

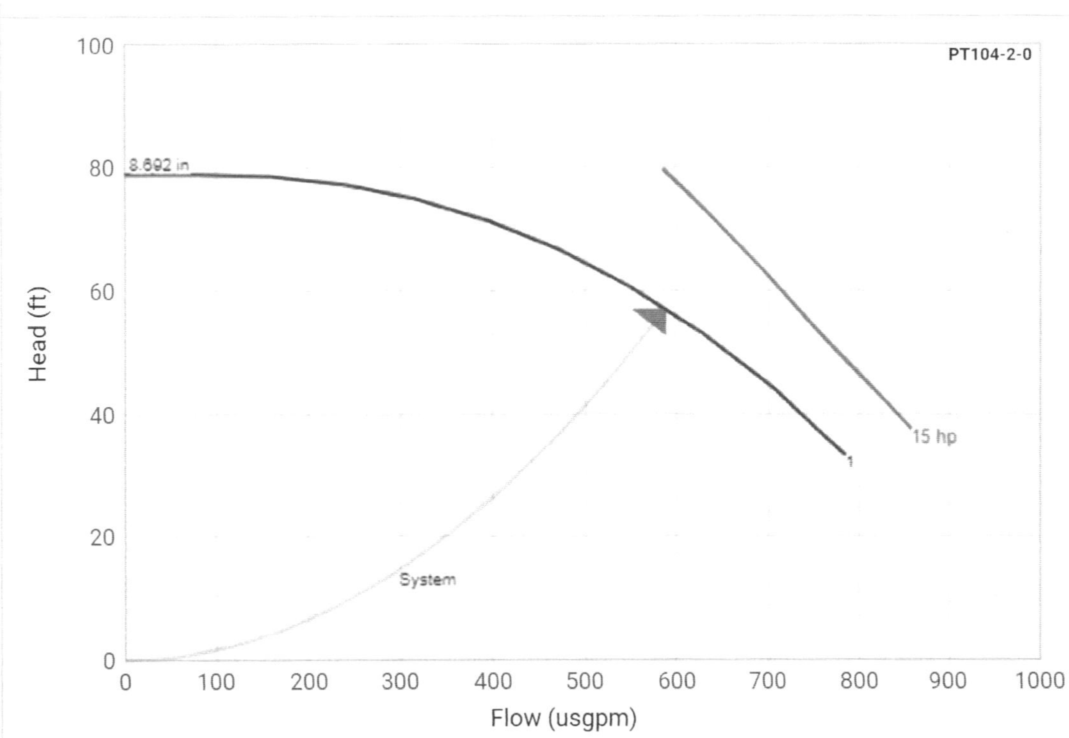

Fig. 2. The operating point of the pump under maximum heat load

Under these conditions, the pump motor has a power of 15 [Hp] or 11,2 [kW], with a speed of 1750 [rpm]. An appropriate control strategy will allow the system to save energy at partial heat loads, reducing the frequency and, therefore, the rotational speed of the pump motor.

3. Pump speed control strategy

Although the project's design engineer can propose several control options for changing the pump rotational speed, in this article we are focusing on changing the pump RPM while maintaining a constant pressure [5]. Assuming an operating condition that requires 300 [gal/min] (22,7 [L/s]). If the centrifugal pump does not have a variable frequency drive (VFD), then the decrease in flow rate is compensated by the increasing of head, the new operating point moves along the pump curve (Fig. 3). Although the braking power is reduced (Table 2), the pump motor remains the same (15 [hp]), switching to 1750 [rpm] thus, the pump efficiency is significantly reduced. When the motor has VFD, it is possible to reduce the frequency in this 4-pole motor and reduce the rotation frequency. This will lead to electricity saving [5].

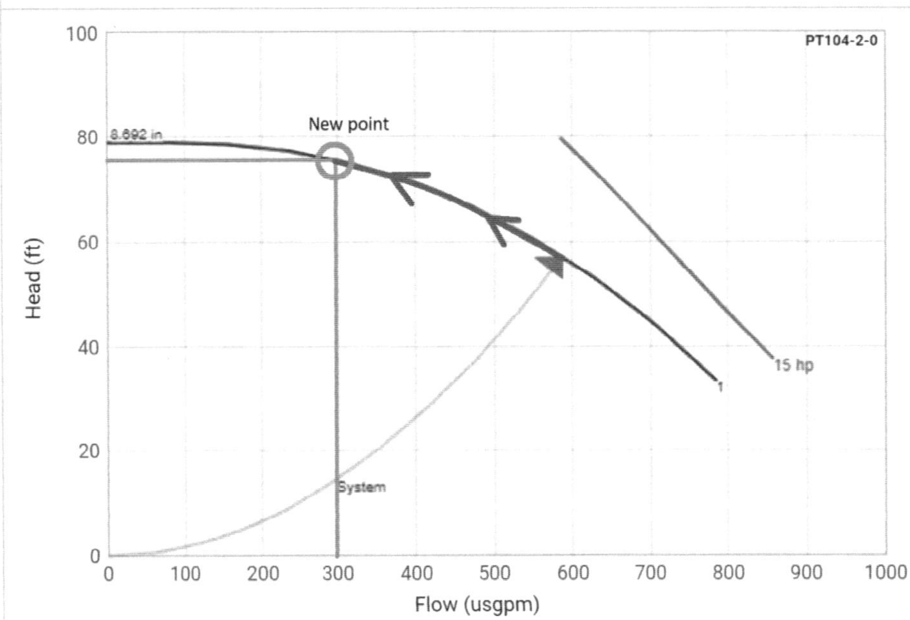

Fig. 3. New pump operation item, engine without VFD

Table 2. Required power

Operating point	Heat load [Btu/hr]	Flow rate [GPM]*	Head[Ft wg]	Bhp [kW]
Design load (max)	2,940,000	588	57	7.60
New point	1,500,000	300	75.87	6.90
Partial load - VFD	1,500,000	300	57	3.87

When the system is operating at partial load, the volumetric flow rate is less than the design condition, for this reason, for a new flow rate of 300 g / min, the operating frequency and, therefore, the pump rotational speed must be reduced, which provides the saving energy. As mentioned above, if the pressure in the main pipe must be maintained for the new requirements for volumetric flow rate, then the known similarity principles for pumps cannot be used, because the system curve is changing. Thus, using the following equation:

$$H_P = A^* \propto^2 + B * Q \propto + C * Q^2$$

Where H_P — is head that must remain unchanged; A, B, C constants can be determined using the initial pump curve (1750 rpm). The alpha constant $\propto$ can be defined by the following equation:

$$\propto = \sqrt{\frac{n}{1750}}$$

Now, using Matlab and using the iterative process, we can determine how many rotations per minute "n" are necessary for H_P to be 57 [ft]. This is shown on the Fig. 4 that at this new rotational speed (1450 rpm) the pressure remains approximately constant, providing pressure on the main branch with the required flow rate. The power can be calculated using the following expression:

$$BHP_{Hp} = \frac{Q * H_P * SG}{3960 * \eta}$$

The specific gravity (SG) of water is equal to 1, and for a constant pressure of 57 [ft] we plot the power behavior depending on different water flows, as can be seen in Fig. 5.

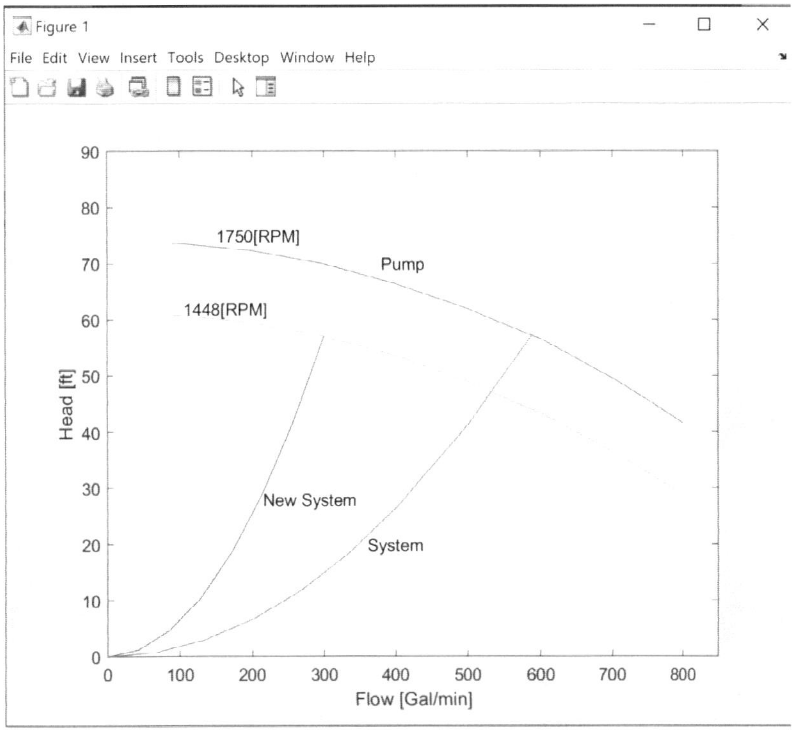

Fig. 4. Required pump behavior under partial load

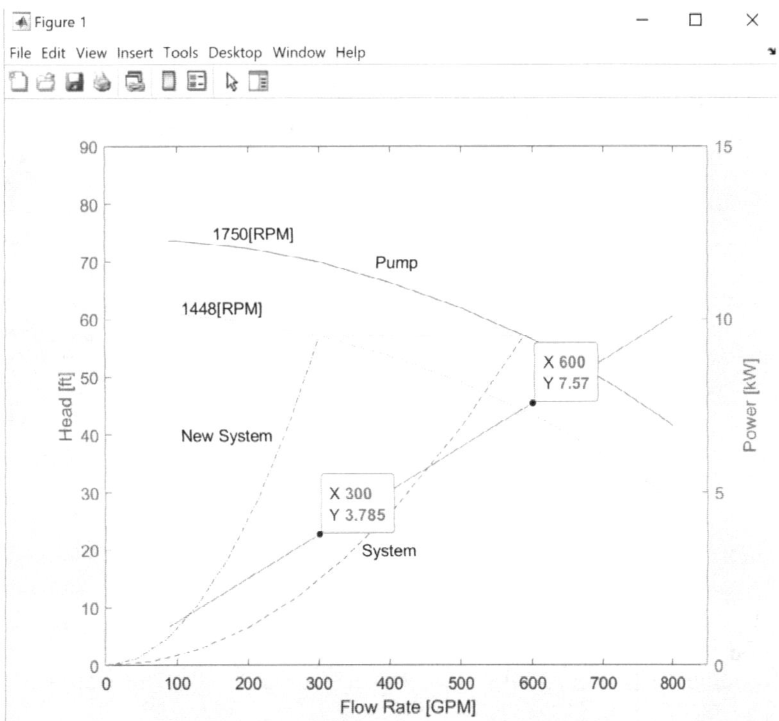

Fig. 5. Power change for a head of 57 [feet]

4. Elements necessary for speed control

To control the rotation speed of the pump, the system must have special devices - sensors that measure variables, such as pressure and flow rate. These sensors send signals to the controller, which can send a control signal to VFD and control the motor speed [6]. The pressure sensor can be used in the pump supply line at the appropriate distance, as recommended by the manufacturer. The choice of this element should take into account the maximum process pressure of the system and the operational temperature of the liquid. The pressure sensor should be a device for measuring the increase in pressure in the main supply pipe.

The flow measurement can use a diaphragm that will measure the pressure differential [7]; this differential will be transmitted to the differential pressure sensor, which will send a signal to the controller. These elements should be used to control the operating flow, which should have a lower limit. If the flow rate sensor is capable to detect a flow equal to or less than the minimum flow rate, the controller must be programmed to either stop the system or not to reduce the motor speed to values when the operating of the pump is unsafe.

Although the HVAC system has more devices [8], for monitoring and controlling all the equipment that form the system, the above elements are basic in order to be able to control the rotational speed of the pump [9]. In the case when the loads of the last three floors do not work, or another similar condition will lead to a decrease in the necessary pressure on the main branch, since there is no need to direct the fluid to these points. Therefore, the project designer should propose several options of controlling [10] that help to reduce the energy consumption of the pump system either by working at constant pressure or by reducing pressure. For all these alternatives, controllers, pressure sensors and other various sensors should be used [11]. This will be set in each of the air handling installations.

5. Conclusions

Controlling the rotational speed of a centrifugal pump using VFD provides energy saving [12]. A detailed analysis of the power, as can be seen from Fig. 5, shows that a decrease in the rotational speed to 1450 [rpm] leads to a decrease in power consumption by up to 50%, providing the necessary flow rate to the required pressure. The project designer must carry out a study of the time periods during which various equipment (supply and exhaust units) are operating, and in accordance with this establish one or more control procedures to provide the use of VFD.

Hydraulic systems with variable flow rate or individual cooled water pumps serving the systems with variable flow rate with engines having the capacity of more than 5 hp (3.7 kW) must have controls and / or devices (such as a variable speed controller) that ensure that the pump motor needs no more than 30% of the design power at 50% of the calculated water flow rate [13].

Using this type of control allows avoiding the unnecessary use of expensive control devices, such as pressure independent characterized control valves known as PICCV [14], these valves maintain a constant pressure inside the supply and exhaust system. The control system [15] maintains a constant pressure in the main pipe, thus, the required pressure inside the coil is guaranteed without requiring this type of valve.

References

[1] The American Society of Heating, Refrigeratingand Air-Conditioning Engineers. (2013). Energy StandardforBuildingsExcept Low-RiseResidential Buildings. ANSI/ASHRAE/IES Standard 90.1-2013. Atlanta, GA 30329.ASHRAE STANDARDS COMMITTEE 2012–2013.

[2] Air-Conditioning, Heating, & Refrigeration Institute. (2015). Performance Rating of Water-chilling and Heat Pump Water-heating Packages Using the Vapor Compression Cycle. AHRI Standard 550/590 (I-P).

[3] A Protopopov et al 2019 IOP Conf. Ser.: Mater. Sci. Eng. 589 012003.

[4] Armstrong Fluid Technology, All Rights Reserved (2015). Adept Select: Online Pump Product Selection Tool Software. Available from: http://armstrongfluidtechnology.com/adept/early-access.html.

IOP Conf. Series: Materials Science and Engineering **963** (2020) 012026 doi:10.1088/1757-899X/963/1/012026

[5] A Martynyuk et al 2019 IOP Conf. Ser.: Mater. Sci. Eng. 589 012029.
[6] E Morozova et al 2019 *IOP Conf. Ser.: Mater. Sci. Eng.* 589 012008.
[7] V Cheremushkin and APolyakov 2019 IOP Conf. Ser.: Mater. Sci. Eng. 589 012001.
[8] Dwyer Instruments International. (2019). Series 673 Pressure Transmitter. Retrieved fromhttp://www.dwyer-inst.com/Product/Pressure/SinglePressure/Transmitters/Series673#ordering
[9] Dwyer Instruments International. (2019). Series 673 Pressure Transmitter. Retrieved fromhttp://www.dwyer-inst.com/Product/Pressure/SinglePressure/Transmitters/Series673#ordering
[10] Dwyer Instruments International. (2019). Series PE & TE Orifice Plate Flowmeters. Retrieved fromhttp://www.dwyer-inst.com/Product/Flow/Flowmeters/OrificePlate/SeriesPE-TE#videos.
[11] A Martynyuk et al 2019 IOP Conf. Ser.: Mater. Sci. Eng. 589 012029.
[12] I Borovik et al 2019 IOP Conf. Ser.: Mater. Sci. Eng. 589 012002.
[13] N Sosnovsky and D Ganieva 2019 *IOP Conf. Ser.: Mater. Sci. Eng.*589 012016
[14] BELIMO —The Pressure Independent Characterized Control Valve (PICCV). https://www.belimo.ru/resources/obshchaya-informatsiya-belimo/
[15] Carel Industries S.p.A CAREL - Sensors and protection devices. https://www.carelusa.com/sensors-and-protection-devices.

IOP Conf. Series: Materials Science and Engineering **963** (2020) 012027 doi:10.1088/1757-899X/963/1/012027

Types of heat exchangers in industry, their advantages and disadvantages, and the study of their parameters

Edreis Edreis[1] and A Petrov[1,2]

[1]Bauman Moscow State Technical University, 5 Second Baumanskaya Street, Moscow, 105005, Russian Federation

[2]E-mail: alexeypetrov@bmstu.ru

Annotation The heat exchanger as the main element in various types of energy complex is widely studied in technological and production processes. Heat exchangers are classified according to their application. Therefore, in this article we will consider the main types of heat exchange equipment, the choice of a specific type, depending on temperature, flow, pressure drop, and so on. The paper also provides a brief description of software for calculating, modeling and designing CFD, the advantages of each of the methods used to study heat transfer and its efficiency.

1. Introduction

A heat exchanger is a device that is used to transfer thermal energy (enthalpy) between two or more environments using a working fluid and a working surface having different temperatures. Heat transfer can occur between a solid surface and a liquid, between solid particles and a liquid, etc.

The heat-exchange (or heat-utilizing) apparatus is one of the most common and important elements of power, utility and technological installations. Any conversion of energy from one type to another, as well as the transfer of energy from one device to another, is accompanied by the transition of a certain part of the energy to heat. Therefore, in almost all machines and apparatuses, heat transfer is important.

Heat exchangers play an important role in technological processes, energy, oil refining, manufacturing, transportation, air conditioning, cryogenic and recovery systems [1]. They also serve as key components of many industrial products available on the market. All heat exchangers can be classified according to various criteria.

2. Fields of applications of heat exchangers:

Heat exchangers are used in many different technological processes. The following roster lists some types of heat exchangers and their main purpose.

- Chiller: a heat exchanger that uses refrigerant to cool a liquid through a vapor compression or absorption refrigeration cycle.
- Condenser: condenses steam or a mixture of vapors with or in the presence of non-condensing gases.
- Cooler: cools a liquid or gas, usually using water.
- Heat exchanger: cools one liquid while heating another.
- Heater: transfers heat to a liquid or gas by contact with a heated surface.

IOP Conf. Series: Materials Science and Engineering **963** (2020) 012027 doi:10.1088/1757-899X/963/1/012027

- Reboiler: generates steam through fractional distillation. Heating occurs with the help of a heating element, as a result of which condensation of steam occurs.
- Thermosiphonreboiler: the natural circulation of boiling liquid is maintained by sufficient pressure.
- Forced circulation reboiler: A pump is used to circulate the fluid in the reboiler.
- Superheater: Heats the steam to a temperature above its boiling point.
- Evaporator: a heat exchanger that evaporates part or all of the liquid flow. [2]

3. Classification of Heat Exchangers:

3.1. «Pipe-in-pipe» heat exchangers

Equipment consisting of two pipes with different diameters inserted one into the other. With the help of clutch couplings, all parts of the pipes are assembled into a coil, which provides the necessary space for the heating and cooling medium. Sections are placed one above the other. The flows are directed counter-currently (towards each other). The cooling agent comes from below, and after heating rises up. The heated steam accumulates from above. After condensation, it goes to the bottom of the heat exchanger. This heat exchange equipment is used in the food industry. Heat exchangers of this design are characterized by a significant heat transfer coefficient and can operate at high pressure. The pipes are cleaned mechanically on level areas. The flow inside the two-pipe heat exchangers can be parallel or counter-current. [3]

Advantages and disadvantages of «pipe-in-pipe» heat exchangers.

1. The main advantages of the device of this design include:

• High flow rate of the coolant: this is achieved through careful selection of water pipes of the desired diameter, which allows the medium to flow freely inside the pipes.

• Ease of maintenance. This property makes it possible to carry out regular cleaning of equipment, which allows to increase the duration of its operation.

• Versatility. In systems, it is permissible to use a coolant both in the liquid and in the vapor phase.

2. The disadvantages of the equipment are:

• Dimensions: due to the large size, difficulties arise during transportation and use of the device. Most of all this refers to individual use, where space is very limited.

• High cost: the price of external pipes that are not involved in heat exchange, and which are connected to the heat exchanger, is quite impressive.

• Difficulties in the design: when choosing this equipment, you must contact the professionals, which is associated with the complexity of the calculation. At the same time, the overall cost of manufacturing and installation work increases.

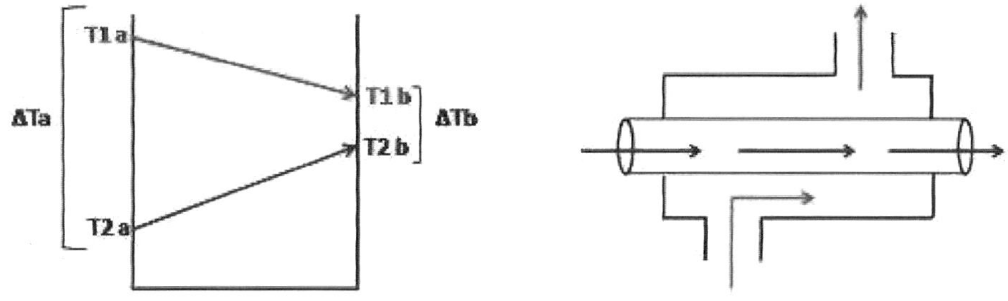

Fig.1: Parallelflow

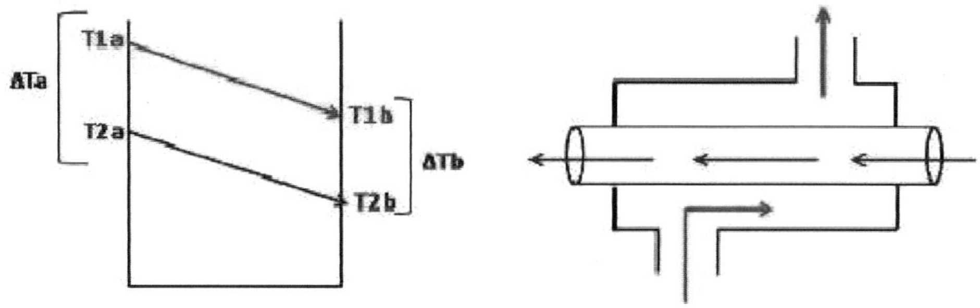

Fig.2: The counter-current flow.

3.2. Shell-and-tube heat exchangers:

The shell-and-tube heat exchanger includes a tubular tank and an integrated tubing section. The heat carriers in the heat exchanger are directed both parallel and towards each other. Shell-and-tube heat exchangers are used in the chemical, food, oil and gas and other fields. They are used as evaporators and condensers. Depending on the operating conditions of the equipment, it is installed in a vertical or horizontal position [4].

In multi-way devices, it is necessary to firmly fix the base and pipe sections. Such modules function even with a small difference in temperature of the working environment. When choosing the material of the heat exchanger, it is necessary to take into account the aggressiveness of the environment. Due to the inaccessibility of the heat exchanger tubes, the formation of corrosion is highly undesirable. Cleaning is carried out exclusively by a chemical method [5].

Advantages and disadvantages of shell-and-tube heat exchangers

1. The advantages of the devices are:

• Internal reliability. Shell-and-tube heat exchangers are more resistant to scale formation, which implies that cleaning should be done less frequently than with other heat exchangers.

• Possibility of power regulation. If necessary, increase or decrease the power, adjust the number of sections, the length and diameter of the pipes.

• Long service life. Shell-and-tube heat exchangers have a long service life.

2. Disadvantages:

• Large dimensions. A heat exchanger weighing 120–150 kilograms and a length of 4 meters can not always be fitted and installed at the facility.

• Vulnerability of the outer part of the case. Tube heat exchangers are made of electric welded pipe. After a short period of work, the outer coating begins to diverge along the seam, leaks appear, as a result of which oxygen begins to be released when the water is heated. This contributes to the development of metal corrosion [6].

• Efficiency. The coefficient is only 70%, which increases energy losses.

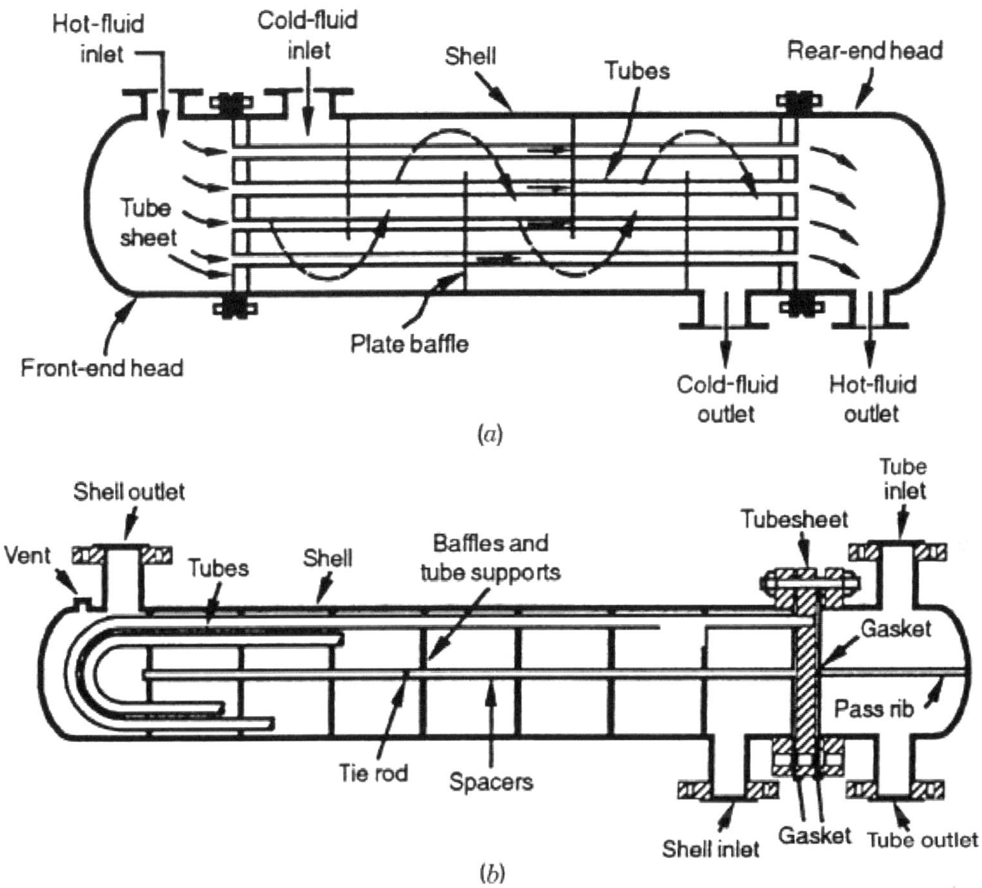

Fig.3: a) shell-and-tube heat exchanger with one outer shell and one pipe passage; b) shell-and-tube heat exchanger with one outer shell and two pipe passages.[7]

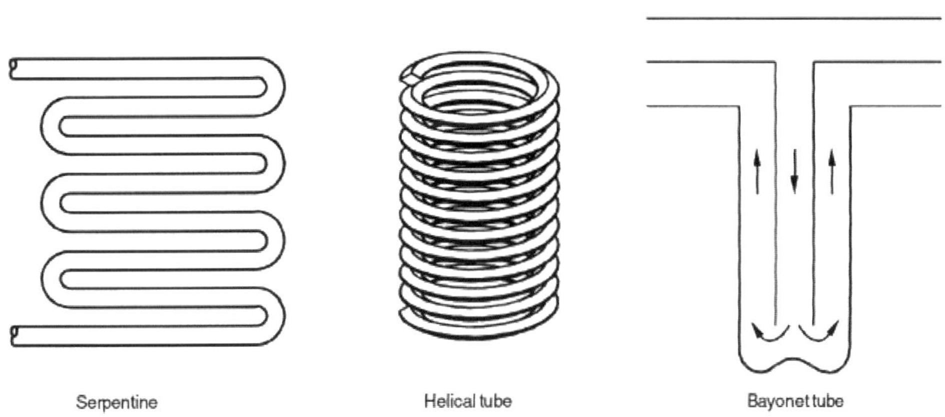

Fig.4: Various configurations of pipes used in shell and tube heat exchangers. [7]

3.3. Plate Heat Exchangers

They consist of a large number of corrugated plates made of stainless steel. They are separated by seals that are installed without the use of adhesive mixtures, but allow tight fit to each other. Gaskets provide absolute tightness and do not allow mixing of media. The direction of flow is counter-current. The power of the heat exchanger is determined by the number of plates installed inside. Service, cleaning and repair of the device is done by disassembling it. Areas of use: housing and public utilities, shipbuilding, metallurgy, oil and gas, pharmaceutical industries and so on. The choice of material of the heat exchanger must be carried out depending on the technological process, the type of coolants in the system, temperature load and pressure. The most universal in application: plate heat exchangers made of stainless steel with copper pipes [8].

Advantages and disadvantages of a plate heat exchanger

1. Theadvantagesare:

• High efficiency. Due to the large area of the heat exchange surface, the efficiency reaches 95%, which is much higher than that of tubular apparatuses.

• Compactness. The device is selected in accordance with the required heat consumption. With a small number of plates, the differences will be less, respectively, with a larger number of plates, the differences will increase.

• Multifunctionality. Plate heat exchangers are used in many areas of life, have a wide range of capacities.

• The cost of the device depends on the number of plates installed in it. There is the possibility of selecting the right number of plates. Repair costs replacing a worn (damaged) plate, and not the entire system.

2. Disadvantages:

• Short service life. Plate heat exchangers are quickly clogged. The maximum service life without cleaning is 3 years.

Fig.5: Plate heat exchanger

3.4. Spiral plate heat exchangers:

Spiral plate heat exchangers are made of two metal plates that are wound on each other. One stream of process fluid enters the heat exchanger through the center and flows from the outside, while the second stream enters from the outside and flows inward. This creates a close to natural backflow [9].

Advantages and disadvantages of a spiral plate heat exchanger:

1. The advantages are:
   - Single flow paths reduce the rate of scale build-up associated with liquids containing solids.
   - Ability to work with two highly polluting liquids.
   - Lack of dead zones for accumulation of solid particles inside the heat exchanger
   - Counterflow.
   - Madeofmanyalloys.
   - Very low pressure drop.
2. Disadvantages:
   - Designs are the intellectual property of companies, a limited number of manufacturers.
   - As a rule, the design is more expensive than other types of heat exchangers.

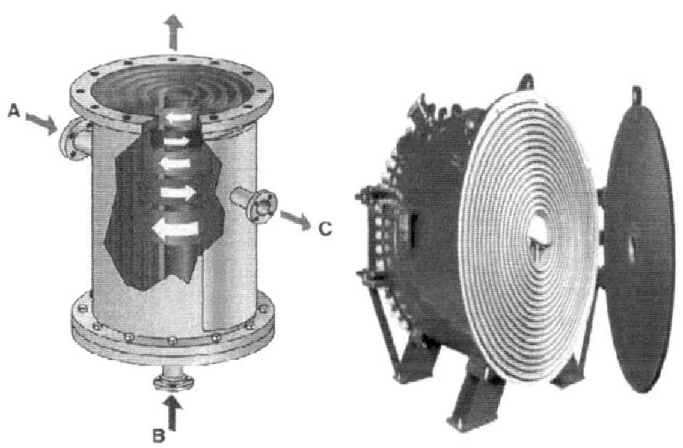

Fig.6: Spiral plate heat exchanger

3.5. Spiral tube heat exchangers:

Spiral tubular heat exchangers are made of spiral pipes. In some cases, the tube is installed inside the bundle to ensure the compactness of the heat exchanger. These heat exchangers are mainly used for small capacities [10].

Advantages and disadvantages of a spiral tube heat exchanger:

1. The advantages are:
   - Compact and inexpensive heat exchanger requiring low power.
   - Canwithstandhighpressures.
2. Disadvantages:
   - Designs are the intellectual property of companies - a limited number of manufacturers.

Fig.7: Spiral tube heat exchanger

3.6. Air-cooled heat exchangers:

Air-cooled heat exchangers use ambient air to cool and condense the working medium. They are usually used in places where there is a shortage of cold water. Air-cooled heat exchangers are commonly used when the temperature at the outlet of the heat exchanger is at least 20° C higher than the ambient temperature. They can be designed for closer temperatures, but often become expensive compared to a combination of a cooling tower and a water-cooled heat exchanger. Air-cooled heat exchangers use electric fans to move air through a series of pipes [11]. Therearetwomainmechanisms:

- Induced draft fans draw air through the tube blocks.
- Fans blow air through a series of pipes.

Air-cooled heat exchangers are expensive compared to water-cooled heat exchangers due to their large size, low air heat transfer coefficient, and structural and electrical requirements. In addition, air-cooler heat exchangers require large surface areas of pipes and must be designed in such a way as to withstand daily and seasonal changes in air temperature [11].

The low heat transfer coefficient associated with the flow around the air of the outer sides of the pipes is partially overcome due to the wide use of finned pipes to increase the outer surface area. Changes in ambient temperature are often controlled by fans with a variable speed or airflow adjustment step. In cold climates, it may be necessary to develop a design with the ability to recirculate air to prevent freezing during operation. Smaller heat exchangers (similar to radiators) are used for small areas of responsibility [12].

Advantages and Disadvantages of Air-cooled Heat Exchangers

1. Theadvantagesare:
   - Do not use water for cooling.
2. Disadvantages:
   - Requires a large area of the site.
   - Highmanufacturingcosts.
   - Ribs may become clogged in dirty environments.
   - Fansmaymakenoise.

Fig.8: Finnedpipes

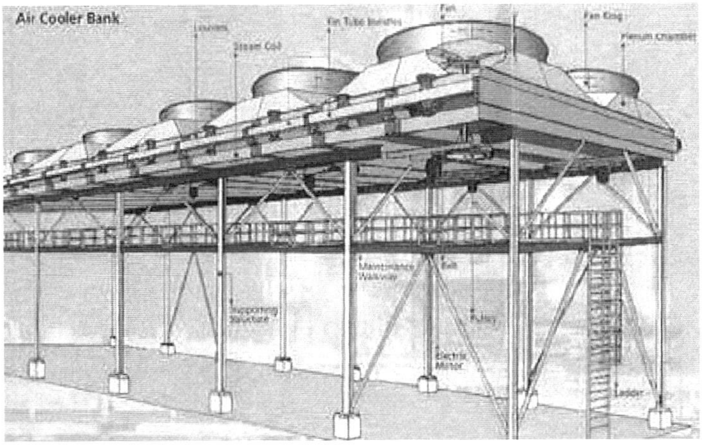

Fig.9: Air-cooled heat exchanger

4. Heat exchanger using the CFD simulation method:

There are two classes of CFD methods: this is the Euler approach and the Lagrange approach. In the Lagrangian approach, a liquid is represented as a set containing a large number of particles with properties such as mass, velocity, and temperature. Then all particles are tracked along the flow, and based on the interactions between the particles, temporary changes in various properties are calculated. In the Euler approach, a liquid is represented using selected volumes in the form of mesh elements whose coordinates are fixed. Observe the flow through these control volumes and calculate the flows to measure the rate of change of properties such as speed and temperature. Methods developed on the basis of the Euler approach are called mesh methods, and methods developed on the basis of the Lagrangian approach are called methods without a particle grid. Both classes of methods have their pros and cons. The meshless feature has one of the biggest advantages for R&D engineers, as mesh is the most time-consuming and complex preprocessing phase for CFD Euler solvers. In recent decades, researchers have developed methods that combine the Lagrange and Euler approaches to make up for the weaknesses of both methods [13–14].

A wider study of heat exchangers can be carried out using specialized software, such as CFD packages, in which variables such as temperature, volumetric or mass flow rate, differential pressure, characteristic of heat transfer in fluid mechanics can be determined. These CFD packages provide ease of displaying areas of influence of pressure drop or temperature using a convenient graphical interface [15].

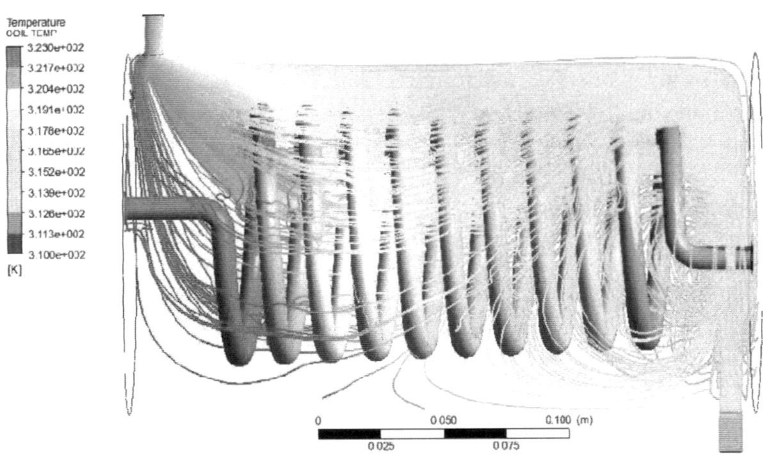

Fig.10: Temperature distribution in the heat exchanger

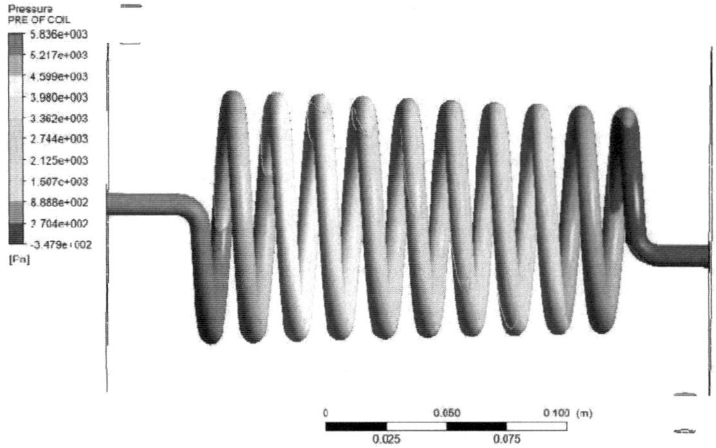

Fig.11: Pressure distribution in the heat exchanger.

5. Conclusion:

In this paper, various types of heat exchangers used in industry, the criteria for the selection of devices depending on power and purpose are considered. Basically, they are made as condensers, evaporators, heaters, depending on the process conditions.

References

[1] V Tkachuk et al 2019 IOP Conf. Ser.: Mater. Sci. Eng. 589 012007.
[2] HEAT EXCHANGERS Prepared by Bob Heaslip KESCO For Queens University CHEE 470 — Fall 2008,p[1–11].
[3] Kays, W. M., and A. L. London, 1998, Compact Heat Exchangers, reprint 3rd ed., Krieger Publishing, Malabar, FL.
[4] Shah, R. K., 1991a, Compact heat exchanger technology and applications, in Heat Exchanger Engineering, Vol. 2, Compact Heat Exchangers: Techniques for Size Reduction, E. A. Foumenyand P. J. Heggs, eds., Ellis Horwood, London, pp.
[5] Evans, F.L., Equipment Design Handbook for Refineries and ChemicalPlants, 2nd Ed., Gulf Publishing Company, Houston, TX (1974)
[6] TEMA, 1999, Standards of TEMA, 8th ed., Tubular Exchanger Manufacturers Association, New York.
[7] Walker, G., 1990, Industrial Heat Exchangers: A Basic Guide, 2nd ed., Hemisphere Publishing, Washington, DC.
[8] K Abramov 2019 IOP Conf. Ser.: Mater. Sci. Eng. 589 012013
[9] STANDARDS OF THE TUBULAR EXCHANGER MANUFACTURERS ASSOCIATION EIGHTH EDITION.
[10] E Morozova et al 2019 IOP Conf. Ser.: Mater. Sci. Eng. 589 012008
[11] V Cheremushkin and APolyakov 2019 IOP Conf. Ser.: Mater. Sci. Eng. 589 012001
[12] V Tkachuk et al 2019 *IOP Conf. Ser.: Mater. Sci. Eng.*589 012007
[13] E Morozova et al 2019 *IOP Conf. Ser.: Mater. Sci. Eng.* 589 012008
[14] N Isaev et al 2019 *IOP Conf. Ser.: Mater. Sci. Eng.*589 012009
[15] MdLokmanHosain, FLUID FLOW AND HEAT TRANSFER SIMULATIONS FOR COMPLEX INDUSTRIAL APPLICATIONS, Printed by E-Print AB, Stockholm, Sweden, 2018.

Mathematical modeling and analysis of the influence of various regulators on the quality of the transient hydraulic drive with throttle control.

A Protopopov[1,2], **A Gontarev**[1] **and Ya Manenak**[1]

[1]Bauman Moscow State Technical University, 5 Second Baumanskaya Street, Moscow, 105005, Russian Federation

[2]E-mail: proforg6@yandex.ru

Abstract. This article provides an analysis of transients in various hydraulic drive regulators. Hydraulic drives are widely used in a number of areas of power engineering and other types of equipment, which is due to a number of advantages, such as: small size and weight with high energy consumption, high speed, high positioning accuracy, a wide range of operating pressures and speeds of executive bodies.

Introduction

Hydraulic actuators [1-5] are widespread in a number of areas of power engineering and other types of equipment, which is due to a number of advantages, such as: small size and weight with high energy intensity, high speed, high positioning accuracy, a wide range of operating pressures and speeds of movement of the executive bodies.

The most widely used electro-hydraulic drives (EHD) with throttle control. In connection with the specifics of the work, their requirements are high requirements for ensuring reliability, stability, optimal values of static and dynamic characteristics.

Static and dynamic characteristics of drives can be improved through the use of various corrective devices (controllers). In this paper, several types of correction devices are considered: the introduction of additional feedbacks on state variables (state controller), the introduction of additional pressure feedback, and the P-controller.

The first type (state controller) is based on the formation of feedback loops according to the state coordinates, which give the servo drive closed in position to the desired location of the roots of the characteristic equation of its linear model(poles of the transfer function).

The introduction of additional pressure feedback makes it possible to increase the damping coefficient of a system with a large inertial load. Significant pressure values in the cavities of the hydraulic actuator allow practically realizing pressure feedback without additional amplification steps.

The use of a P-controller is associated with a control method based on the law of proportional control, in which the characteristics of the output signal are proportional to the characteristics of the input signal.

In this work, the influence of the proposed regulators on the quality of the transient EHD process was also studied, and their comparative analysis was carried out.

Mathematical model of hydraulic drive with throttle control

To compile a mathematical model, we consider the dynamics of individual elements of the hydraulic drive.

The equation of motion of the piston of a loaded hydraulic cylinder:

$$p_н F_c - \left(P_{mp}\right)_c - c_{cв}\left(y - y_m\right) = m_p \frac{d^2 y}{dt^2}$$

where $p_н = p_1 - p_2$ — pressure difference in the cavities of the hydraulic cylinder under the action of the load.

The equation of the load on the rod of the hydraulic cylinder:

$$m\frac{d^2 y_m}{dt^2} + k_{mp}\frac{dy_m}{dt} + \left(c_{cв} + c_н\right) y_m = c_{cв} y$$

In real hydraulic drive systems, there are often non-linearities that distort the characteristics of drive elements. In order to avoid excessive complication of the mathematical model, we consider only the basic nonlinearities that can affect the dynamics of the process.

The fluid flow rate Q_s is determined by the flow-differential characteristic of the spool, which in the general case non-linearly depends on the movement of the spool x_s and the pressure difference $p_н$ in the hydraulic cylinder cavities:

$$Q_s = Q_s\left(x_s, p_н\right) = k_s^{'} \cdot x_s \cdot \sqrt{\frac{p_p - p_{cl} - p_н \cdot sign\left(x_s\right)}{2}}$$

Where $k_s^{'} = \mu_s \cdot \pi \cdot d_s \cdot k_p \cdot \sqrt{\frac{2}{\rho}}$ — specific (referred to the unit of movement of the spool) conductivity of the windows of the spool valve.

In this work, to optimize the parameters, we used the criterion for the product of time t and absolute error $\left|e(t)\right| - ITAE$ (Integral of Time multiplied by Absolute value of Error). It has found widespread application in practice by providing a compromise between the overshoot value and the transition time. Mathematically, it is calculated as follows:

$$I = \int_0^T t \cdot \left|e(t)\right| dt$$

Since integration to infinity is practically impossible, the criterion consists in choosing a value of T large enough so that $e(t)$ for $t > T$ is negligible.

Mathematical Modeling Results

The simulation was carried out in the MATLAB Simulink software package based on the classical linear system model. The following structural diagrams were created.

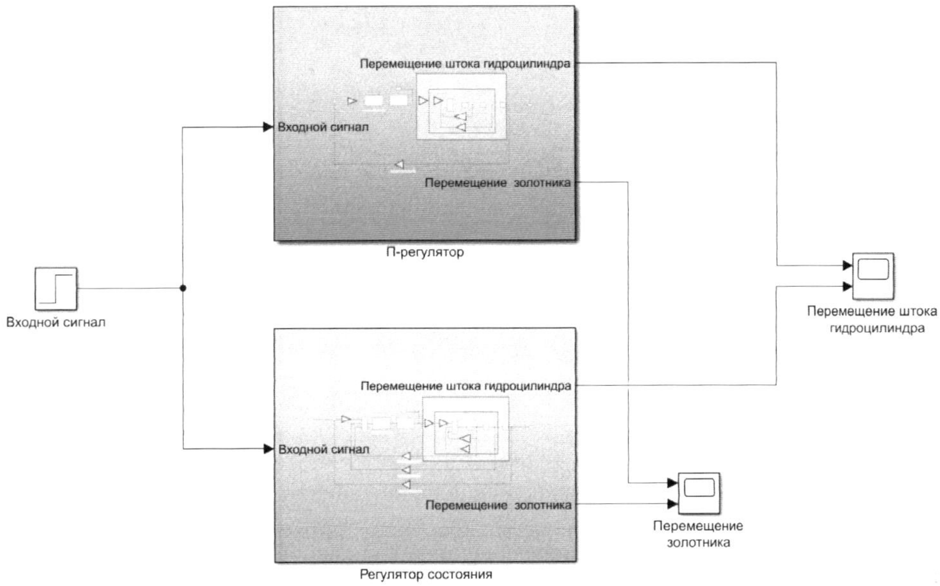

Fig.1. General view of the structural scheme for comparing the P-controller and the state controller.

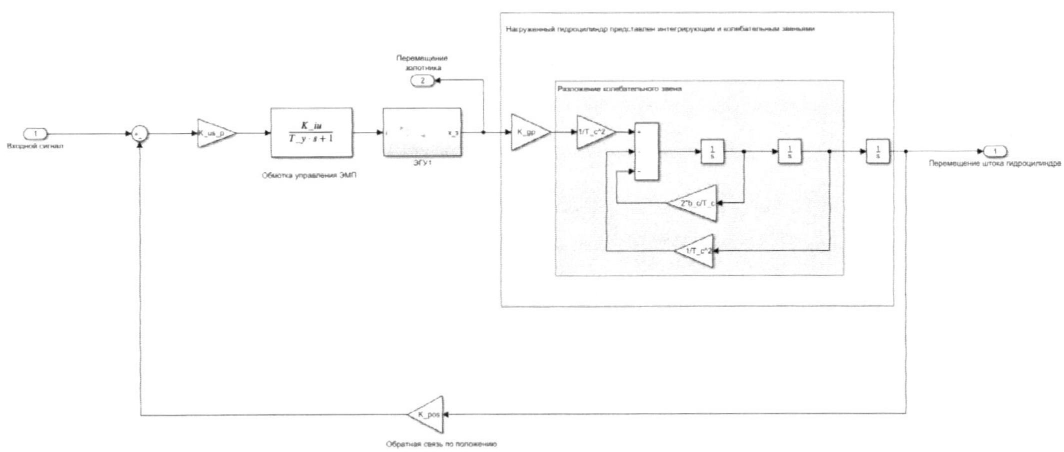

Fig. 2. Block diagram of a linear model with feedback on the position of the cylinder head (P-regulator).

IOP Conf. Series: Materials Science and Engineering **963** (2020) 012028　doi:10.1088/1757-899X/963/1/012028

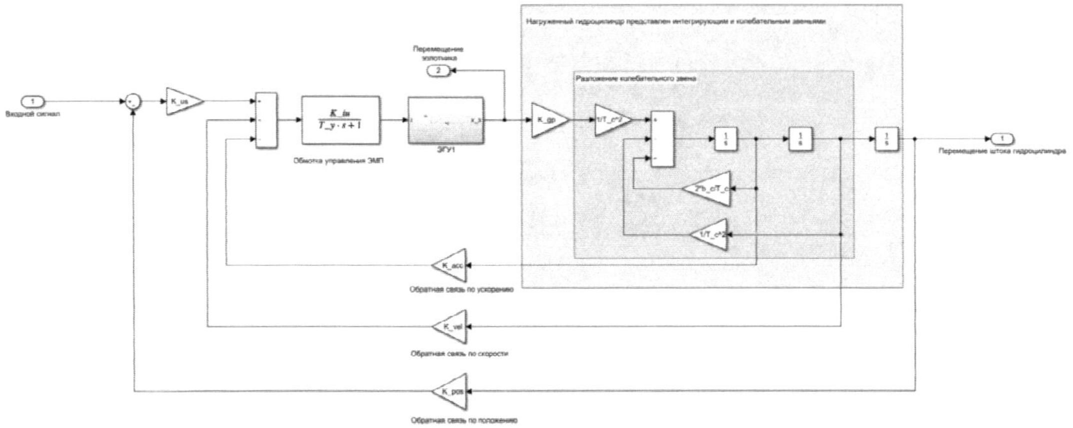

Fig. 3. Structural diagram of a linear model with feedback on acceleration, speed and position of the cylinder head (state controller).

The following transient graphs were obtained:

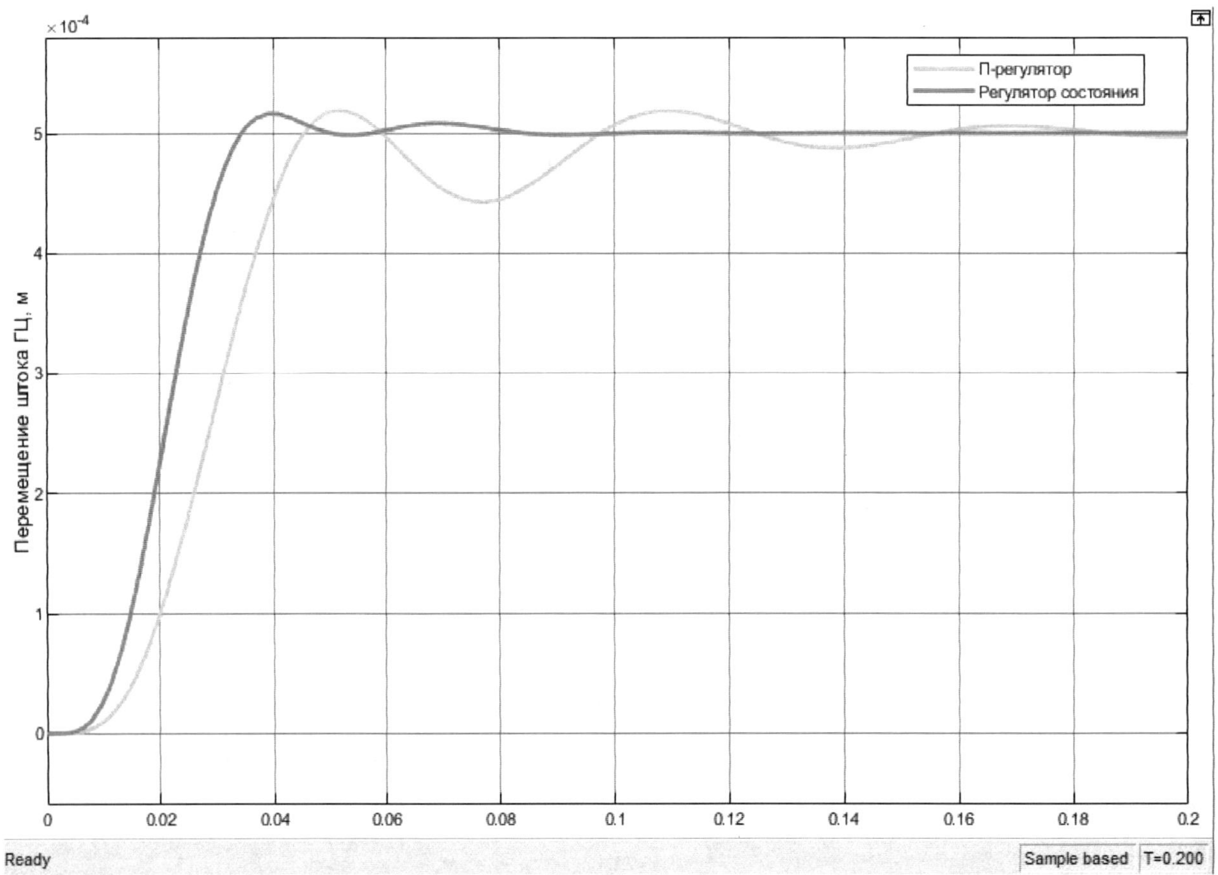

Fig. 4. Transient system with a P-controller (yellow) and with a state controller (blue).

Criteria for assessing the quality of transients are shown in table 1.

IOP Conf. Series: Materials Science and Engineering **963** (2020) 012028 doi:10.1088/1757-899X/963/1/012028

Table 1

	Dynamic Error	Transient Time
Type of controller	σ, %	$t_{t.p.}, c$
P-controller	3,81	0,145
Statecontroller	3,28	0,0445

Further, taking into account in the mathematical model that the flow-differential characteristic of the hydraulic drive is non-linear, we study the effect of various types of corrective devices on the transient process (TP) of the system (Fig. 5).

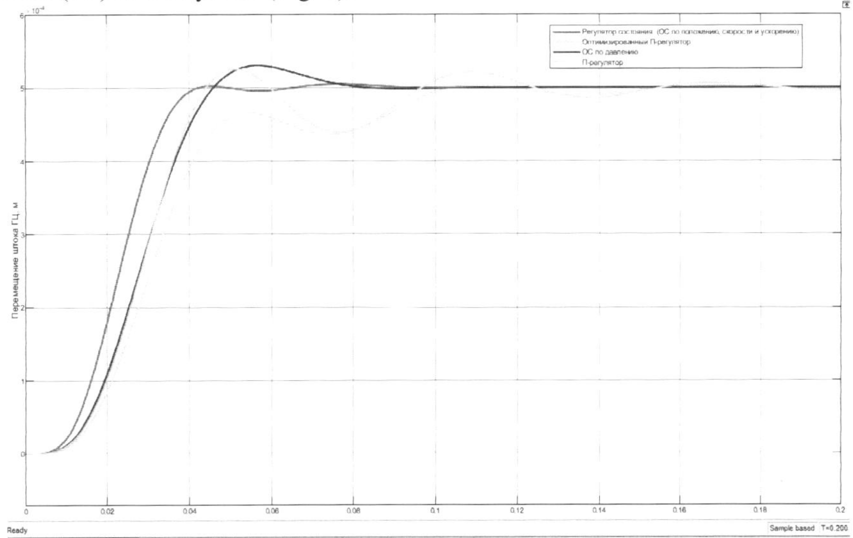

Fig. 5. Transients of systems with various types of corrective devices.

Red color shows the TP of the system with a state controller;

Yellow color indicates the TP of the system with position feedback (P-controller);

The blue color shows the TP of the system with the optimized value of the gain (regulation coefficient) of the P-controller;

The blue color shows the TP of the system with feedback (F) by pressure;

The numerical values of the results of mathematical modeling are presented in table 2.

Table 2

	Dynamic error	Transient time	Controlcoefficient (amplification factor)
Type of controller	σ, %	$t_{t.p.}, c$	K_{yc}
Status controller	1,01	0,0396	3,0071
P-controller	5,2	0,144	1,39
Optimized P-controller	0,559	0,0986	1,1738
Pressure differential feedback	6,0	0,071	1,58

IOP Conf. Series: Materials Science and Engineering **963** (2020) 012028 doi:10.1088/1757-899X/963/1/012028

Figure 6 and 7 shows the structural diagram of the ITAE criterion. Using this criterion optimal F coefficient were obtained for speed ($6{,}6711$), acceleration ($0{,}07552$) and pressure drop ($2\cdot10^{-8}$).

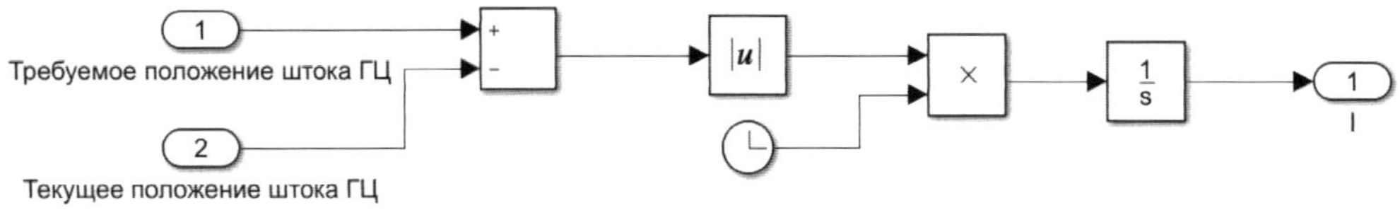

Fig. 6. ITAE calculation block diagram

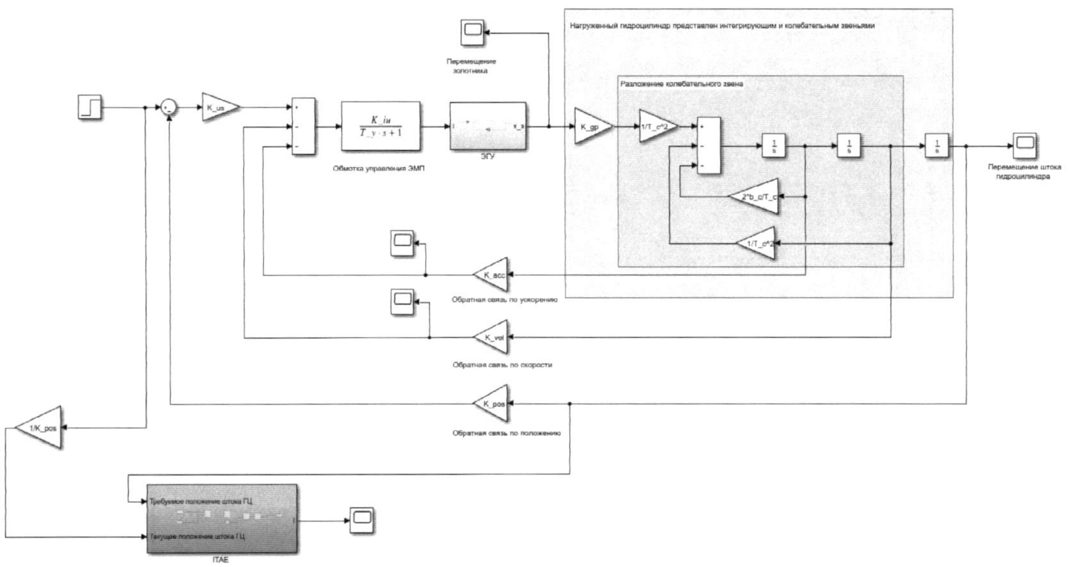

Fig. 7. Block diagram of a hydraulic actuator with ITAE calculation

Conclusion

During the work, an analysis of three types of regulators for electro-hydraulic drives was carried out, their dynamic characteristics were obtained.

It was found that the introduction of additional pressure feedback can significantly reduce the time of the transient process and its oscillation, but at the same time it has the largest dynamic error.

The use of a P-controller as a correcting device (without optimization) has significant drawbacks in the form of a large dynamic error and the longest transient time. Optimization based on the integral ITAE criterion contributes to a significant decrease in the dynamic error — by 9,3 times, but the transition time does not change significantly.

The best EHD quality indicators are provided by the state regulator. When this regulator is connected to the drive circuit

References

[1] S Semenov 2020 *IOP Conf. Ser.: Mater. Sci. Eng.* **779** 012043
[2] Li Jianhui and N Sosnowski 2020 *IOP Conf. Ser.: Mater. Sci. Eng.* **779** 012035
[3] V Pilgunov and K Efremova 2020 *IOP Conf. Ser.: Mater. Sci. Eng.* **779** 012032
[4] B Kulakov and D Kulakov 2020 *IOP Conf. Ser.: Mater. Sci. Eng.* **779** 012027
[5] V Bozrov et al 2020 *IOP Conf. Ser.: Mater. Sci. Eng.* **779** 012025

Investigation of the dependence of the axial force acting on the rotor of a vertical centrifugal pump on the area of the discharge holes of the impeller

A Protopopov[1,2], N Zhestkiy[1], S Godorozha[1], M Kalugina[1], A Mukhlaeva[1] and A Zaitseva[1]

[1]Bauman Moscow State Technical University, 5 Second Baumanskaya Street, Moscow, 105005, Russian Federation

[2]E-mail: proforg6@yandex.ru

Abstract. The article describes a vertical centrifugal pump with a bearing Assembly in the upper part of the rotor. It is shown that for certain values of design parameters, the influence of the so-called discharge holes in the impellers can significantly affect the axial force acting on the rotor, and for certain parameters, for example, with a large weight of the rotor, this influence is insignificant and the manufacture of such impellers is impractical.

Introduction

The issues of modeling working processes in centrifugal pumps are widely illustrated in the literature [1]–[10]. Of particular interest are publications using modern modeling tools [11]–[15]. In a number of constructions of modern centrifugal pumps, discharge holes are used at the impellers.

These holes are used to reduce bearing loads. In this article, the task is to estimate the effect of the area of these holes on the total axial force acting on the rotor.

Methods

The most loaded bearing is the one that perceives the total axial force, because in vertical pumps it is usually maximum and radial ball bearings (see figure 1) do not accept it well.For determination this axial force, it is necessary to consider the rotor equilibrium.

The total axial response perceived by the most loaded bearing from the rotor equilibrium condition is:

$$P_{pod} = G - P_{\omega} - P_D + P_o \quad (1)$$

where G — the weight of the rotor,P_{ω} — axial force arising from the rotation of the impeller, P_D — axial force arising from the rotation of the working fluid in the impeller,Po — the force due to pressure in the upper part of the impeller from the discharge holes.Let's considerthe last force in more detail.

In fig. 3: R_{u1} and $R_{u2}$are the Radius of the 1st and 2nd groove seals, R_2 is the outer radius of the impeller, R_{vx} is the radius of the entrance to the impeller, R_{vt} is the radius of the upper bushing, Δ is the radial clearancein the groove seals, it is assumed to be equal for both groove seals in this mathematical model, $p_{u2}$is the pressure in front of the upper groove seal, Δp_{u2} is the pressure drop on the upper groove seal, Δp_o is the pressure drop on the discharge hole. Take the pressure at the pump inlet as 0.

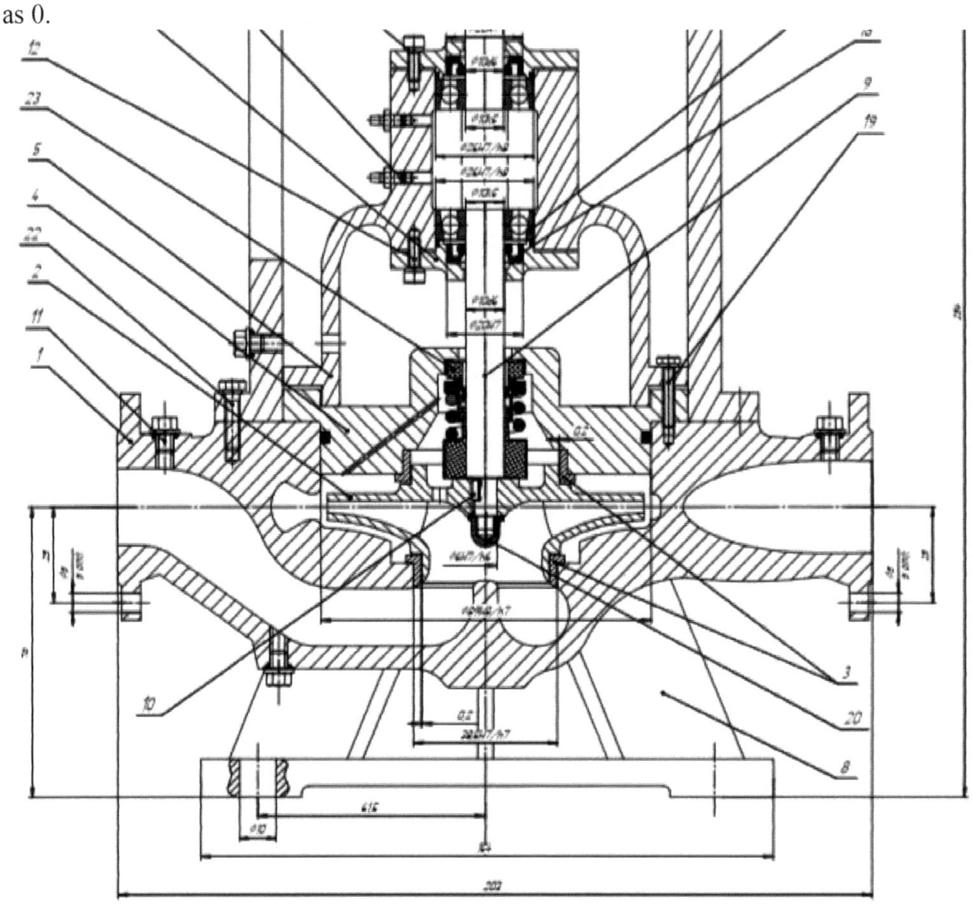

Fig. 1. Typical construction of a vertical centrifugal pump.

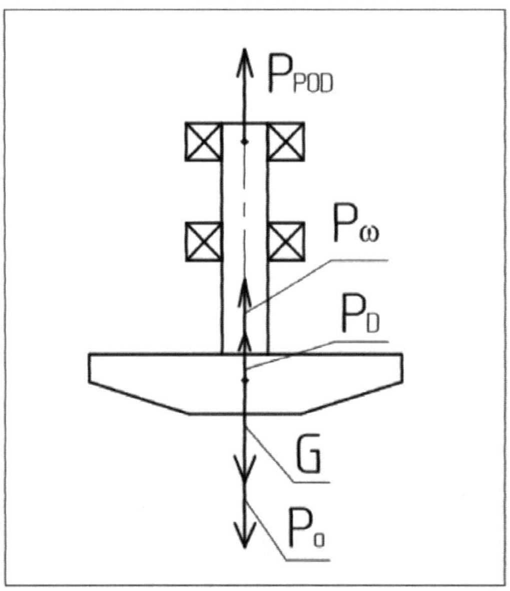

Fig. 2. Plan of forces acting on the rotor in the vertical direction

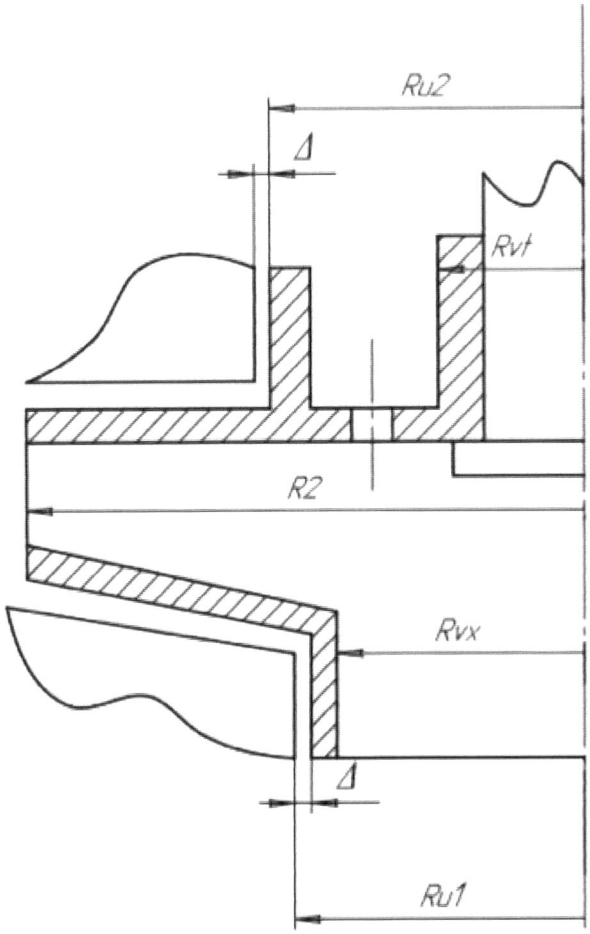

Fig. 3. Basic diagram of the impeller with discharge holes and two groove seals.

Then the Po force will be equal to:

$$P_o = \Delta p_o \cdot \left(R_{u2}^2 - R_{vt}^2\right) \cdot \pi \tag{2}$$

For determination Δp_o, consider the pressure change on the back of the impeller. The pressure before the groove seal can be found from theory of similarity of centrifugal pumps:

$$p_{u2} = \rho \cdot g \cdot H_n \cdot \left(\frac{R_{u2}}{R_2}\right)^2 \tag{3}$$

This pressure is divided into two components:

$$p_{u2} = \Delta p_o + \Delta p_{u2} \tag{4}$$

From Bernoulli's relation orifice discharge is equal to:

$$Q_{u2} = \mu \cdot f_{u2} \cdot \sqrt{\frac{2 \cdot \Delta p_{u2}}{\rho}} \tag{5}$$

where$f_{u2}$is the area of the groove seal, and μ is the discharge coefficient.
From (5) weget:

$$\Delta p_{u2} = \left(\frac{Q_{u2}}{\mu \cdot f_{u2}}\right)^2 \cdot \frac{\rho}{2} \tag{6}$$

Similarly for the next section:

$$\Delta p_o = \left(\frac{Q_{u2}}{\mu \cdot f_o}\right)^2 \cdot \frac{\rho}{2} \tag{7}$$

wherefo is the area of the discharge hole. This is the variable quantity for us in this study. Now we can consider the other power factors in equation (1). The weight of the rotor is calculated using the following formula:

$$G = m \cdot g \tag{8}$$

where m is the rotary mass and g is the acceleration of free falling.
Axial force arising from the rotation of the impeller [16]:

$$P_\omega = \pi \cdot \left(R_{u1}^2 - R_{u2}^2\right) \cdot \rho \cdot g \cdot H_n - \frac{\pi \cdot \rho \cdot \omega^2}{8} \cdot \left[R_{u1}^2 - R_{u2}^2\right] \cdot \left[R_2^2 - 0.5 \cdot \left[R_{u1}^2 + R_{u2}^2\right]\right] \tag{9}$$

whereHn is the pump head.
Axial force arising from the rotation of the working fluid in the impeller [16]:

$$P_D = \rho \cdot Q_n \cdot \left(\frac{Q_n}{\pi \cdot R_{vx}^2}\right) \tag{10}$$

where$Q_n$is the pump flow. We introduce the following auxiliary designation.

$$A = G - P_\omega - P_D \tag{11}$$

$$B = \left(R_{u2}^2 - R_{vt}^2\right) \cdot \pi \tag{12}$$

$$D = \rho \cdot g \cdot H_n \cdot \left(\frac{R_{u2}}{R_2}\right)^2 \tag{13}$$

Then from the equations (3), (4), (6), (7) it turns out:

$$D = Q_{u2}^2 \cdot \left[E + \frac{I}{f_o^2}\right] \tag{14}$$

From which:

$$Q_{u2}^2 = \frac{D}{\left[E + \frac{I}{f_o^2}\right]} \tag{15}$$

Substitute in (7):

$$\Delta p_o = \frac{D}{\left[1 + \frac{f_{u2}^2}{f_o^2}\right]} \tag{16}$$

IOP Conf. Series: Materials Science and Engineering **963** (2020) 012029 doi:10.1088/1757-899X/963/1/012029

Substitute (16), (12) and (11) in (1) and get:

$$P_{pod} = A + \frac{D \cdot B}{\left[1 + \frac{f_{u2}^2}{f_o^2}\right]} \tag{17}$$

where A, B, and D are constants if we analyze equations (11)-(15). Thus, the mathematical model is ready.

Results

If you substitute the numbers for a typical vertical pump in formula (17), you can get the following dependence of the axial force on the area of the discharge hole:

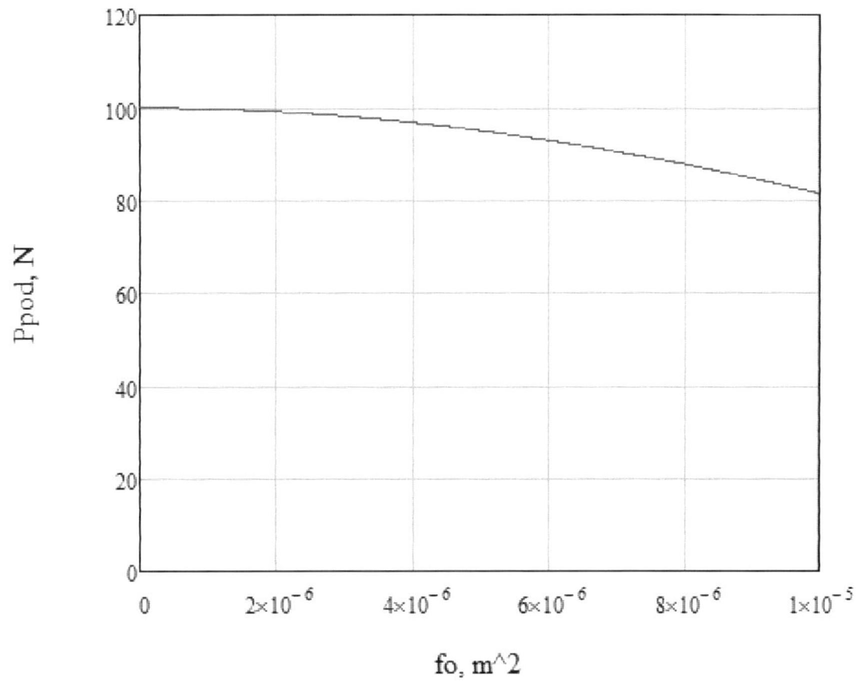

Fig. 4. The dependence of the axial reaction of the most loaded bearing on the area of the discharge hole.

Therefore, the axial force depends on the area of the discharge hole.

However, if this model is substituted, for example, a centrifugal pump with a heavy rotor, the dependence does not look so obvious:

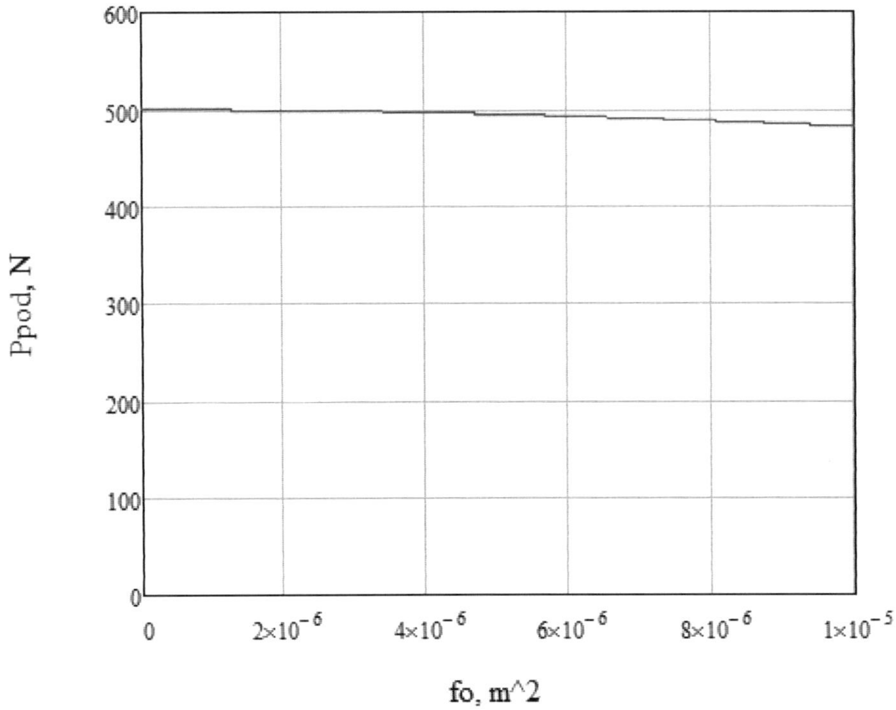

Fig. 5. The dependence of the axial reaction of the most loaded bearing on the area of the discharge hole in the case of a heavy rotor.

Conclusions

A mathematical model is obtained that makes it relatively easy to evaluate the feasibility of using impeller discharge holes for the case of vertical centrifugal pumps with ball bearings, two impeller groove sealsthat are not equal in diameter, and one discharge hole at a radius below the upper grooveseal. For the convenience of using the obtained mathematical model, auxiliary constants (11)–(15) were introduced, the values of the design parameters in which did not change during the study. The formula (19) is obtained, which allows you to quickly estimate the effect of using the discharge hole on the value of the axial reaction in the bearing, which perceives the total axial force on the rotor.

References

[1] V Cheremushkin and APolyakov 2019 *IOP Conf. Ser.: Mater. Sci. Eng.***589** 012001
[2] N Isaev et al 2019 *IOP Conf. Ser.: Mater. Sci. Eng.***589** 012009
[3] K Abramov 2019 *IOP Conf. Ser.: Mater. Sci. Eng.***589** 012013
[4] A Boyarshinova et al 2019 *IOP Conf. Ser.: Mater. Sci. Eng.***589** 012014
[5] M Saprykina and V Lomakin 2019 *IOP Conf. Ser.: Mater. Sci. Eng.***589** 012017
[6] A Boyarshinova et al 2019 *IOP Conf. Ser.: Mater. Sci. Eng.***589** 012028
[7] A Martynyuk et al 2019 *IOP Conf. Ser.: Mater. Sci. Eng.***589** 012029
[8] Shin, D. -., Morimoto, Y., Sawayama, J., Miura, S., & Takeuchi, S. (2019). Centrifuge-based step emulsification device for simple and fast generation of monodispersepicoliter droplets. *Sensors and Actuators, B: Chemical, 301* doi:10.1016/j.snb.2019.127164
[9] Si, Q., Bois, G., Liao, M., Zhang, H., Cui, Q., & Yuan, S. (2019). A comparative study on centrifugal pump designs and two-phase flow characteristic under inlet gas entrainment conditions. *Energies, 13*(1) doi:10.3390/en13010065

[10] Lai, F., Wang, Y., Ei-Shahat, S. A., Li, G., & Zhu, X. (2019). Numerical study of solid particle erosion in a centrifugal pump for liquid-solid flow. *Journal of Fluids Engineering, Transactions of the ASME, 141*(12) doi:10.1115/1.4043580

[11] Jia, X. -., Cui, B. -., Zhu, Z. -., & Zhang, Y. -. (2019). Experimental investigation of pressure fluctuations on inner wall of a centrifugal pump. *International Journal of Turbo and Jet Engines, 36*(4), 401-410. doi:10.1515/tjj-2016-0078

[12] Guo, C., Wang, J., &Gao, M. (2019). A numerical study on the distribution and evolution characteristics of an acoustic field in the time domain of a centrifugal pump based on powell vortex sound theory. *Applied Sciences (Switzerland), 9*(23) doi:10.3390/app9235018

[13] Wang, L., Lu, J., Liao, W., Zhao, Y., & Wang, W. (2019). Numerical simulation of the tip leakage vortex characteristics in a semi-open centrifugal pump. *Applied Sciences (Switzerland), 9*(23) doi:10.3390/app9235244

[14] Kim, H., Posa, A., Nerg, J., Heikkinen, J., &Sopanen, J. T. (2019). Analysis of electromagnetic excitations in an integrated centrifugal pump and permanent magnet synchronous motor.*IEEE Transactions on Energy Conversion,34*(4), 1759–1768. doi:10.1109/TEC.2019.2935785

[15] Guo, C., Gao, M., Wang, J., Shi, Y., & He, S. (2019). The effect of blade outlet angle on the acoustic field distribution characteristics of a centrifugal pump based on powell vortex sound theory. *Applied Acoustics, 155*, 297-308. doi:10.1016/j.apacoust.2019.05.031

[16] Cherkassky, V. M. Pumps, fans, compressors / V. M. Cherkassky. — Moscow: Energiya, 1977. — 416 p.

IOP Conf. Series: Materials Science and Engineering **963** (2020) 012030 doi:10.1088/1757-899X/963/1/012030

Method for increasing the damping of an electro-hydraulic drive system of anthropomorphic walking robots

S Semenov[1,2] and M Ryabinin[1]

[1]Bauman Moscow State Technical University, 5 Second Baumanskaya Street, Moscow, 105005, Russian Federation

[2]E-mail: a1e10@rambler.ru

Annotation. Due to the fact that anthropomorphic walking robots (AWR) are becoming one of the classes of extreme robotics tools, the task of creating drive systems that simultaneously have high indicators for a set of parameters arises. In particular, the problem arises of preventing the occurrence of self-oscillations in conditions where servo drives with high quality factor experience strong dynamic mutual influence through the actuator. According to the totality of indicators for AWR designed for operation in extreme conditions, electro-hydraulic servo drives are preferable. Features of the dynamics of the ASH actuators and the electro-hydraulic drives used on them force us to look for new ways to correct their dynamic properties. This article presents a method of reducing the propensity of a drive system to generate self-oscillations by correcting hydraulic drives using acceleration feedbacks. It uses an estimate of the relative accelerations of the robot links during movement using the signals from pressure sensors installed in the actuator hydraulic motor. The calculations are made using the equations of dynamics of the AWR actuator, taking into account the temporarily imposed external connections. The method allowed to significantly increase the dynamic accuracy of the drive system of the prototype AWR developed by MSTU N.E. Bauman.

Introduction

The fact that anthropomorphic walking robots (AWRs) are becoming one of the classes of extreme robotics is explained, among other things, by the fact that most of the emergency situations (accidents, catastrophes, fires), to eliminate the consequences of which robots have to be used, occur on artificial objects. Robots have to operate in an impenetrable environment, and in many cases it is an environment adapted for humans: narrow corridors, steep stairs, cabins and other rooms of various equipment.

To use AWR as a means of replacing a person when working in an environment adapted for a person, it is advisable that the kinematics (Fig. 1) and the weight and size parameters of the robot be comparable to human ones. This allows him to win in such conditions in efficiency in comparison with other means of extreme robotics [1, 2, 3]. If possible, it is advisable to reduce the mass of the robot. In the event of it falling on a person, the robot should not seriously injure him. It is also desirable that two people were able to lift the robot, so its mass should not exceed 100 kilograms.

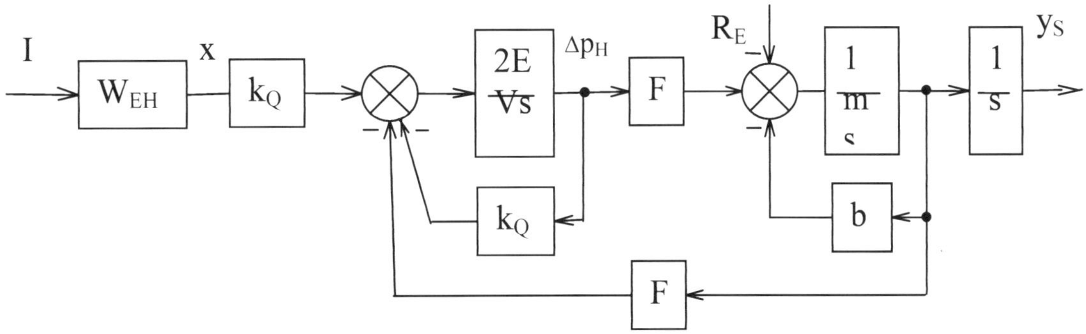

I_c — control current, x_s — the displacement of the spool, Δp_{HM} — the pressure difference in the cavitiesHM, R_E — external load, y_s — the movement of the rod HM, W_{EHV} — transfer function of an electrohydraulic amplifier, k_{Qx}, k_{Qp} — the linearization coefficient of the expenditure flow and pressure drop characteristics of the spool, E — volume modulus of elasticity of a liquid, V — the volume of cavities HM, F — piston area HM, m — given to the stock HM weight,b — coefficient of viscous friction.

Fig. 1. Block diagram of an open electrohydraulic drive.

The executive system of extreme AWR works in severe external conditions, a set of conflicting requirements is presented to it.

1) Rigid weight and size restrictions on the parameters of the drives; a large number of drives (up to 30 or more) must be arranged in dimensions commensurate with human ones.

2) Limited energy resources; power and cable management seriously complicates the work of the robot in the emergency zone, although in some cases this is justified. At the same time, weight and size restrictions do not allow the use of a powerful on-board power source with a large supply of energy.

3) A large range of operating temperatures, since it is impossible to predict in advance the place and time of use of the AWR. For example, it is possible to use AWR in a fire.

4) Aggressive external environment; extreme AWR may be required to act during chemical and possibly radioactive contamination, and possibly under water.

5) The impact of vibration and shock; AWR can be delivered to the work site by various vehicles that are not always adapted for this. When acting near the object of work, there is a danger of collapse of structural elements of buildings and structures, falling from a height.

Under these conditions, the ASR executive system must perform complex and diverse tasks. Moving on a surface with a simple relief in marching mode, using dynamic walking or running, the robot should approach the object of work with a sufficiently high speed. In the rubble or in the presence of complex obstacles, a transition to static walking or climbing is possible (moving with the additional help of manipulators). The same modes of movement are advisable when performing work operations with objects of work.

When manipulating a work object, the executive system must move the grippers of the manipulators or the tools used in a predetermined way, or act on the work object with a given force, providing a given static and dynamic accuracy. As well as in the shutdown mode (the degree of mobility equipped with a drive turns into a simple hinge) or vice versa locks (the degree of mobility is "frozen"). Moreover, in the shutdown and lock modes, it is desirable to ensure minimum energy consumption or its absence.

An important role in choosing the type of drive for a robot designed for operation in extreme situations is played by specific power. Its minimum value can be estimated on the basis of the similarity of overall dimensions and characteristic work operations of a robot and a person.

It is generally accepted that an average person in terms of physical capabilities can develop a power of 100 W for a long time, a limited time of 300 W, and a peak maximum power of 1.2 to 2 kW [4]. It is clear that these data are very arbitrary, since it does not stipulate what kind of work is performed, there is no binding to specific muscles. However, we can say for sure that continuous power is determined by the capabilities of the human energy system, and above all the cardiovascular component. When it is mentioned about peak power, they usually mean work done by the leg muscles (running, cycling).

Based on this, it can made an estimate of the total power of the main muscles of a person having their own analog drives in AWH. If it is assumed that the estimated movement is performed simultaneously by three sagittal degrees of one leg, then the total power of the corresponding muscles is approximately 0.1 of the total power of all muscles. Then the total (installed) peak power of the muscles of the average person is approximately from 12 to 20 kW. This power corresponds to the work of the muscles in the middle of their working characteristic, known as the Hill-Abbot Curve [4]. Such a concave characteristic is quite rare in the most common drives in technology. Therefore, for example, throttle hydraulic drives with the same maximum force and speed as the muscles, should have almost three times the maximum power, and electric drives with a linear inclined characteristic - two and a half or more.

According to the author's estimates, made on the basis of experience with a prototype AWR created at MSTU. N.E.Bauman [5, 6], the total mass of the drives can be approximately 20% of the total mass of the robot. Therefore, in order for the ASR to have working capabilities comparable to at least a person average in physical development, the specific power of hydraulic drives should be more than 1.8 kW / kg, and electric drives together with gearboxes should be more than 1.3 kW / kg.

In addition to power, the drives must provide comparable idle speeds and braking moments. For example, it can given the following approximate estimate: the greatest braking torque is required to extend the knee joint (approximately 400 N · m); the least braking torque is required to rotate the wrist around the longitudinal axis forearm (approximately 20 N · m). The limiting angular velocities have a much smaller spread and are approximately 10 r / s in the most loaded joints, and 20 r / s in the least loaded.

It is extremely important to ensure sufficient dynamic accuracy of the drive system. Based on the analysis of the operation of the foot drives of the aforementioned AWR prototype model in the process of dynamic walking, the author obtained an estimate of the minimum required drive bandwidth of 10 Hz, and it should be ensured under conditions of strong interference through the robot's actuator.

Also, high energy efficiency and ease of control are required from the drives.

Based on the set of requirements, electro-hydraulic drives are best suited for the ASR executive system. Electro-hydraulic drives have a number of specific features, some of which are especially manifested in the complex AWR multi-link executive system. In particular, much more acute than for a single drive, the problem arises of ensuring high dynamic accuracy, and in conjunction with the problem of eliminating the risk of self-oscillations. Methods for solving this problem for electro-hydraulic drive systems also have their own specifics. One of the possible methods, which has shown its effectiveness in practice, is discussed below.

Methods

Executive systems are not only walking, but many other modern robots are complex controllable multi-link spatial mechanisms with many drives and a tree-like kinematic structure. They can have several tens of degrees of mobility. Drives installed in the degrees of mobility of such robots form a single multi-connected system. They experience significant interference through the actuator and a common energy source of limited power, therefore, it is advisable to control the drives with the help of a multi-connected controller. The modern theory of automatic control has many methods for the synthesis of such controllers, however, when creating systems for controlling the movement of walking robots, a lot of problems arise. In particular, the well-known "curse of dimensionality" makes the synthesis of regulators for systems with so many drives quite complicated and time-consuming even if the structure and parameters of the system are constant.

In AWR, the configuration of the actuator is constantly changing during movement. External additional mechanical bonds appear and disappear imposed on the links of the mechanism, primarily on the feet. This is accompanied by impacts, the parameters of which can be predicted only very approximately. External loads and mass inertia characteristics of the actuator are changing. Therefore, it is necessary to continuously adapt the controller settings in real time, and in conditions when not all system parameters are available for direct measurement, and the corresponding signals from the sensors can be very noisy.

The task of controlling such a complex multidimensional, high-order, non-stationary drive system in practice is still difficult to solve. Therefore, a popular method for the synthesis of drive control systems, in which the drives are initially considered independent and calculated based on the worst combination of reduced masses (moments of inertia) and static loads. Further, the correction of the obtained solution is carried out according to the simulation results, taking into account the mutual influence of the drives, external communications and other operating features. In reality, this means a significant (often an order of magnitude) decrease in the quality factor of drives in terms of speed. The system in this case is not optimal, but it often allows us to solve practical problems.

In the case of AWR, this approach is applicable only when creating robots of very small sizes, usually toys. With an increase in the geometric dimensions of the robot, the temporal characteristics of motion, subject to similar conditions, also increase. The forces that mainly determine the movement of a walking robot are the forces of weight and inertia. Thus, subject to geometric similarity of actuators and kinematic similarity of movements, dynamic similarity is achieved when the Froude numbers are equal

$$Fr = \frac{v^2}{gl},$$

Where v — robot speed,

g — acceleration of gravity,

l — some characteristic size of the robot, for example, its height.

Therefore, the temporal parameters of the movement, for example, the step period, increase in proportion to the square root of the linear size (for dynamically similar movements)

$$T \sim \sqrt{l}$$

For example, a small anthropomorphic robot with a height of 0.3 m must walk almost 2.5 times more often to perform a movement similar to the normal walking of a person with a height of 1.8 m. Hence, many people have the illusion of ease of controlling the robot's dynamic walking (toys can!). In fact, if the walking periods are equal, the dynamic walking of a person can correspond to the static walking of a robot toy with relatively large feet, the implementation of which is fundamentally simpler.

As the robot's growth increases, the time characteristics of its Executive system also change. The robot's mass, all other things being equal (true for similar robots), is proportional to the cube of its linear size. Then the moments of inertia that characterize the Executive mechanism change proportionally to the fifth power of the linear size. Time constants of hydraulic drives, taking into account the volume of hydraulic cylinder cavities, their geometry and reduced moments of inertia, are proportional to the linear size of the robot

$$T_{dr} \sim l$$

It follows that when the linear dimensions of the robot change, the ratio of time constants characterizing the dynamic properties of the drives and the temporal characteristics of the movement changes in proportion to the square root of the characteristic linear size. For example, if the same materials used proportionally increase all linear dimensions of the robot twice, the time constants of the drives will increase 20.5 times more than the temporal characteristics of the movements. This leads to the fact that, firstly, with an increase in the size of the robot, the dynamic accuracy of working out the required movements decreases, and secondly, the time constants of the drives approach the time constants of the stabilization loop of the robot according to the signals of the orientation system, which reduces the quality of its operation.

This leads to the conclusion that you should not increase the linear dimensions of the AWR unnecessarily. You should also reduce the robot's weight by using light construction materials. This trend can be seen in the works of most AWR developers. The growth of most recently developed AWR is less than 2 m. Based on the experience gained in working with the layout model of the machine created with the participation of the author, we can recommend the robot's height in the range from 1.5 to 1.7 m. At the same time, the speed and dynamic accuracy of the robot's drive system are subject to strict, difficult-to-meet requirements.

The block diagram corresponding to the linearized equations of the mathematical model of a typical single-channel electro-hydraulic drive with throttle control and a symmetrical hydraulic cylinder is shown in Fig. 2. Modern electro-hydraulic amplifiers (EHAs) have high speed, due to which, in most cases, executive hydraulic motors (HM) together with mechanisms connected to the output link have a decisive influence on the dynamics of the drives. In this case, the damping is determined by the coefficients k_{Qp} and b (the coefficient of viscous friction in the hydraulic motor in total with the reduced total coefficient of viscous friction in the actuator). Moreover, k_{Qp} is proportional to the displacement of the spool of the hydraulic amplifier, so when positioning the drive, the damping is reduced, and in the case of small leaks through the closed spool slits, it is mainly determined by friction in the gaskets of the HM. However, the friction parameters in the seals are not stable and depend on many factors that change during the operation of the drive: the pressure drop on the seals, the temperature of the working fluid, the condition of the working surfaces, and the degree of aging of the seal materials. In addition, to improve the accuracy of the drives, friction is minimized, which reduces damping.Even if we consider the idealized situation and when analyzing the dynamics of the drive, consider all other degrees of mobility frozen, when changing the configuration of the robot, the mass reduced to the main rod of the main cylinder and the load on the drive change. Accordingly, the frequency characteristics of the drive change. As a result, a certain region of the probable location of the top of the resonant peak of the drive on the LAH plane is formed (Fig.2).

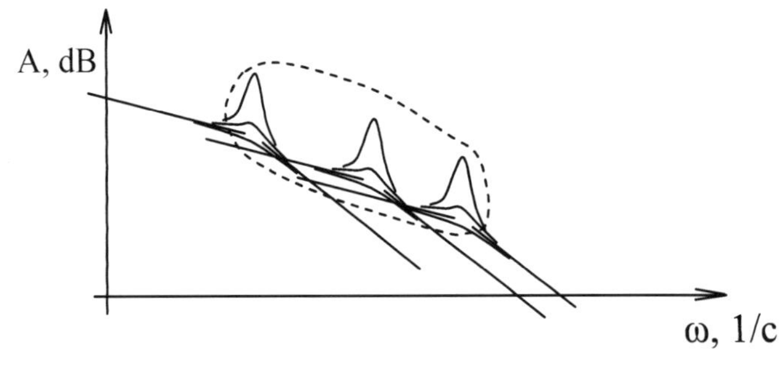

Fig. 2.LAFR view of the linear model of the power part of the hydraulic drive.

The probability of the appearance of high peaks in the mid-frequency region of the LAH forces either to reduce the speed factor of the drive or to find dynamic correction methods to obtain the necessary phase and amplitude reserves. Introduction of damping speed feedbacks for hydraulic drives is ineffective. In aviation, for example, in a similar situation, hydro mechanical correction of dynamic pressure is used. A similar result can be achieved using electric feedback to accelerate the output link of the drive, measured relative to its body. However, in the case when the hydraulic motor is mounted on a moving base, for example, on a moving link of the AWR mechanism, this acceleration is difficult to measure. In systems with one degree of mobility in practice, as a rule, instead of acceleration, an estimate of its magnitude is used using signals from pressure sensors installed in cavities This is easy to do, since in systems with one degree of mobility, the acceleration of the output link is proportional to the pressure drop up to the influence of external loads. That is, to assess the magnitude of the acceleration, it is necessary to convert the pressure drop on the hydraulic motor into force (moment) at its output link and divide by the mass (moment of inertia) reduced to the output link. Moreover, the obtained value will depend not only on acceleration, but also on the reduced output link of the external load hydraulic motor. But this problem can be solved using the circumstance that for correcting the properties of the drive, the mid-frequency part of the spectrum of the received signal is required, and the influence of the external load mainly affects the low-frequency part of the spectrum. A correction device (high-pass filter) with a transfer function of the form is installed in the damping feedback loop

$$W = \frac{Ts}{Ts+1}.$$

Its time constant is chosen so that it reduces the influence of low-frequency vibrations and eliminates the influence of the constant component associated with external load, but leaves the signal unchanged in the mid-frequency region that determines the stability and dynamic properties of the drive. Figure 4 shows the corresponding block diagram, where k_aF/m is the feedback coefficient for dynamic pressure, and k_a is the transmission coefficient for acceleration

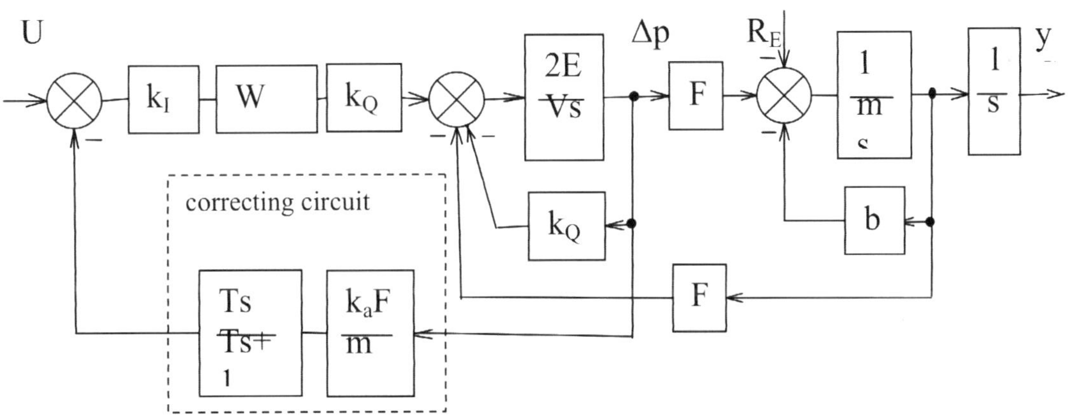

Fig. 3. Block diagram of an open drive with dynamic pressure feedback.

Using the method of inverse characteristics, it can be shown that as a result, the drive LAH will lie no higher than the ABC line (Fig. 5), which is a reflection relative to the horizontal axis of the LAH correction chain. Thus, the resonant peak, regardless of its location, will be "cut off".

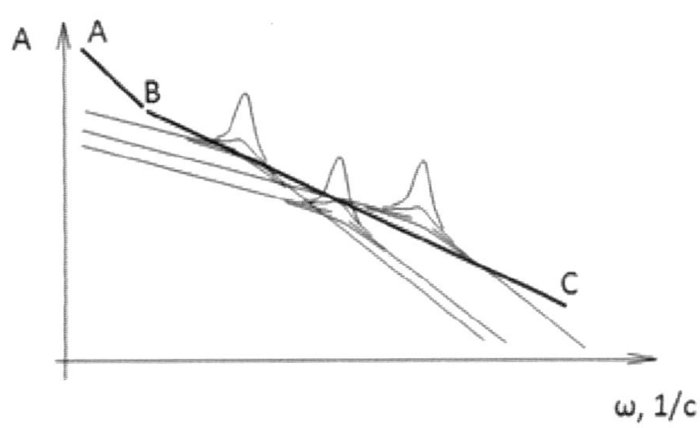

Fig. 4. Correction of frequency characteristics of an open drive using dynamic pressure feedback.

However, this correction method is not directly applicable to multi-stage systems with a strong interaction of the drives through the actuator, which includes robots. In the case of AWH, additional difficulties are caused by the tree-like nature of the actuator, its lack of attachment to a fixed base, and the variability of external relations. The methodology of mathematical modeling of the dynamics of the executive system of AWR, worked out during the creation of a prototype AWR in MSTU. N.E. Bauman, described in [5, 6, 7, 8]. The method involves the use of a "fake" actuator to the absolute coordinate system of a fictitious kinematic chain, consisting of weightless links connected by three translational and three rotational degrees of mobility with zero hinge moments. As a result, the dynamics of the actuator of the robot is described by the following block-matrix equation:

$$\begin{pmatrix} I_{II} & I_{IR} & -J_I^T \\ I_{IR}^T & I_{RR} & -J_R^T \\ -J_I & -J_R & 0 \end{pmatrix} \begin{pmatrix} \ddot{q}_I \\ \ddot{q}_R \\ R \end{pmatrix} + \begin{pmatrix} H_I \\ H_R \\ -P \end{pmatrix} = \begin{pmatrix} 0 \\ \tau \\ 0 \end{pmatrix}$$

Where I_{II}, I_{IR}, I_{RR} — components of the inertia matrix of the actuator of the robot;

J_I, J_R— components of the Jacobi matrix, which is part of the equations of relations superimposed on the actuator (for example, on the supporting foot), and expressed relative to accelerations; this matrix also establishes the relationship between the reactions of superimposed bonds and the components of the hinge moments caused by these reactions;

P— a column matrix constituting the right-hand side of the equations of relations expressed with respect to accelerations;

R — a column matrix containing a vector of reactions of superimposed external relations;

$\ddot{q}_I, \ddot{q}_R$—components of a matrix-column containing a vector of generalized accelerations of the mechanism;

H_I, H_R — components of the matrix-column of reduced external, centrifugal and Coriolis forces;

τ — vector of hinge moments (forces) developed by drives.

In this equation, the submatrix indicated by the index I refers to the fictitious kinematic chain, and the index R refers to the real links of the mechanism. An efficient procedure for calculating matrix elements of this equation adapted to tree mechanisms is described in [5, 7]. The Jacobi Matrix J has as many rows as there are additional external links imposed on the Executive mechanism. For example, if the robot is stable on one leg, and translational and rotational connections are superimposed on all three axes, then the matrix J contains 6 rows, and if the robot is standing on two legs, then 12 rows.

Since the elements of the Ni matrices depend on the current values of generalized speeds, these values are small at low speeds, and even more so when stopping in place, and in some cases they can be ignored. Ignoring also the external forces applied to the mechanism, with their predominantly low-frequency spectrum (except for coupling reactions), the equations of dynamics of the robot's Executive mechanism can be written in simplified form:

$$\begin{cases} I_{IR}\ddot{q}_I + I_{IR}\ddot{q}_R - J_I^T R = 0 \\ I_{IR}^T\ddot{q}_I + I_{RR}\ddot{q}_R - J_R^T R = \tau \\ J_I\ddot{q}_I + J_R\ddot{q}_R = 0 \end{cases}$$

At low speeds and when the robot is stopped, the spool in the electrohydraulic drive amplifiers are at zero or offset by a small amount, which directs the coefficient to zero k_{Qp}in a linear mathematical model of the drive, and exacerbates the problem of providing sufficient damping of the drives to prevent self-oscillation.

The author suggests using the above simplified form of the equation to determine the approximate values of relative accelerations in the degrees of mobility of the robot according to the measured in the process of movement of the robot arm torques (forces) and loads on the robot's feet. Hinge moments (forces) are determined during the robot's movement by the measured values of pressure in the cavities of hydraulic motors, just as in the case of single-channel hydraulic drives. Forces and moments on the robot's feet are measured during movement using six-component force-moment sensors installed on its feet. The obtained values of τ and R Allow us to calculate approximate values of generalized accelerations. To do this, the above system of equations is proposed to be represented as a matrix expression:

$$\ddot{q}_R \approx (I_{RR} - I_{IR}^T I_{II}^{-1} I_{IR})^{-1}(\tau + (J_R^T - I_{IR}^T I_{II}^{-1} J_I^T)R)$$

The acceleration estimates obtained from it can be used to dampen the robot's drive system in the same way as it is done in single-channel electrohydraulic tracking drives. Accordingly, to reduce the influence of external forces that load the drives, as well as in single-channel hydraulic drives, it is needed to use high-frequency filters.

One of the advantages of this method is that there is no need to change the calculation algorithm when changing the movement phases corresponding to the change of the robot's support legs. In this expression, it can be included equations of connections that are imposed on both feet at once. They will be activated when the corresponding reactions appear, measured by sensors installed on the robot's feet.

Conclusion

This method was used in the control system of the prototype ASR model and allowed to increase the quality factor of its speed by 3 ... 10 times. Due to this, the dynamic accuracy of the drives has increased to a level sufficient to implement the dynamic walking of the robot.

In the Figure 6 shows experimental curves illustrating the effect of the depth of damping feedback on the quality of the drives. During the experiment, the robot stood on its right foot.

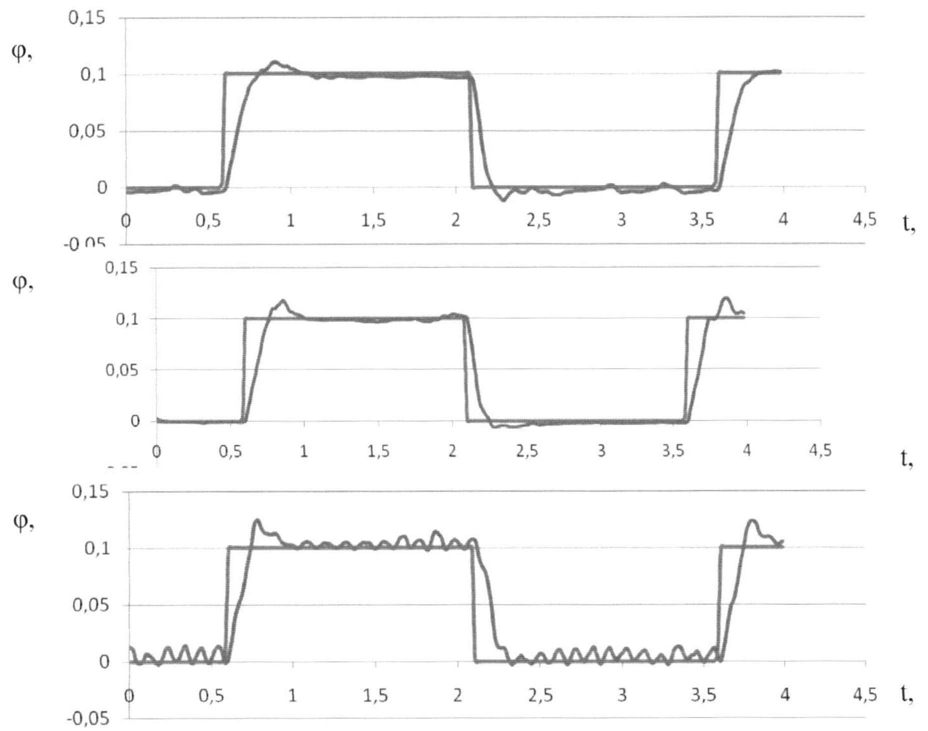

Fig. 5. Transient processes in the knee drive of the AWR at various settings of feedback on dynamic pressure.

A rectangular signal with an amplitude of 0.05 rad was applied to the input of the knee drive of the raised left leg. The input action and response of the drive were recorded with three damping feedback settings. The average process was obtained with damping bond coefficients close to optimal. The upper one — with an increase in the damping feedback coefficients in all drives by 5 times, and the lower one — with their decrease by 4 times. This experience illustrates the existence of an optimal setting and the impossibility of a system functioning with a given speed factor without additional damping. It can be stated the stability of the system to change settings over a wide range.

References

[1] Filimonov, A.B., Filimonov, N.B., Barashkov, A.A. Construction of Potential Fields for the Local Navigation of Mobile Robots (2019) Optoelectronics, Instrumentation and Data Processing, 55 (4), pp. 371–375.
[2] Tovarnov, M.S., Bykov, N.V. A Mathematical Model of the Locomotion Mechanism of a Mobile Track Robot with the Magnetic-Tape Principle of Wall Climbing (2019) Journal of Machinery Manufacture and Reliability, 48 (3), pp. 250–258.
[3] Kozov, A.V., Volosatova, T.M., Vukolov, A.Y. Structural obstacle recognition method and its application in elevated terrain objects search (2018) 2018 International Russian Automation Conference, RusAutoCon 2018, статья № 8501765 .
[4] Utkin B. V. Biomechanics of physical exercises. –Moscow: Nauka 1989.
[5] Semenov S.E., Kovalchuk A. K., Kulakov D. B. Mathematical modeling of mechanisms with a tree-like kinematic structure // Drive equipment. –M.: 2010 N2.
[6] Kulakov, D.B., Semenov, S.E., Kulakov, B.B., Shcherbachev, P.V., Tarasov, O.I. Hydraulic Bipedal Robots Locomotion Mathematical Modeling (2015) Procedia Engineering, 106, pp. 62–70.
[7] Semenov, S.E. Mathematical modeling of the electro-hydraulic actuation systems of the machines with tree-like kinematic structure (2015) Proceedings of 2015 International Conference on Fluid Power and Mechatronics, FPM 2015 article № 7337184, pp. 583–592.
[8] Yaskevich, N. Recursive algorithm for simulation of single chain manipulator dynamics (2014) MESA 2014 - 10th IEEE/ASME International Conference on Mechatronic and Embedded Systems and Applications, Conference Proceedings, article № 6935553.

IOP Conf. Series: Materials Science and Engineering **963** (2020) 012031 doi:10.1088/1757-899X/963/1/012031

Research of composition porosity based on 3d-printed frames and impregnated with epoxy resin

Y Lopatina[1,2,3] and A Filippova[2]

[1]Federal State Budgetary Scientific Institution "Federal Scientific Agricultural Engineering Center VIM"
[2]Bauman Moscow State Technical University, 5 Second Baumanskaya Street, Moscow, 105005, Russian Federation

[3]E-mail: lopatina.julia@yandex.ru

Annotation. The paper discusses a method of increasing the strength of 3D-printed structures made by the fused deposition modelling process, which consists in vacuum impregnation of printed products in a compound based on epoxy resin. Studied the number of pores in samples obtained by this method depending on the percentage of 3D printing filling and direction of layers relative to the axis of the samples. The porosity of the samples was determined by comparing the theoretical ideal density with the experimental value determined using the hydrostatic weighing method. As a result, it was found that the porosity of the samples lays in the range from 17 to 18.5%. It was also found that with increasing percentage of filling the porosity of the samples increases. It has been suggested that this dependence is associated with a complication of the processes of air release from samples during the impregnation.

Introduction

In recent years, 3D printing technology is gaining more and more popularity. This fact is related to a number of advantages in comparison with traditional technologies, such as the ability of creating complex shaped products without special auxiliary tools (injection molds or press forms), in many cases lower cost of equipment, high speed of manufacturing parts due to the lack of a large number of technological operations, as well as the absence of the need for readjustment of equipment when changing the nomenclature [1]–[5]. One of the most common 3D printing technologies today is the FDM technology (fused deposition modeling, Figure 1). The stages of work look like this: polymer thread (filament) comes into the extruder where it melts; extruder moves along a given directory, forming a part layer; the layer is cured by cooling; cured layer shifts down; the extruder gives shape to a new layer, fusing it on top of the previous one, etc.

Despite the obvious advantages, 3D printing currently finds its main application for creating dummies that do not carry loads during operation. The creation of machine parts, which are able to work in real constructions under loads, is difficult due to the low mechanical properties of materials used for 3D printing. The strength of structures seeks to increase by printing continuous fiber reinforced composites [6], [7], as well as the creation of superengineering plastics for 3D printing

[8]–[15]. However, these solutions require the use of expensive equipment for the printing process. The cost of the materials themselves is also very high.

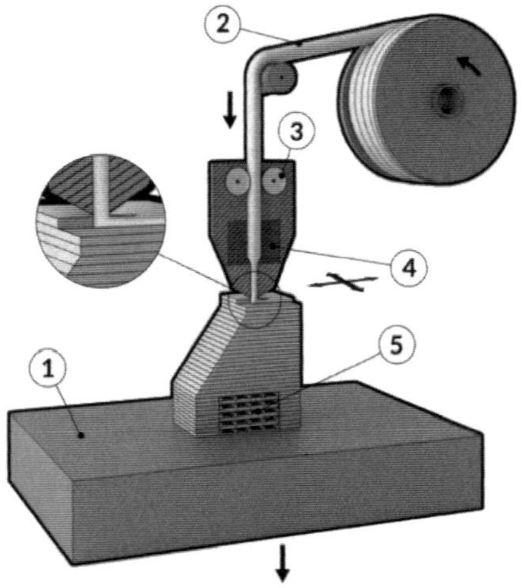

Figure 1. Scheme of 3D printing using FDM technology: 1 — desktop; 2 — polymer thread (filament); 3 — feed rollers; 4 — heating element; 5 — product

One of the possible ways for increasing the strength properties of 3D-printed parts is its impregnation in a vacuum chamber in the thermosetting resins (Figure 2). The method is related to the vacuum infusion technology used to create woven composite [16]–[20]. Thermosetting resins have high strength properties compared to standard general-purpose plastics that are widely used for 3D printing. The main problem inherent in all similar methods of creating composites by impregnation is the residual porosity, which impairs mechanical properties. This work is devoted to the study of the influence of various 3D printing parameters of construction for the porosity of composite material, which is obtained in the process of impregnation.

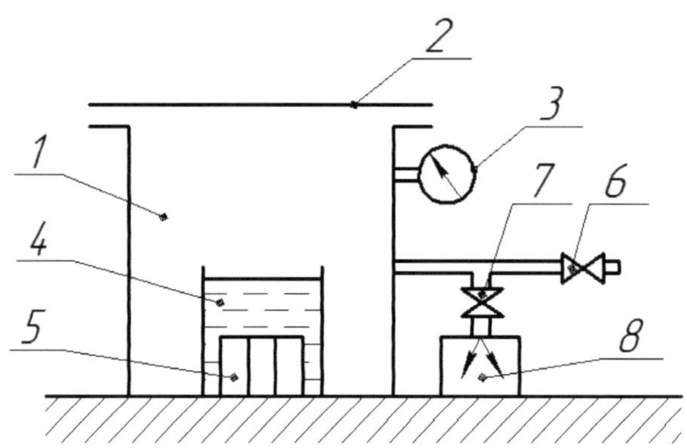

Figure 2. Vacuum system for impregnation: 1 — chamber; 2 — a transparent cover; 3 — manometer; 4 — polymer compound; 5 — impregnated part; 6 — exhaust valve; 7 — shutoff valve; 8 — vacuum pump

Materials and methods

In this work, porosity was studied on impregnated samples intended for further tensile and toughness tests (Figure 3, a). Samples were printed on a Picasso 3D Designer XPro 3D printer using FDM technology with different percentages (20, 33 and 50%, Figure 3, b), as well as with different lay directions — along and across the axis of the samples (Figure 3, c). Printing material is a polylactide (PLA). The impregnation was carried out using the vacuum system shown in Figure 2. As an impregnation material was used epoxy diane resin with 20% of epoxy groups called ED-20 with a polyethylene polyamine as a curative. Exposure in vacuum was 30 minutes. After the impregnation, the samples were cured at a room temperature for 24 hours.

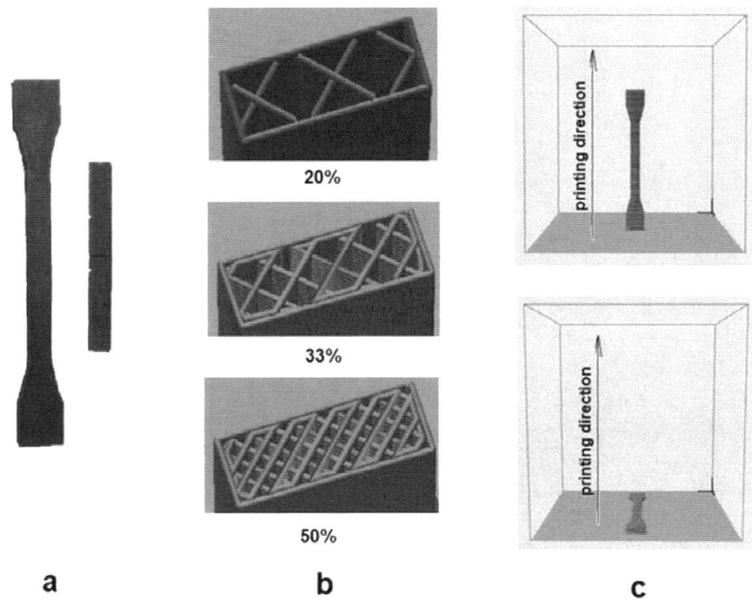

Figure 3. Parameters of the testing samples: a — shape (on the left are standard samples for tensile tests, on the right — for impact tests); b — percentage of filling; c — print direction

The porosity of the samples was estimated by comparing the theoretical ideal density of the samples ρ_t with the real determined by hydrostatic weighing ρ_r:

$$P = \frac{(\rho_t - \rho_r)}{\rho_t} \cdot 100\%$$

The theoretical density of the samples was determined as follows:

$$\rho_t = \rho_{PLA} \cdot V_{PLA} + \rho_{ED-20} \cdot V_{ED-20}$$

where $\rho_{PLA} = 1.24\ g/cm^3$ — printing material density; $\rho_{ED-20} = 1.21\ g/cm^3$ — cured compound density; V_{PLA}, V_{ED-20} — volume fractions of material parts, $V_{PLA} + V_{ED-20} = 1$. The amount of plastic V_{PLA} used for printing for each filling was determined using a Polygon X 3D printing slicer. The volume of epoxy resin for each samples filling was calculated as the difference between the volume of plastic required to print a sample with 100% filling and the volume of plastic required to print a frame of a certain filling.

The real density of the samples was estimated by hydrostatic weighing according to GOST 15139–69. The essence of the method is to compare the mass of the sample, measured in air, with the mass of the sample, measured by immersion in a working fluid of known density, in this study distilled water. Weight is determined using accurate scales with an error of not more than 0.001 g. The density of the sample is determined as follows:

$$\rho_r = \frac{m_a}{m_w} \rho_w$$

where m_a — sample weight in the air; m_w — sample weight in the working fluid; $\rho_w = 1\text{г}/cm^3$ — density of the working fluid (distilled water).

Results and discussion

The results of the study of the porosity of the samples are presented in Figure 4. At each point, the average value of porosity calculated for a group of three samples is given. The confidence interval according to the calculation results ranged from 0.29 to 0.34. The porosity in the samples intended for tensile testing varies in the range from 17.5 to 18.5%, in the samples intended for impact testing — from 17.1 to 18.5%.

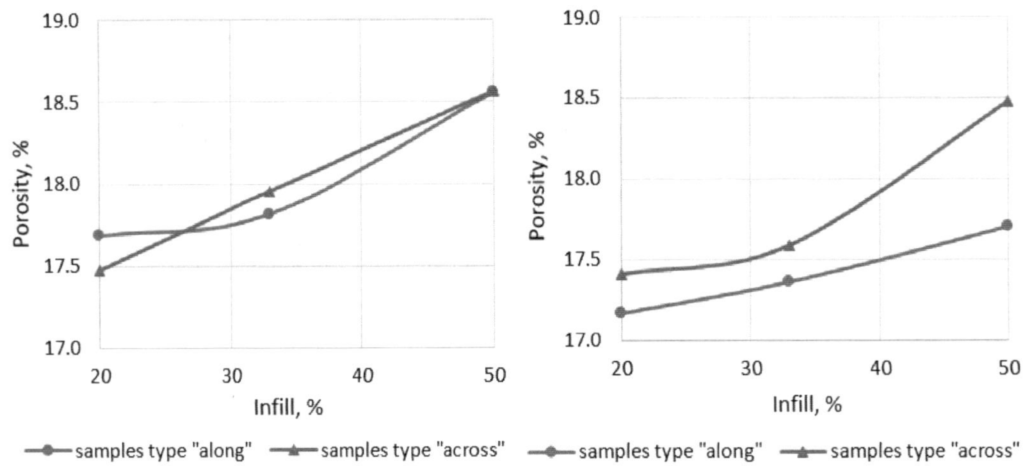

a)b)

Figure 4. Porosity of the tested samples:
a— samples for tensile testing; b — samples for impact testing.

It can be seen that in all cases the value of porosity increases with an increase in the percentage of filling of samples in 3D printing. Moreover, the porosity of the samples intended for tensile testing has very similar values regardless of the orientation of the layers relative to the axis of the sample, while the porosity of the samples for impact tests differs: for samples of the "across" type it turned out to be slightly higher than for samples of the "along" type. In general, the porosity of the tensile test specimens is slightly higher than the porosity of the impact test specimens.

An increase in porosity with an increase in the percentage of filling can be explained by the fact that in the presence of a denser mesh inside the sample it is more difficult for air to go outside during vacuum impregnation. A slightly higher level of porosity in the samples for tensile tests compared with samples for impact tests can be explained by the fact that the former have a slightly larger volume compared to the latter, which means that the likelihood of an air bubble is higher.

Conclusion

As a result of the studies, the dependence of the porosity of the samples of composites based on 3D-printed frames impregnated with epoxy resin on the percentage of 3D printing filling was revealed. The higher the sample filling percentage, the higher the porosity was observed, which is probably can be explained by the fact that with a denser internal structure, it is more difficult for air to leave the sample during the impregnation process. On average, the porosity was about 18% and had a small variation depending on the parameters of the manufacture of the sample (within ± 0.5%). This level of porosity is large enough and can significantly affect the mechanical properties of the samples during testing, and therefore it is recommended to adjust the impregnation process to improve quality.

Acknowledgement

The reported study was funded by RFBR, project number 19-38-90037.

References

[1] Katkar R.A., Taft R.M., Grant G.T. 3D Volume Rendering and 3D Printing (Additive Manufacturing). Dent Clin North Am. 2018 Jul;62(3):393–402. doi: 10.1016/j.cden.2018.03.003.
[2] Rinaldi, M., Ghidini, T., Cecchini, F., Brandao, A., Nanni, F. Additive layer manufacturing of poly (ether ether ketone) via FDM // Composites Part B: Engineering. 2018. Volume 145. pp. 162–172
[3] Lifton V., Lifton G., Simon S. Options for additive rapid prototyping methods (3D printing) in MEMS / Rapid Prototyping Journal. 2014. V.20. No. 5, 403–412 pp.
[4] Stansbury J., Idacavage M. 3D Printing with polymers: Challenges among expending options and opportunities / Dental Materials.2016. V.32. 54–64 pp.
[5] Zukas V., Zukas J. An Introduction to 3D Printing / Published and Distributed by First Edition Design Publishing. 2015. 41 pp.
[6] Azarov, A.V., Antonov, F.K.,Golubev, M.V., Khaziev, A.R., Ushanov, S.A. Composite 3D printing for the small size unmanned aerial vehicle structure // Composites Part B: Engineering. 2019. Volume 169. pp. 157–163. DOI: 10.1016/j.compositesb.2019.03.073
[7] Mirzaev M.A., Smirnov I.G, Dyshekov A.I., Slavkina V.E., Shereuzhev M.A. and Aliev S.A. Design of the solar panel frame of the spacecraft by 3D printing with composite materials. IOP Conference Series: Materials Science and Engineering. 2020. V. 747. P. 1–5. DOI:10.1088/1757-899X/747/1/012009
[8] A. Sviridov, Iu. Lopatina, Iu. Kurganova. 3D-printed polyether ether ketone samples mechanical properties estimation. IOP Conf. Series: Materials Science and Engineering. 2019. V. 589. Number of article 012021. doi:10.1088/1757-899X/589/1/012021
[9] Singh, S., Prakash, C., Ramakrishna, S. 3D printing of polyether-ether-ketone for bio-medical applications // European Polymer Journal. 2019. Volume 114. pp. 234–248. DOI: 10.1016/j.eurpolymj.2019.02.035
[10] Geng, P., Zhao, J., Wu, W., Ye, W., Wang, Y., Wang, S., Zhang, S. Effects of extrusion speed and printing speed on the 3D printing stability of extruded PEEK filament // Journal of Manufacturing Processes. 2019. Volume 37. pp. 266–273. DOI: 10.1016/j.jmapro.2018.11.023
[11] Stepashkin A., Chukov, D.I., Senatov, F.S., Salimon, A.I., Korsunsky, A.M., Kaloshkin, S.D. 3D-printed PEEK-carbon fiber (CF) composites: Structure and thermal properties // Composites Science and Technology. 2018. Volume 164. pp. 319–326. DOI: 10.1016/j.compscitech.2018.05.032
[12] Basgul, C., Yu, T., Macdonald, D.W., Siskey, R., Marcolongo, M., Kurtz, S.M. Structure-property relationships for 3D-printed PEEK intervertebral lumbar cages produced using fused filament fabrication // Journal of Materials Research. 2018. Volume 33, Issue 14. pp. 2040–2051. DOI: 10.1557/jmr.2018.178
[13] Hoskins, T.J., Dearn, K.D., Kukureka, S.N. Mechanical performance of PEEK produced by additive manufacturing // Polymer Testing. 2018. Volume 70. pp. 511–519. DOI: 10.1016/j.polymertesting.2018.08.008
[14] Rinaldi, M., Ghidini, T., Cecchini, F., Brandao, A., Nanni, F. Additive layer manufacturing of poly (ether ether ketone) via FDM // Composites Part B: Engineering. 2018. Volume 145. pp. 162–172. DOI: 10.1016/j.compositesb.2018.03.029
[15] Arif, M.F., Kumar, S., Varadarajan, K.M., Cantwell, W.J. Performance of biocompatible PEEK processed by fused deposition additive manufacturing // Materials and Design. 2018. Volume 146. pp. 249–259. DOI: 10.1016/j.matdes.2018.03.015
[16] Kobets L.P., Borodulin A.S., Malysheva G.V. Study of microcapillary impregnation of carbon fibres by epoxy binders. Fibre chemistry. DOI: 10.1007/s10692-017-9788-8

[17] Belov P.A., Borodulin A.S., Kobets L.P., Malysheva G.V. Kinetics of fiber impregnation by a binder gradient generalization of Navier–Stokes–Darcy equations. Polymer Science. Series D. 2016. T. 9. № 2. C. 205–208. DOI: 10.1134/S1995421216020039

[18] Dumansky, A.M., Alimov, M.A., Terekhin, A.V. Experiment- and computation-based identification of mechanical properties of fiber reinforced polymer composites // Journal of Physics: Conference Series. 2019. Volume 1158, Issue 2. DOI: 10.1088/1742-6596/1158/2/022037

[19] Borodulin A.S., Malysheva G.V., Romanova I.K. Optimization of rheological properties of binders used in vacuum assisted resin transfer molding of fiberglass. Polymer Science. Series D. 2015. T. 8. № 4. C. 300–303. DOI: 10.1134/S1995421215040036

[20] Gorodetskii, M.A., Nelyub, V.A., Malysheva, G.V., Shaulov, A.Y., Berlin, A.A. Technology of Forming and the Properties of Reinforced Composites Based on an Inorganic Binder // Russian Metallurgy (Metally). 2018. Volume 2018, Issue 13. pp. 1195–1198. DOI: 10.1134/S0036029518130074

IOP Conf. Series: Materials Science and Engineering **963** (2020) 012032 doi:10.1088/1757-899X/963/1/012032

Study of tribotechnical characteristics of 3D printed abs plastic samples

V Slavkina[1] and Yu Lopatina[1,2,3]

[1]Federal State Budgetary Scientific Institution "Federal Scientific Agricultural Engineering Center VIM", 1st Institutskyproezd 5, Moscow, Russia
[2]Bauman Moscow State Technical University, 5 Second Baumanskaya Street, Moscow, 105005, Russian Federation

[3]E-mail: lopatina.julia@yandex.ru

Abstract. This work is devoted to research of tribotechnical characteristics of 3D printed samples from ABS plastic. The results of determination of friction coefficient and depth of counterbody introduction into the sample surface on Tribometer TRB-S-DE-0000 are presented. The article also presents the advantages of using 3D printing for repairing machine parts and describes the causes of gear failure.

Introduction

The profitability of expensive equipment is largely determined by the useful life of components of nodes and machine parts which often quickly become unusable. Due to systematic breaks for the replacement of faulty parts decreases productivity and the quality of goods produced by this technique. At the same time, the costs of producing new parts and paying for repair workers will increase. All this slows down the work of the enterprise, destroys key elements necessary for automation of production and operational processes. With such dynamics, the losses of the enterprise will be very large [1]–[3]. The fact that nowadays a lot of manufacturers have to buy spare parts abroad also leads to certain difficulties: shortage of spare parts, their high price, long delivery times. Many manufacturers of machinery to reduce the cost of parts and metal consumption for their production replace the material with plastic. Such parts fail much faster which again leads to downtime in production [4],[5].

The majority of machine parts fail due to wear and tear [6],[7]. Wear leads to destruction and separation of solids from the surface,as well as accumulation of deformation during friction. Friction can be defined as a process that occurs when objects interact with each other in their relative motion. The causes of friction are the attraction of molecules of mutual nature in the contacting surfaces at the point of their contact as well as mechanical interaction of irregularities of friction surfaces [8]. The key indicator of friction is the coefficient of friction, which varies depending on the material of the interacting objects. But this is not the only indicator that should be taken into account it is equally important to measure material wear. When estimating the wear, it is necessary to take into account the changes in mass, parameters of the surface and shape of the original part resulting from the destruction of the upper layers during friction [9],[10].

Gear which in operation are subjected to dynamic and cyclic loads leading to surface destruction, are an example of parts that often fail due to friction wear.The extremely constant gear ratio determines the smooth running of the gears. However operating and technical errors (deformation) exclude the possibility of a constant gear ratio. Dynamic loads such as shocks also have an undesirable effect. The above-mentioned errors do not allow the load to be distributed evenly over the width of the toothed rim which leads to the appearance of the angle of misalignment as a result of which the tension in some points will increase [11]–[13].As a result, a defect such as wear of the working profile of teeth can often be found on gears of any kind. It is not accepted to correct such defect and usually it is replaced with a new part. The exception is if the defect does not exceed the permissible limits. The permissible wear of gear teeth is given in Table 1 [14],[15]. If there is a need to replace one gear wheel it is also necessary to replace the conjugated one the defect of which is still acceptable because the necessary contact can only be achieved in this way. Usually gear wheels working in a pair are produced on the same machine to achieve better meshing effect. Also, it is necessary to consider that at replacement only one defective wheel can increase noise level [16]–[18].

Table 1. Permissible wear of gear teeth

Working Mode	Circular speed, m/s	Limit wear to nominal tooth thickness along initial circle, %		
		Type of repair		
		Small	Medium	Capital
Power transmission in one direction with shockless load	under 2	20	15	10
	2…5	15	10	6
	Over 5	10	7	5
Reverse gear with impact load	under 2	15	10	5
	2…5	10	5	5

Reducing material and time costs in manufacturing spare parts in particular plastic gears can be achieved by using 3D printing technology. The most commonly used 3D printing technology today is fusion deposition modeling (FDM). Fusion deposition modeling is shown in Figure 1.

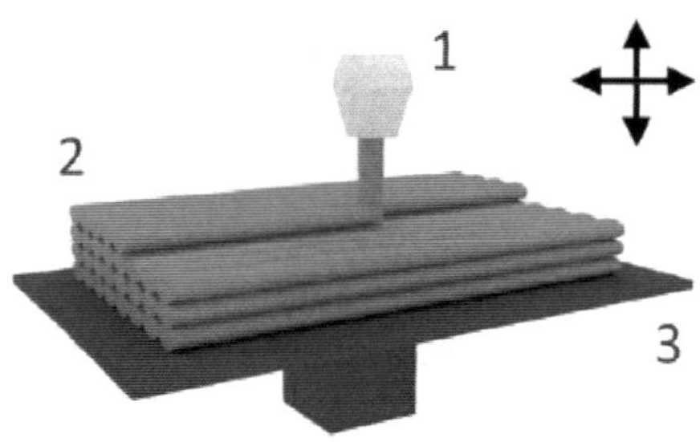

Figure 1. Melting method simulation: 1 — controlled mobile extruder feeding molten plastic through a nozzle; 2 — molten material; 3 — controlled mobile platform.

The FDM printing method uses a polymer thread called filament to produce the part. The extruder melts the incoming filament forming a layer of the part on the desktop. The layer is formed according to the contour of the part and the scheme specified by the special program. As the heat dissipates the molten polymer layer solidifies and then goes down together with the worktable under the next layer. Then the actions are repeated [19]–[23]. The main problem for wide application of additive technologies is a weak study of physical and mechanical characteristics of used materials in particular their friction characteristics. It should also be noted that tribotechnical characteristics of a part depend to a large extent on the quality of its surface and at 3D-printing roughness of formed surfaces can vary in very wide limits.

Materials and methods

A tribometer TRB-S-DE-0000 for measuring the sliding friction coefficient in contact between two mutually movable surfaces has been used to carry out the research. The device is a desktop unit consisting of a friction machine with a measuring device and a computer (figure 2, a). The investigated sample in the form of a flat disk is established on a platform in a 6- cam holder. The counterbody is a ball of 6 mm diameter made of steel 100Cr6 fixed immovably in the rod holder (figure 2, b). The counterbody was placed on the sample at a distance of 1–2 mm from the axis of rotation of the platform.

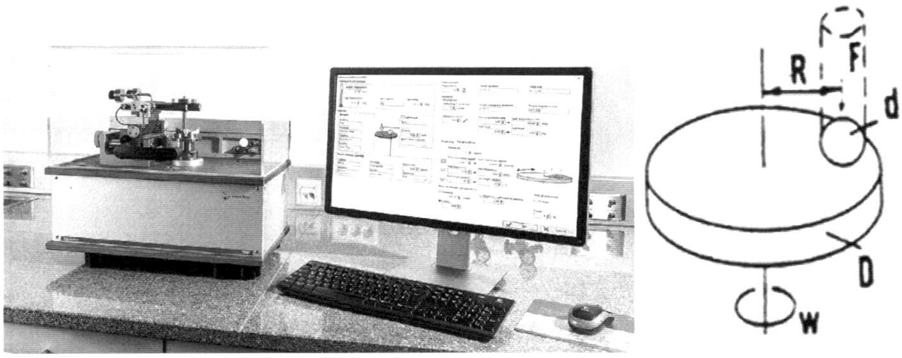

a) b)

Figure 2. Tribometer TRB-S-DE-0000 (a) and friction scheme used (b)

The tests were carried out at a load of 20 N, the speed of relative sliding — 0.20 m/s, at a distance of 4500 m and ambient air temperature — 23–25 ° C, the path of friction was 500 m. During the test the friction coefficient and the depth of penetration of the control sample into the body were continuously measured. The sliding of the counterbody relative to the sample is ensured by rotation of the platform with the sample.

Acrylonitrile butadiene styrene (ABS) is one of the most common materials used for 3D printing due to its high tensile strength and hardness [24] as well as its high impact toughness. At the same time products based on it have stable dimensions and chemical resistance. The main advantages also include ease of processing and low cost. In order to exclude shrinkage of the obtained parts it is necessary to observe the temperature regime during printing [25],[26]. In the Table 2 characteristics of ABS plastic used for 3D printing on FDM technology are presented.

Table 2.ABS plastic characteristics for 3D printing

Extrusion temperature, °C	210–245
Operating temperature of products, °C	–40 ÷ +80
Tensile strength, MPa	26–47
Rockwell hardness	R118

In this study, the samples were flat discs with a diameter of 30 mm and a thickness of 5 mm printed on a Picasso 3D Designer XPro printer. The printing parameters are presented in Table 3. The coefficient of friction was measured on two different surfaces of the sample: the ribbed one and the opposite smooth one which was pressed to the desktop surface during printing.

IOP Conf. Series: Materials Science and Engineering **963** (2020) 012032 doi:10.1088/1757-899X/963/1/012032

Table 3. Options for printing ABS plastic samples on the Picasso 3D Designer XPro 3D printer

Nozzle diameter, mm	0,3
Print speed, mm/s	100
Height of print layer, mm	0,2
Thickness of printing, mm	0,15
Print temperature, ° C	250

Results and discussion

The data obtained from the tests have been processed and presented in graphs. During the tests two indicators were recorded:the coefficient of friction and the depth of penetration of the counterbody into the surface of the analyzed sample. The data obtained from the tests of ribbed and smooth samples presented in graphs (figure 3). In both cases a similar character of friction coefficient behavior is observed: at the beginning of the test the coefficient of friction is reduced, then there is area where the value is practically unchanged. The first section is called the run-in section. After running in, the coefficient of friction corresponds to the minimum possible for a given pair of friction.

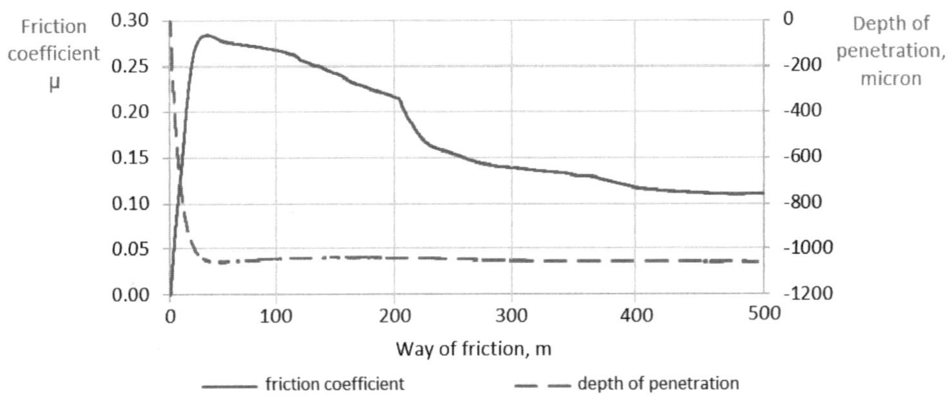

a)

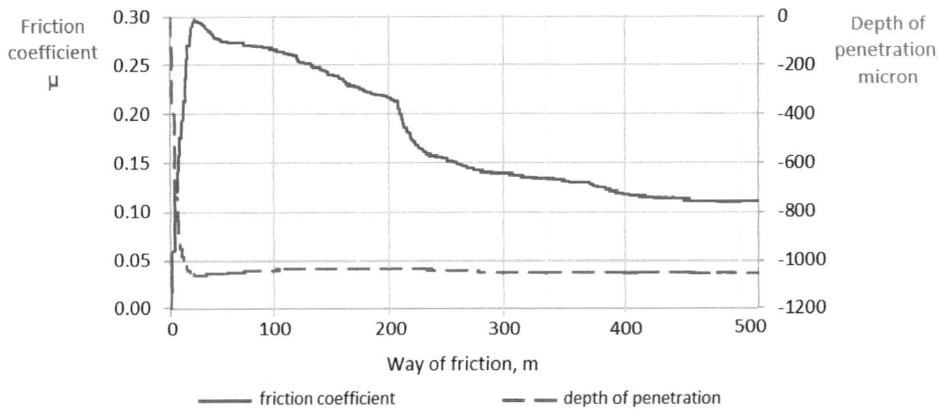

b)

Figure 3. Dependence of the friction coefficient and the depth of counterbody penetration into sample surface from the friction path: a) a sample with a ribbed surface (№ 2); b) a sample with a smooth surface (№ 4)

The tribotechnical parameters of the specimens determined as a result of the test are given in Table 4. The running-in distance varied in the range from 234 to 412 m, and the coefficient of friction during the tests ranged from 0.11 to 0.37. After running-in, the average coefficient of friction for samples with a ribbed and smooth surface practically does not differ (0.20 and 0.24, respectively).

Table 4. Values of coefficients of friction of ABS plastic obtained on tribometer TRB-S-DE-0000

Way of friction	Surface	Sample №	Run-in distance, m	Coefficient of friction			Medium value
				min	max	Medium in normal operating mode	
500 m	Ribbed	1	412	0.11	0.27	0.19	0.20
		2	234	0.16	0.25	0.20	
		3	257	0.17	0.29	0.23	
	Smooth	4	387	0.24	0.37	0.30	0.24
		5	400	0.11	0.29	0.18	

Conclusion

The study showed that the surface quality does not significantly affect the coefficient of friction of 3D-printed samples from ABS. In all cases, the coefficient of friction in normal operation (after running in) was on average about 0.2. Thus, studies have shown that spare parts for friction machines can be produced using FDM 3D printing technology. The frictional characteristics of these parts will be close to those for the molded plastic with a smooth surface.

References

[1] Kandeva M., Kalitcin Z.H., Kornic P. Ecology for Renovation of Automotive Air Filters / Journal of Enviromental Protection and Ecology. 2017. V.18. №2. 641–651 pp.
[2] Reznic S.V., Prosuntsov P.V., Michailovskii K.V. Prediction of Thermophysical and Thermomechanical Characteristics of Porous Carbon-Ceramic Composite Materials of the Heat Shield of Aerospace Craft / Journal of Engineering Physics and Thermophysics. 2015. V.88. №3. 594–601 pp.
[3] Eitherton J.R., Myers J.R. Agricultural Machine-Related Death / American Journal of Public Health. 1991. V.81. №6. 766–768 pp.
[4] Buyanov I.A., Vdovin D.S. Development of a Design Procedure and Preform Piercing Technology for Carbon-Composite Manufacturing / Polymer Science: Series D. 2017. V.10. №2. 103–105 pp.
[5] Signhal N., Levchenko I., Huang S., Xu L., Potrivitu G.S., Cherkun O., Bazaka K., Xu S., Fang J.
[6] Denisenko N.I., Maslyuk V.A., Yakovenko R.V. Reinforced Powder Chromium Steels and their Use for Hardening of Feed Mill Hammers / Powder Mettallurgy and Metal Ceramics. 2019. V. 57. № 11–12. 740–746 pp.
[7] Lavrov N.A., Kiemov S.N., Kryzhanovskii V.K. Tribotechnical Properties of Composite Materials Based on Epoxy Polymers / Polymer Science: Series D. 2019. V.12. №2. 182–185 pp.
[8] Kobernic N.V., Mikheev R.S., Kalashnokov I.E., Kobeleva L.I., Bolotova L.K. Tribological Properties of Babbitt Alloy Coating Modified with Carbon Nanotubes / Inorganic Materials: Applied Research. 2017. V.8. №3. 428–433 pp.
[9] Fernandes C.M., Martins R.C., Seabra J.H. Coefficient of Friction Equation for Gears Based on a Modified Hersey Parameter / Tribology International. 2016. V.101. 204–217 pp.
[10] Li S., Chu Y. Application Research of Automatic Control Technology in Agriculture Machinery / Revista De La Facultad De Agronomia. 2019. V.36. №4. 1106–1115 pp.

[11] Medelyaev I.A. Wear of Machine Parts with Mixed Lubrication / Russian Engineering Research. 2018.V.38. 956–61 p.

[12] Gorodetskii, M.A., Nelyub, V.A., Malysheva, G.V., Shaulov, A.Y., Berlin, A.A. Technology of Forming and the Properties of Reinforced Composites Based on an Inorganic Binder // Russian Metallurgy (Metally). 2018. Volume 2018, Issue 13. 21–25 pp

[13] Borodulin A.S., Kalinnikov A.N., Beshtoev B.Z., Kharaev A.M., Bazheva R.C., Kvashin V.A. Sunthesis and Performance Characteristics of Superstructure Polyethers /Key Engineering Materials. 2019. V.816. 307–311 pp.

[14] Polyakov S.A., Kuksenova L.I., Alekseeva M.S. Methodologycal Foundations for Choosing the Materials for Gears According to Wear Resistance Criteria / Journal of Machinery Manufacture and Reability. 2019. V.48. № 4. 328–335 pp.

[15] PolezhaevYu.V., Reznic S.V. Advanced Thermal Technologies and Materials / High Temperature. 1998. V.36. №3. 503–506 pp.

[16] Stepashkin A., Chukov, D.I., Senatov, F.S., Salimon, A.I., Korsunsky, A.M., Kaloshkin, S.D. 3D-printed PEEK-carbon fiber (CF) composites: Structure and thermal properties // Composites Science and Technology. 2018. Volume 164. pp. 319–326.

[17] Trifonova O. M., Panova Y.A., Akhmetov I.G. A Study of the Effect of Maleic Anhydride on the Properties of Acrylonitrile Butadiene Sterene Plastic / International Polymer Science ans Technology. 2018. V.45. №6. 251–256 pp.

[18] Lifton V., Lifton G., Simon S. Options for additive rapid prototyping methods (3D printing) in MEMS / Rapid Prototyping Journal. 2014. V.20. No.5, 403–412 pp.

[19] Zukas V., Zukas J. An Introduction to 3D Printing / Published and Distributed by First Edition Design Publishing. 2015. 41 pp.

[20] Stansbury J., Idacavage M. 3D Printing with polymers: Challenges among expending options and opportunities / Dental Materials.2016. V.32. 54–64 pp.

[21] Rinaldi, M., Ghidini, T., Cecchini, F., Brandao, A., Nanni, F. Additive layer manufacturing of poly (ether ether ketone) via FDM // Composites Part B: Engineering. 2018. Volume 145. pp. 162–172

[22] Hoskins, T.J., Dearn, K.D., Kukureka, S.N. Mechanical performance of PEEK produced by additive manufacturing // Polymer Testing. 2018. Volume 70. pp. 511–519

[23] Tuan D., Kashani A., Imbalzano G., Nguyen K., Hui D. Additive manufacturing (3D printing): A review of materials, methods, application and challenges / Composites Part B: Engineering. 2018. V. 143. P. 172–196

[24] Olivera S., Muralidhara H., Venkatesh K., Gopalakrishna K., Vivek C. Plating on acrylonitrile-butadiene-styrene (ABS) plastic: a review / Journal of Materials Science. 2016. V. 51. P. 3657–3674

[25] Letcher T., Rankouhi B., Javadpour S. Experimental Study of Mechanical Properties of Additively Manufactured ABS Plastic as a Function of Layer Parameters / Proceedings Paper. 2016. V.5. 14–22 pp.

[26] Armilotta A. Tolerance Analysis of Gear by Static Analogy / Mechanism and Machine Theory. 2019. V.135. 65–80 pp.

IOP Conf. Series: Materials Science and Engineering **963** (2020) 012033 doi:10.1088/1757-899X/963/1/012033

Propeller optimization for underwater propulsion

V Cheremushkin[1,2] and A Polaykov[1]

[1]Bauman Moscow State Technical University, 5 Second Baumanskaya Street, Moscow, 105005, Russian Federation

[2]E-mail: vcheremushkin@bmstu.ru

Annotation. This article is devoted to the optimization of an underwater propulsion propeller and the identification of the main dependencies in the first stages of optimization. A mathematical model and a parameterized 3d model with indication of variable parameters, as well as optimization criteria, are presented. The used method of numerical hydrodynamic modeling is described. The obtained characteristic dependencies are presented in the form of graphs.

1. Introduction.

The screw propeller is an extremely widely used device, relatively simple and quite effective, used primarily as a watercraft propulsion system [1]. Currently, there are new types of units that use a screw propeller, and serve various purposes. This is also a diving device that allows a person to develop a significant speed under water using a relatively small device.

The possibilities for optimizing propellers have not yet been exhausted, especially with the use of computational hydrodynamics [2, 3]. This work is devoted to the initial stage of optimizing the propeller for underwater propulsion for diving according to various criteria in order to determine which geometric parameters and how they are affected.

2. Mathematical model and methods.

For optimization, a parameterized 3d model of the screw propeller is used. An example of this model is shown in figure 1.

Parameters that changed during the optimization process:

- Inlet blade width
- blade length
- Inlet angles
- Outlet angles
- Change the diameter of the sleeve along the blade (linear)

The search for the optimal solution was carried out in two settings with a restriction in the magnitude of the moment, respectively, and a power of 1 Nm:

- Maximizing axial force on the screw
- Maximizing efficiency and axial force, with a larger linear weight of efficiency in the objective function.

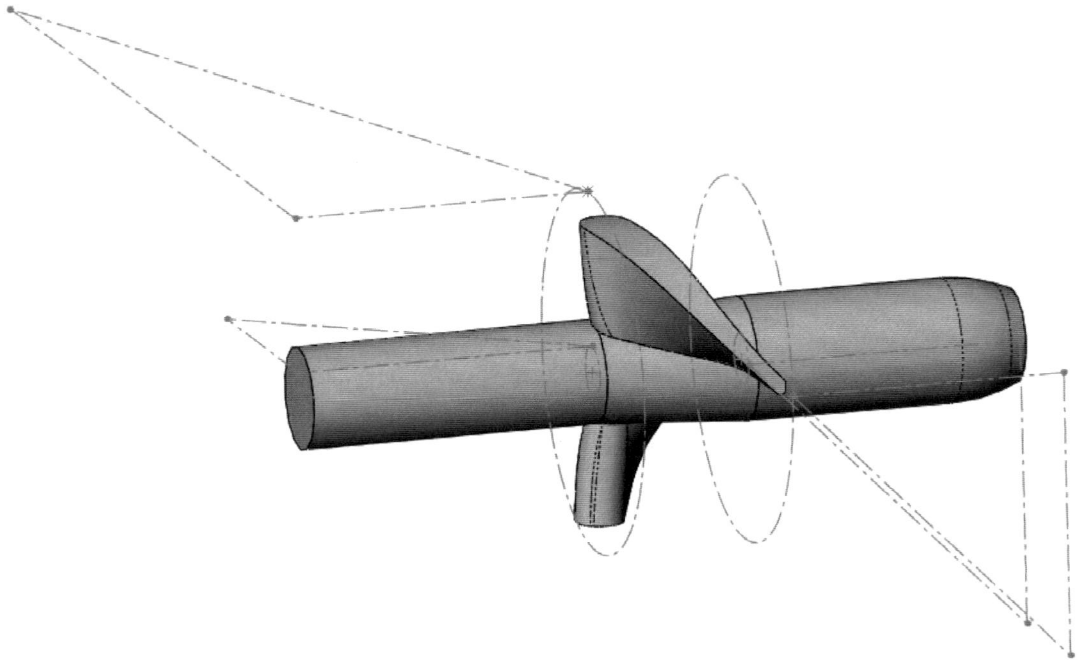

Fig. 1. Parameterized 3d model.

Efficiency is a value determined by the ratio of axial force to moment.

To determine the hydraulic characteristics of the propeller, the method of numerical hydrodynamic modeling was used, based on the solution of discrete analogues of the equations of dynamics of a continuous viscous medium. For the case of an incompressible fluid model, are used the following equations [4, 5, 6]:

Mass conservation equation (continuity equation)

$$\frac{\partial \tilde{u}_j}{\partial x_j} = 0,$$

$\tilde{u}_j$ — the averaged value of the fluid velocity in the projection on the j-th axis (j=1,2,3);

Equation of change in momentum (Reynolds averaging) in an unsteady setting

$$\rho\left[\frac{\partial U_i}{\partial t} + U_j\frac{\partial U_i}{\partial x_j}\right] = -\frac{\partial P}{\partial x_i} + \frac{\partial}{\partial x_i}\left[T_{ij}^{(v)} - \rho u_i u_j\right],$$

Where U, P — averaged speed and pressure;

$\tilde{T}_{ij}^{(v)} = 2\mu\tilde{s}_{ij}$ —Viscous stress tensor for incompressible fluid;

$\tilde{s}_{ij} = \frac{1}{2}\left[\frac{\partial \tilde{u}_i}{\partial x_j} + \frac{\partial \tilde{u}_j}{\partial x_i}\right]$ — Instant strain rate tensor;

$\rho u_i u_j$ — Reynolds stresses.

ρ — liquid density;

μ — dynamic coefficient of liquid viscosity.

Reynolds stress was modeled using the k-ω SST turbulence model, which gives good results when calculating inseparable flows [7, 8, 9, 10].

3. Results and analysis.

The most obvious results of the simulation are the dependences of the axial force and efficiency on the length of the blade, shown in figure 2

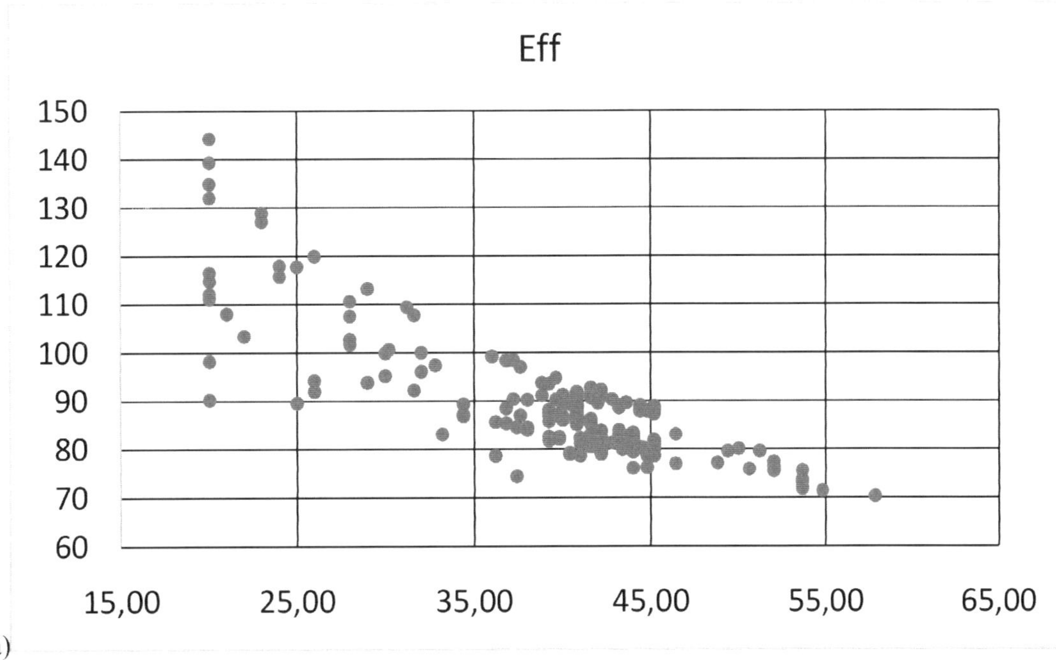

a)

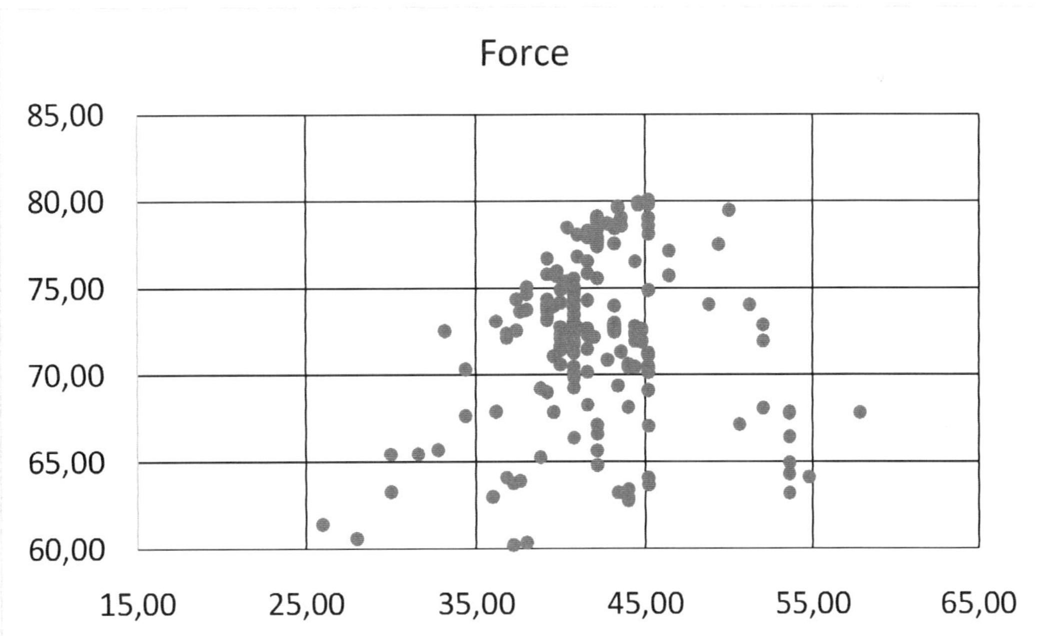

б)

Fig. 2. Depending on the length of the blade: a) efficiency; b) axial force.

It should be noted that these graphs are given taking into account the power limitations (With an increase in the length of more than 45 mm, the axial force can continue to grow, but with an excess of power).

In the given ranges of parameter changes, the achievable screw propeller efficiency tends to decrease clearly, the achievable axial force value, on the contrary, tends to increase. The model, taking into account the specified weights of optimization criteria, has the following characteristics: efficiency — 88.9, axial force — 72.9 N, length — 40.8 mm.

Conclusion

Thus, as a result of the work done, it was shown that for the considered screw propeller, the main parameter affecting the optimization criteria is the blade length. At the same time, the axial force and efficiency have a different tendency depending on the length, and the correct choice of weight coefficients is an important part of the problem of optimizing the screw.

References

[1] Antonenko S. V. Ship engines: a textbook. Far Eastern state technical University. — Vladivostok: Publishing house, 2007. — 126 p.

[2] Kovacevic A., Stosic N., Mujic E., Smith I.K., Gueratto D., Extending the role of computational fluid dynamics in screw machines. Proceedings of the institution of Mechanical Engineers, Vol 225, Issue 2, Pp 83–97.DOI: 10.1177/0954408910397586.

[3] Ding H., Jiang Y., CFD simulation of a screw compressor with oil injection. IOP Conference series: Materials Science and Engineering, Vol 232, Issue 1. DOI: 10.1088/1757-899X/232/1/012020.

[4] Ma, X.,Zhang, R., Xiao, C.Application of disk friction pump in principal axis seal of hydroelectric generator. Shuili Fadian Xuebao/Journal of Hydroelectric Engineering, Vol 25, Issue 1, 2006, Pp 124-126+113.

[5] Lomakin, V., Cheremushkin, V.,Chaburko, P. Investigation of vortex and hysteresis effects in the inlet device of a centrifugal pump. 2018 Global Fluid Power Society PhD Symposium, GFPS 2018, 2018, №8472374. DOI: 10.1109/GFPS.2018.8472374

[6] Lomakin, V.O., Chaburko, P.S., Kuleshova, M.S. Multi-criteria Optimization of the Flow of a Centrifugal Pump on Energy and Vibroacoustic Characteristics // Procedia Engineering. Vol. 176, 2017, pp 476–482. DOI: 10.1016/j.proeng 2017.02.347.

[7] Lomakin, V.O. Investigation of two-phase flow in axial-centrifugal impeller by hydrodynamic modeling methods. Proceedings of 2015 International Conference on Fluid Power and Mechatronics, FPM 2015, 2015, № 7337302, Pp 1204-1206. DOI: 10.1109/FPM.2015.7337302

[8] Gouskov, A.M., Lomakin, V.O., Banin, E.P.,Kuleshova, M.S. Minimization of Hemolysis and Improvement of the Hydrodynamic Efficiency of a Circulatory Support Pump by Optimizing the Pump Flowpath. Biomedical Engineering, Vol 51, 2017, Pp 229–233. DOI: 10.1007/s10527-017-9720-9

[9] V. Cheremushkin and A. Polyakov 2019 Optimization of the output device of a disk pump for high viscous fluid. IOP Conf. Series: Materials Science and Engineering 589 (2019). doi:10.1088/1757-899X/589/1/012001

[10] Cheremushkin V.A., Petrov A.I., Chaburko P.S. Using Stator Blades in Accessory Tracts of Hermetic Pumps // Mashiny I Ustanovki: Proektirovanie, Razrabotka I Ekspluataciya. 2017, no. 2, pp 1–12

Development of an algorithm for determining the main quality indicators of the transition process on the example of a LUDV hydraulic system

S Godorozha[1], A Protopopov[1] and M Ryabinin[1,2]

[1]Bauman Moscow State Technical University, 5 Second Baumanskaya Street, Moscow, 105005, Russian Federation

[2]E-mail: ryabinin@bmstu.ru

Abstract. When optimizing hydraulic systems, various algorithms are used, which are based on the control quality indicators selected as target functions. This is why there is a need to automate the calculation of these indicators. The article considers the main indicators of the quality of the transition process, a mathematical model of the LUDV hydraulic system is compiled and its calculation is performed in two mathematical modeling packages in order to confirm the accuracy of the initial data. An algorithm for determining the main quality indicators of the transition process has been developed, which makes it possible to simplify and speed up the optimization process of the considered LUDV-hydraulic system.

Introduction

In the literature [1]–[14] when optimizing hydraulic systems, various algorithms are used that rely on control quality indicators selected as target functions. Therefore, there is a need to automate the calculation of these indicators.

According to sources [15]–[16], the main indicators (assessments) of the quality of the transition process can be divided into two groups:

direct assessment;

indirect (integral) estimates.

Direct estimates include (Fig. 1):

transition time t_p;

time to reach the first maximum t_m;

time of first approval t_i;

the maximum dynamic error

$$\delta_{max} = x_{max} - x_{\infty};$$

maximum overshoot

$$\sigma_{max} = \frac{x_{max} - x_{\infty}}{x_{\infty}} \cdot 100\%,$$

and the maximum dynamic error δ_{max} and maximum overshoot σ_{max} together, they determine the assessment of " oscillation».

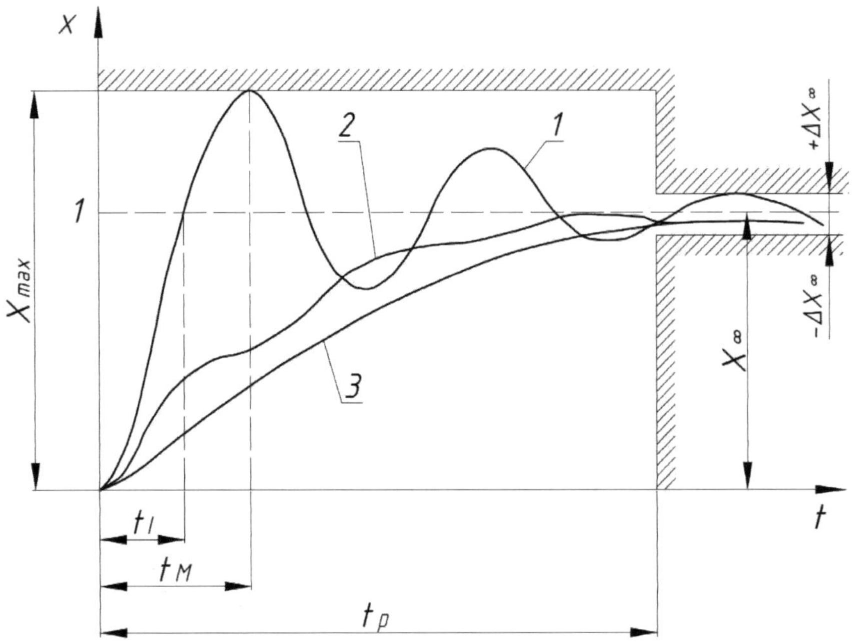

Figure 1. Main types of transients
1 — oscillatory; 2 — monotonic; 3 — aperiodic

Among indirect (integral) estimates, the most widely used ones are:

- linearintegralestimation I_1

$$I_1 = \int_0^\infty \delta(t)\,dt,$$

where $\delta(t)$ — the deviation of the controlled variable $X(t)$ from the set value X_∞,

$$\delta(t) = X_\infty - X(t);$$

- quadraticintegralestimation I_2

$$I_2 = \int_0^\infty \delta^2(t)\,dt;$$

- ITAE (Integral of Time and Absolute Error) I_3

$$ITAE = \int_0^\infty t \cdot |\delta(t)|\,dt;$$

- modifiedquadraticestimation I_4

$$I_4 = \int_0^\infty t \cdot \delta^2(t)\,dt.$$

These estimates are indirect because their absolute value is not related to numerical indicators of the quality of the transition process. They can serve as a generalized indicator when comparing two or more variants of systems: the variant in which the value of the integral estimate is less is considered the best.

The mathematical model of the system assumes a certain type of transition process of the controlled variable. The integral indicator is selected in such a way that it better characterizes the quality of the system. In other words, there are recommendations for selecting integral indicators for each type of transition process.

Let's look at some of the recommendations: the linear integral estimate I_1 can only be used for obviously non-oscillatory (aperiodic) processes. For oscillatory transients, the I_2 estimate is used. However, the choice of variable parameters at the minimum I_2 usually gives an oscillatory transition process with a rather large oscillation. In cases where such a process is unacceptable, we proceed to the integral criteria I_3 and I_4.

Mathematical model of the system

As an example, let's consider the definition of the main indicators of the quality of regulation on the drive of two hydraulic cylinders with a LUDV hydraulic system (Fig. 2).

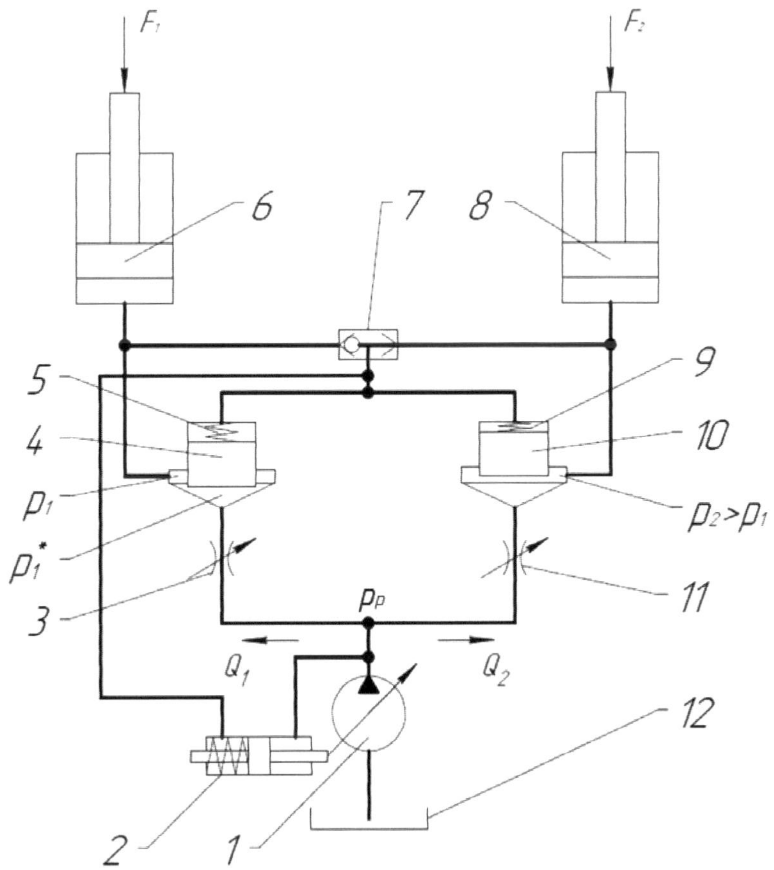

Fig. 2. Schematic diagram of a hydraulic drive with two hydraulic cylinders
1 — adjustable pump; 2 — flow regulator for differential pressure; 3, 11 — control spool; 4, 10 — pressure compensator; 5, 9 — stabilizing spring of low rigidity; 6, 8 — Executive hydraulic cylinder; 7 — valve " or»; $p_p - p_1^* = p_p - p_2 = \Delta p_{regulator}$; $p_{LS} = p_{max} = p_2$.

The operation of the system can be described by the following equations:

$$
\begin{cases}
Q_{fg} = Q_{dg} + A_{cp} \cdot \dfrac{dy_{cp}}{dt} + \dfrac{W_c}{B} \cdot \dfrac{dp_c}{dt}; \\
Q_p = Q_1 + Q_2 + \dfrac{W_p}{B} \cdot \dfrac{dp_p}{dt}; \\
p_p \cdot a_s = p_{LS} \cdot a_s + C_s \cdot \left(x_s + x_{so}\right) + m_s \cdot \dfrac{d^2 x_s}{dt^2} + K_s \cdot \dfrac{dx_s}{dt}; \\
p_c \cdot A_{cp} = p_p \cdot A_r + C_{cp} \cdot \left(y_0 + y_{cp}\right) + M_{cp} \cdot \dfrac{d^2 y_{cp}}{dt^2} + K_{cp} \cdot \dfrac{dy_{cp}}{dt}; \\
\left(p_{cs1} - p_T\right) \cdot a_{s1} = C_{s1} \cdot \left(x_{s1o} + x_{s1}\right) + m_{s1} \cdot \dfrac{d^2 x_{s1}}{dt^2} + K_{s1} \cdot \dfrac{dx_{s1}}{dt}; \\
\left(p_{cs2} - p_T\right) \cdot a_{s2} = C_{s2} \cdot \left(x_{s2o} + x_{s2}\right) + m_{s2} \cdot \dfrac{d^2 x_{s2}}{dt^2} + K_{s2} \cdot \dfrac{dx_{s2}}{dt}; \\
\left(p_1^* - p_{LS}\right) \cdot a_{c1} = C_{c1} \cdot \left(x_{c1} + x_{co}\right) + m_{c1} \cdot \dfrac{d^2 x_{c1}}{dt^2} + K_{c1} \cdot \dfrac{dx_{c1}}{dt}; \\
\left(p_2^* - p_{LS}\right) \cdot a_{c2} = C_{c2} \cdot \left(x_{c2} + x_{co}\right) + m_{c2} \cdot \dfrac{d^2 x_{c2}}{dt^2} + K_{c2} \cdot \dfrac{dx_{c2}}{dt}; \\
Q_1 = \mu \cdot x_{c1} \cdot b_{c\Sigma} \cdot \sqrt{\dfrac{2}{\rho}} \cdot \sqrt{\left(p_1^* - p_1\right)} + \dfrac{W_{c1}}{B} \cdot \dfrac{dp_1^*}{dt} + a_{c1} \cdot \dfrac{dx_{c1}}{dt}; \\
Q_2 = \mu \cdot x_{c2} \cdot b_{c\Sigma} \cdot \sqrt{\dfrac{2}{\rho}} \cdot \sqrt{\left(p_2^* - p_2\right)} + \dfrac{W_{c2}}{B} \cdot \dfrac{dp_2^*}{dt} + a_{c2} \cdot \dfrac{dx_{c2}}{dt}; \\
p_1 \cdot A_{p1} = F_1(t) + M_1 \cdot \dfrac{d^2 y_1}{dt^2} + K_{p1} \cdot \dfrac{dy_1}{dt} + 0,05 \cdot p_1 \cdot A_{p1}; \\
p_2 \cdot A_{p2} = F_2(t) + M_2 \cdot \dfrac{d^2 y_2}{dt^2} + K_{p2} \cdot \dfrac{dy_2}{dt} + 0,05 \cdot p_2 \cdot A_{p2}; \\
\mu \cdot x_{c1} \cdot b_{c\Sigma} \cdot \sqrt{\dfrac{2}{\rho}} \cdot \sqrt{\left(p_1^* - p_1\right)} = \dfrac{W_1}{B} \cdot \dfrac{dp_1}{dt} + A_{p1} \cdot \dfrac{dy_1}{dt}; \\
\mu \cdot x_{c2} \cdot b_{c\Sigma} \cdot \sqrt{\dfrac{2}{\rho}} \cdot \sqrt{\left(p_2^* - p_2\right)} = \dfrac{W_2}{B} \cdot \dfrac{dp_2}{dt} + A_{p2} \cdot \dfrac{dy_2}{dt}.
\end{cases}
$$

where Q_{fg} — flow rate of the working fluid through the filling slot TSR of the flow controller, m^3/s (Fig. 3);

Q_{dg} — working fluid flow through the drain slot TSR flow regulator, m^3/s;

A_{cp} — area of the piston cavity of the control cylinder, m^2;

y_{cp} — moving the control piston of the regulator, m;

W_c — volume of the cavity in the control line of the controller, m^3;

B — interval reduced volumetric modulus of elasticity of a liquid, $B = 10^9\ Pa$;

p_c — pressure in the control line of the regulator, Pa;

Q_p — pump flow, m^3 / c;

Q_1, Q_2 — the cost of the working fluid to consumers, m^3 / c;

W_p — volume of the pressure line cavity, m^3;

p_p — pump pressure, Pa;

p_{LS} — line pressure LS, $p_{LS} = \max(p_1; p_2)$, p_1, p_2 — pressure caused by the load on the hydraulic cylinder rods, Pa;

a_s — the area of the end face of the spool, m^2

C_s — spring stiffness of the regulator spool, N/m;

x_{s0} – preload of the regulator spool spring, m;

m_s — the mass of the spool, kg;

K_s — coefficient of viscous friction of the regulator spool, N·c/m;

x_s — offset of the regulator spool relative to the neutral position, m;

A_r — area of the control cylinder rod cavity, m^2;

C_{cp} — spring stiffness of the control piston, N/m;

y_0 — preload of the control piston spring, m;

M_{cp} — the reduced mass of the piston, kg;

K_{cp} — coefficient of viscous friction of the control piston, N·c/m;

p_{cs1}, p_{cs2} — pressure control the test section, Pa;

p_T — pressure in the drain line, Pa;

a_{s1}, a_{s2} — area of the spool end of the working section, m^2;

C_{s1}, C_{s2} – stiffness of the return spring of the spool of the working section, N/m;

x_{s10}, x_{s20} — preload of the spool return spring, m;

x_{s1}, x_{s2} — displacement of the spool of the working section from the neutral position, m;

m_{s1}, m_{s2} — weight of the spool of the working section, kg;

K_{s1}, K_{s2} — coefficient of viscous friction of the spool of the working section, N·c/m;

p_1^*, p_2^* — pressure in front of the compensator, Pa;

a_{c1}, a_{c2} — the area of the end face of the compensator, m^2;

C_{c1}, C_{c2} — compensator spring stiffness, N/m;

x_{c1}, x_{c2} — offset of the compensator from the neutral position, m;

x_{c0} — preload of the compensator spring, m;

m_{c1}, m_{c2} — compensator weight, kg;

K_{c1}, K_{c2} — coefficient of viscous friction of the compensator, N·c/m;

μ — flow coefficient;

$b_{c\Sigma}$ — total width of the compensator's throttling windows, m;

ρ — the density of working fluid, kg/m^3;

W_{c1}, W_{c2} — volume of the cavity in front of the compensator, m^3;

A_{p1}, A_{p2} — hydraulic cylinder piston area, m^2;

$F_1(t), F_2(t)$ — force on the hydraulic cylinder rod, N;

M_1, M_2 — reduced weight of the hydraulic cylinder piston, kg;

y_1, y_2 — displacement of the hydraulic cylinder from the extremely retracted position, m;

K_{p1}, K_{p2} — the coefficient of viscous friction of the hydraulic cylinder piston (the calculation of the coefficient has a number of specific features and is not considered in this paper), N·c/m;

W_1, W_2 — volume of the cavity in front of the hydraulic cylinder, m^3.

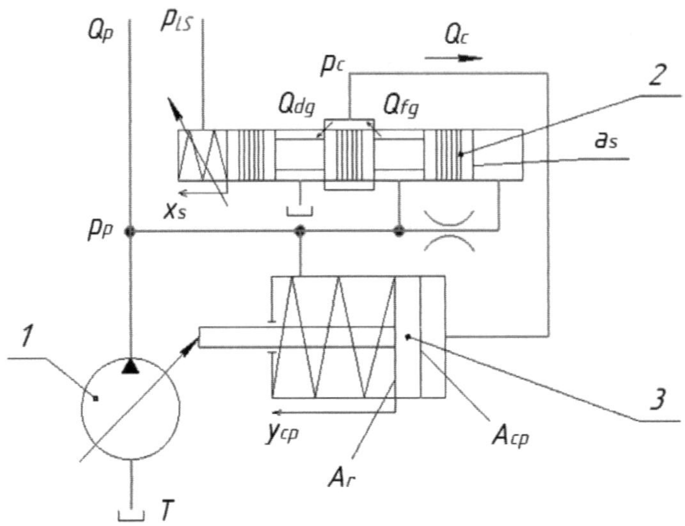

Fig. 3. Pump regulator diagram

Result of calculation. Data analysis

To confirm the accuracy of the source data, the system of differential equations describing the operation of the drive was solved in the VisualStudio integrated development environment in C++ and in the MatLAB/Simulink structural modeling package. The Runge-Kutta method was used to solve both cases.

The input step effect in the calculation is the area of the filling slots of the spool distributors. The output signal is the speed of the hydraulic cylinder rods. Disturbing effects are forces on the rods of hydraulic cylinders.

The calculation results are presented as graphs (Fig. 4, 5). It is worth noting that the Runge-Kutta method does not allow solving the system of nonlinear differential equations described above with a time step of more than 10^{-7}s.

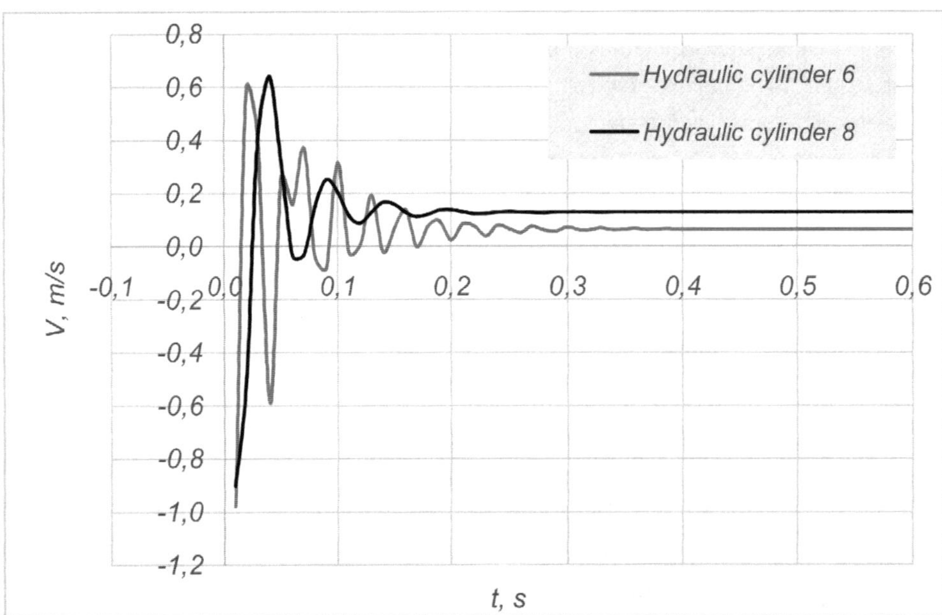

Fig. 4. Graph of changes in the speed of the hydraulic cylinder rod over time (VisualStudio)

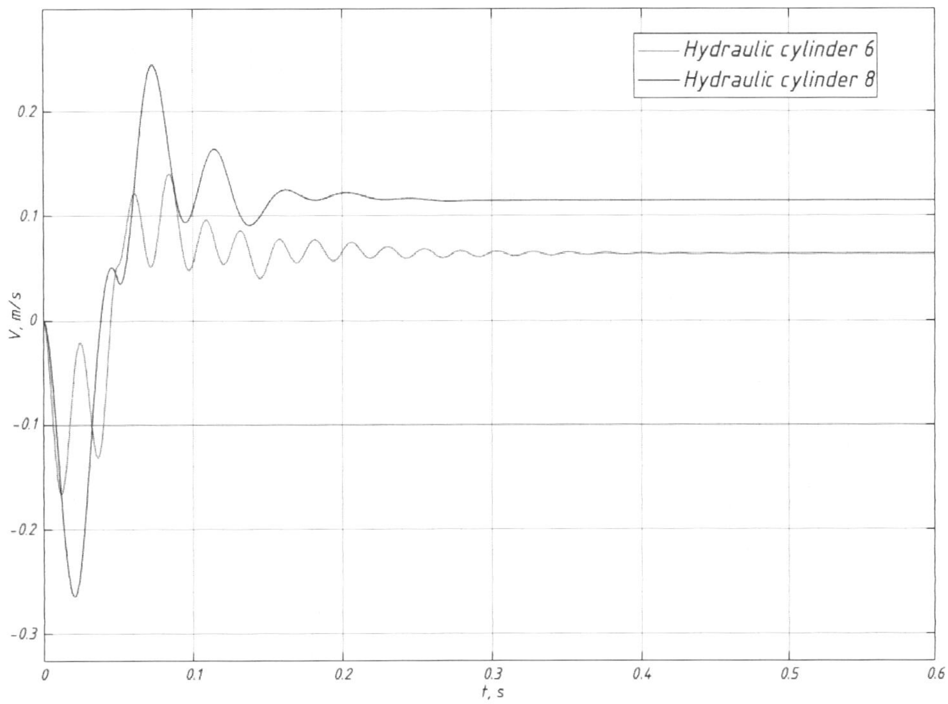

Fig. 5. Graph of changes in the speed of the hydraulic cylinder rod over time (MatLAB/Simulink)

The dependencies shown in figures 4 and 5 are almost identical: they have the same indicators and type of transition process.

Processing of calculation results. Defining the main indicators of transition quality

So, the dependences obtained from the calculation have the form of an oscillatory transition process. In accordance with the recommendations presented above, it is advisable to choose an integral assessment of ITAE.

The definite integral in the expression for the indicator ITAE is calculated using the Simpson formula for a uniform grid, which has high accuracy in comparison with other numerical methods for calculating the definite integral.

The results of calculating the above indicators for the previously obtained dependences of the speed of the hydraulic cylinder rod on time (Fig. 4) are shown in figures 6-9, where z[1] is the speed of the hydraulic cylinder rod 6, z[2] is the speed of the hydraulic cylinder rod 8.

```
Transient time for z[i]:

z[1] : 0.330003
z[2] : 0.200002
```

Fig. 6. During the transition process

```
Maximum dynamic error for z[i]:

z[1] : 0,524638 (при t = 0,020000)
z[2] : 0,513465 (при t = 0,040000)
```

Fig. 7. Maximum dynamic error

```
Maximum overshoot for z[i], :

z[1] : 815,768819 (при t = 0,020000)
z[2] : 401,435332 (при t = 0,040000)
```

Fig. 8. Maximum overshoot

```
ITAE score for z[i], :

z[1] : 0,002624
z[2] : 0,001305
```

Fig. 9. ITAE indicator

Conclusion

The system calculation results obtained in various mathematical modeling packages (VisualStudio and MatLAB/Simulink) are identical. The developed algorithm for determining the main quality indicators of the transition process will optimize the components included in the system to improve the quality of regulation.

References

[1] N Sosnovsky and D Ganieva 2019 *IOP Conf. Ser.: Mater. Sci. Eng.***589** 012016
[2] I Kolodin and M Ryabinin 2019 *IOP Conf. Ser.: Mater. Sci. Eng.***589** 012018
[3] V Brusov et al 2020 IOP Conf. Ser.: Mater. Sci. Eng. 779 012020
[4] Y Vul et al 2020 IOP Conf. Ser.: Mater. Sci. Eng. 779 012021
[5] B Kulakov and D Kulakov 2020 IOP Conf. Ser.: Mater. Sci. Eng. 779 012027
[6] D Vdovin et al 2019 IOP Conf. Ser.: Mater. Sci. Eng. 589 012025
[7] B Kulakov and D Kulakov 2019 IOP Conf. Ser.: Mater. Sci. Eng. 589 012036
[8] Wang, X., Tong, S., Ge, J., & Zhang, J. (2014). Study on dynamic performance of EFMC system in hydraulic excavator. ZhongguoJixieGongcheng/China Mechanical Engineering, 25(15), 2030-2037. doi:10.3969/j.issn.1004-132X.2014.15.009
[9] Ai, C., Lin, J., Kong, X., & Zhao, X. (2016). Design and research of new large flow three-way pressure compensation valves. GaojishuTongxin/Chinese High Technology Letters, 26(5), 498-503. doi:10.3772/j.issn.1002-0470.2016.05.010
[10] Yang, X., Zhao, X., Ai, C., Peng, G., Wang, Z., Dou, J., . . . Li, Y. (2016). Design and research of new large flow three-way pressure compensation valve. Paper presented at the AUS 2016 - 2016 IEEE/CSAA International Conference on Aircraft Utility Systems, 967-972. doi:10.1109/AUS.2016.7748194
[11] Polishchuk, M., Suyazov, M., &Opashnyansky, M. (2020). Study on numerical analysis of dynamic parameters of mobile walking robot. Journal of Mechanical Engineering and Sciences, 14(1), 6380-6392. doi:10.15282/jmes.14.1.2020.14.0499
[12] Zhang, C. (2020). PD plus dynamic pressure feedback control for a direct drive stewart manipulator. Energies, 13(5) doi:10.3390/en13051125
[13] Zhao, B., Guo, W., Ge, L., Hao, Y., &Quan, L. (2020). Experiment study on operation characteristics of new flow self-balancing pump controlled asymmetric hydraulic cylinder. JixieGongchengXuebao/Journal of Mechanical Engineering, 56(8), 257–264. doi:10.3901/JME.2020.08.257
[14] Wang, X., Ge, L., Zhao, B., Hao, Y., Quan, L., & Mu, X. (2020). Energy efficiency characteristics of cable shovel lifting system driven by hydraulic-electric hybrid system. NongyeJixieXuebao/Transactions of the Chinese Society for Agricultural Machinery, 51(4), 418–426. doi:10.6041/j.issn.1000-1298.2020.04.049

[15] Voronov A.A., Titov V.K., Novogranov B.N. Fundamentals of the theory of automatic regulation and control. Textbook for universities. M. "Higher School", 1977
[16] Popov D. N. Dynamics and regulation of hydraulic and pneumatic systems. Textbook for engineering universities. M., "Engineering", 1976. — 424 p.

IOP Conf. Series: Materials Science and Engineering **963** (2020) 012035 doi:10.1088/1757-899X/963/1/012035

Development Prospects of High-Vacuum Mechanical Pumps Using Magnetic Bearing Support

A Ochkov[1,2] and M Hamadi[1]

[1]Bauman Moscow State Technical University, 5 Second Baumanskaya Street, Moscow, 105005, Russian Federation

[2]E-mail: aochkov@bmstu.ru

Abstract. The analysis of existing high-vacuum mechanical pumps with magnetic suspensions is carried out. The schemes and algorithms used for calculations for permanent magnetic suspensions (PMS) and active magnetic suspensions (AMS) are presented. Based on the results of the analysis, the relevance of the use of magnetic suspensions in high-vacuum mechanical pumps was evaluated, the development prospects of high-vacuum mechanical pumps with magnetic suspensions were noted.

INTRODUCTION

Nowadays, mechanical vacuum pumps are increasingly being used, the operating principle of which is based on the movement of gas from the pumped volume due to the mechanical movement of pump parts. [1–5] High-vacuum pumps of this type include: molecular, turbomolecular (hereinafter TMP), and molecular-viscous vacuum pumps.

For most mechanical vacuum pumps, a rotating rotor is an important part of the pumpingprocess. To ensure the rotation of the rotor, supports are needed that will take the load arising in this process. Rolling bearings or plain bearings are usually used as such supports, but often they cannot cope with the extreme requirements, for example, work under vacuum conditionsor durability at high speeds. Such arising problems can be solved by using an active magnetic suspension (hereinafter AMS) or a permanent magnetic suspension (hereinafter PMS) as bearing support.

OPERATING PRINCIPLE OF AMS

The operating principle of AMS is based on the creation of the necessary effect of magnetic forces from the support site on the suspended body (Fig.1). A displacement sensor measures the displacement of a suspended body from a given equilibrium position. The measurement signal is processed by the regulator. A power amplifier, powered by an external power source, converts this signal into a control current in the electromagnetwinding, which induces magnetic adhesion in such a way that the disturbed equilibrium is restored. The stability of such a magnetic suspension depends on the selected control law.

Thus, we get that structurally AMS consists of two main parts (Fig. 2):

- electromechanicalpart;
- electroniccontrolsystem.

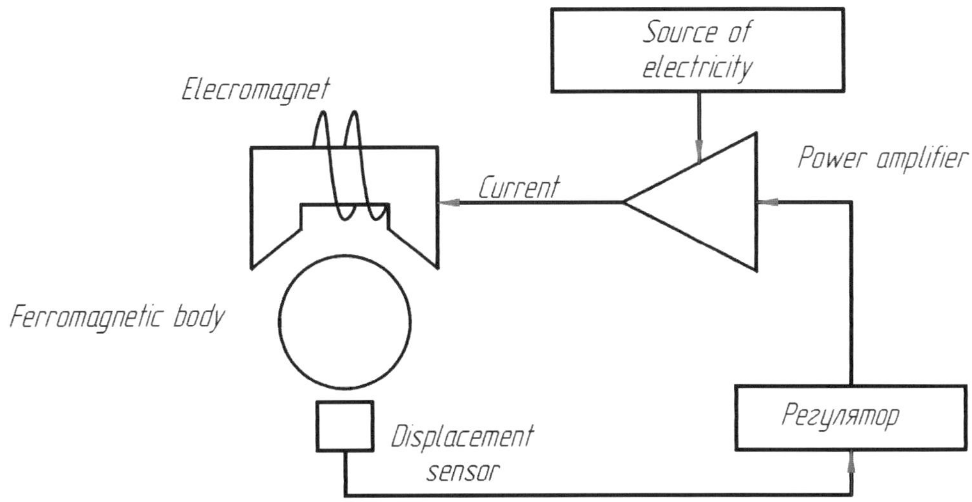

FIGURE.1. Operating principle and functional parts of AMS.

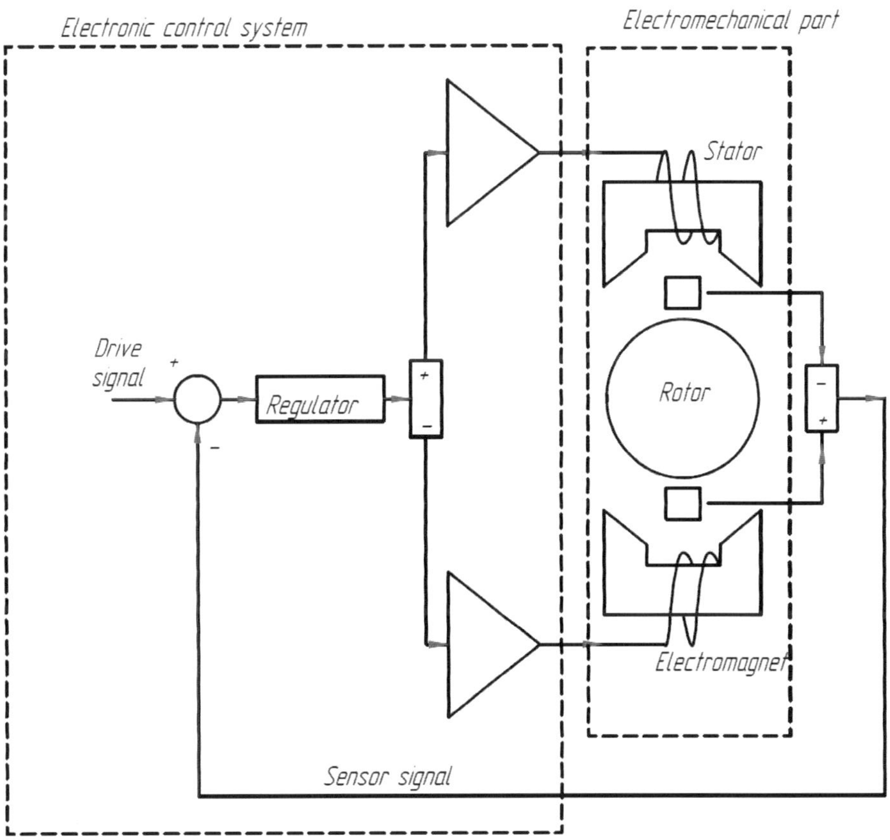

FIGURE.2. Structure diagram of AMS.

The electromechanical part includes a rotor suspended in a magnetic field, electromagnets fixed to the stator, and rotor position sensors. The rotor and stator do not interact mechanically. In case the suspension is disconnected, or in case of an emergency failure in the control system, the rotor rests on auxiliary bearings. These bearings are often rolling bearings, they are installed with a gap, which excludes their rotation when the AMS is connected.

The electronic control system consists of a regulator and a power amplifier. Its main functions are to ensure a stable position of the rotor and its axial alignment in the gap. This is done by processing signals from the rotor position sensors. Structurally, the electronic control system is an electronic unit connected by cables with magnetic bearings and a power supply.

OPERATING PRINCIPLE OF PMS

In magnetic suspensions of this type, permanent magnets or DC electromagnets are used to maintain the rotor in a suspended state, i.e. there is no function of regulating the magnetic field in the suspension. It is customary to make the suspended body partially or completely from a ferromagnetic material; it can also carry permanent magnets.The suspension of the body is carried out using magnetic forces of repulsion and adhesion.

Here is the scheme of radial magnetic bearing on permanent magnets with possible rotor shifts from the equilibrium position (Fig. 3).

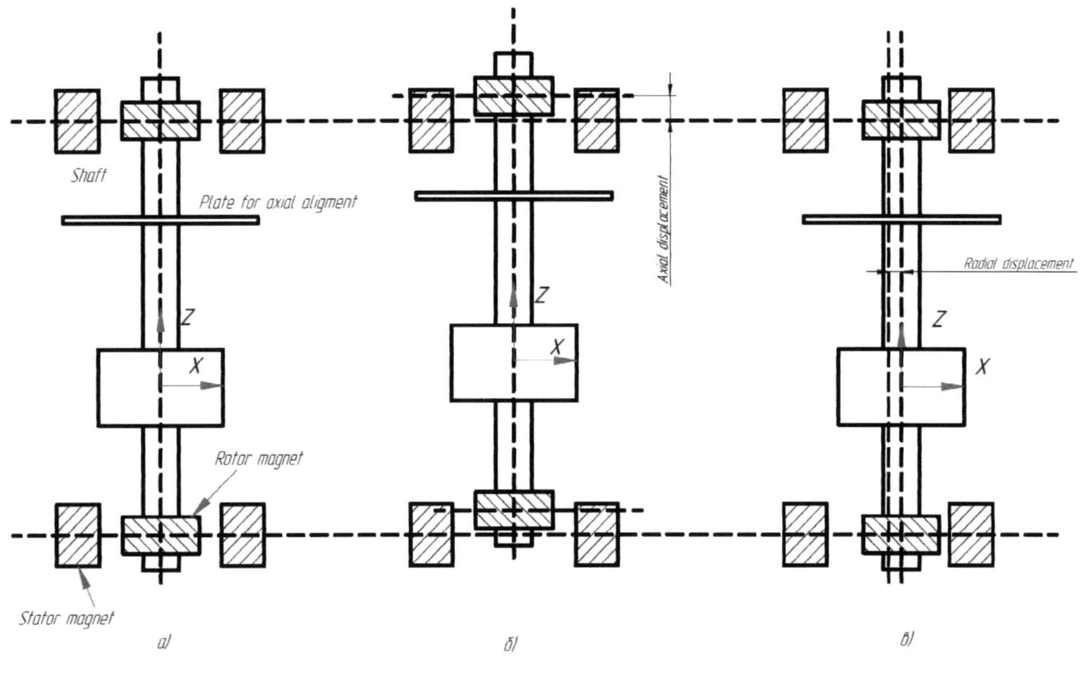

FIGURE. 3. Radial magnetic bearing circuit with possible rotor shifts from equilibrium position: a) no deviation; b) z-axis deviation; c) x-axis deviation.

In this design, the axial stability of the rotor is ensured by a plate that is put on the shaft. The rotor magnets interact with the stator magnets to create magnetic forces that hold the rotor in suspension.

ANALYSIS OF EXISTING CALCULATION ALGORITHMS TO DETERMINE THE MAIN CHARACTERISTICS OF MAGNETIC BEARINGS

In works [6–18], algorithms for calculating the main parameters of a radial magnetic bearing in AMS are considered. The presented algorithms are similar to each other; in general, their content can be shown in the block diagram (Fig. 4).

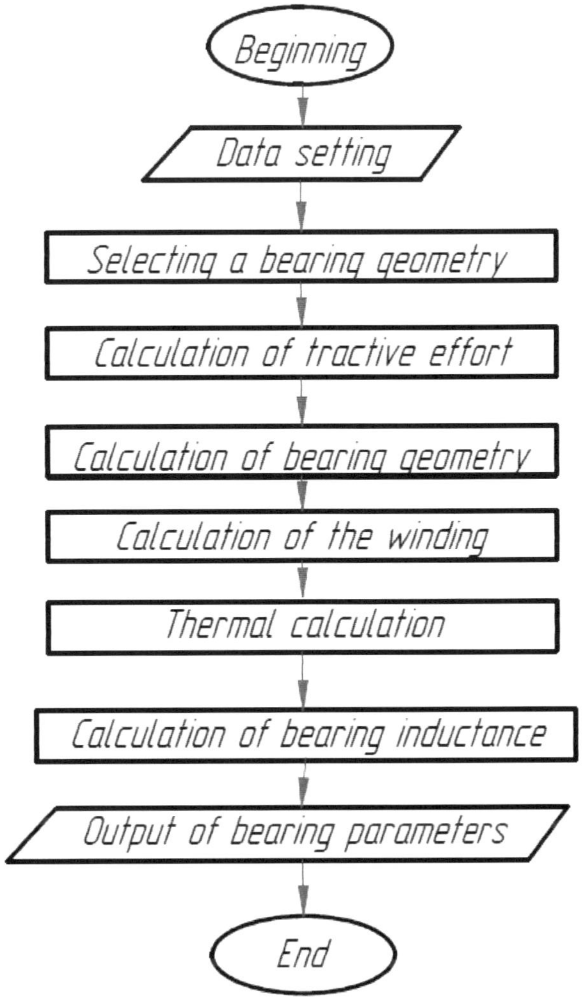

FIGURE. 4. Block diagram of the general algorithm of AMS parameter determination.

The first step is to calculate the tractive effort. This is such a force that can be developed for a long time without overheating the coil. To do this, it is necessary to determine the magnetic force of adhesion acting on the element of the area of the ferromagnetic body in a uniform magnetic field with induction B in the gap:

$$dF = \frac{B^2}{2 \cdot \mu_0} \cdot dA, (0.1)$$

where $\mu_0 = 4\pi * 10^{-7} \frac{H}{m}$ — is themagnetic constant.

The general formula for the tractive effort is

$$F = \frac{B^2}{2 \cdot \mu_0} \cdot \frac{p \cdot a \cdot l \cdot d}{4} \cdot \sin(\frac{\Delta\alpha}{2}), (0.2)$$

where$a$ — is the coefficient of pole number, p — is the polenumber, $\Delta\alpha$ — is the pole angle.

Then the geometric characteristics of the bearing are calculated. This point is necessary to obtain the smallest required volume occupied by the bearing, in which the maximum tractive effort is realized.ThevalueofthebearingcapacityF_{max}(tractive effort in this case) must be greater than the static load of the bearings Q, which is often known precisely or approximately. In calculating, one should take into account the influence of dynamic loads; for this, under standard operating conditions, a safety factor is introduced, varying in the range from 1.5 to 2:

$$\frac{F_{\max}}{Q} = 1{,}5...2. \quad (0.3)$$

The initial problem in calculating the bearing geometry is as follows: let the outer diameter D and the pack length l be given. Let the gap δ and the permissible current density j_{max}be also given. It is required to find the trunnion diameter d and the pole width t at which the tractive effort and induction in the gap will be maximum.

The solution to this problem begins with determining the magnetic induction in the gap. It is determined from Ampere's law for a magnetic circuit. To simplify the task, we neglect the magnetic resistance of steel:

$$j_{\max} \cdot k_{Cu} \cdot A = \frac{B}{\mu_0} \cdot 2 \cdot \delta, \quad (0.4)$$

where k_{Cu} — is the filling area of groove A with copper, A— is the area occupied by the winding inthe groove:

$$A = (\frac{D}{2} - \frac{d}{2} - t) \cdot (\frac{\pi d}{p} - t), \quad (0.5)$$

where$p$ — is the pole number.

Hence follows the expression for the induction in the gap:

$$B(d,t) = C_B \cdot (\frac{D}{2} - \frac{d}{2} - t) \cdot (\frac{\pi d}{p} - t), \quad (0.6)$$

And also, the expression for the tractive effort:

$$F(d,t) = C_F C_B{}^2 \cdot (\frac{D}{2} - \frac{d}{2} - t)^2 \cdot (\frac{\pi d}{p} - t)^2 \cdot t, \quad (0.7)$$

where$C_F$andC_B — constants, are defined by ratios:

$$C_F = \frac{p \cdot a \cdot l}{8 \cdot \mu_0}; \quad (0.8)$$

$$C_B = \frac{\mu_0 \cdot k_{Cu} \cdot j_{\max}}{2 \cdot \delta}. \quad (0.9)$$

The problem of calculating the geometry of a radial active magnetic bearing is mathematically reduced to a problem for a conditional extremum: it is necessary to find the maximum of the function $F(d,t)$with the additional condition $B=B_{max}$, which can be written as an equation:

$$\varphi(d,t) = (\frac{D}{2} - \frac{d}{2} - t) \cdot (\frac{\pi d}{p} - t) - \frac{B_{\max}}{C_B}, \quad (0.10)$$

The method for solving such a problem is known. The function $\Phi(d,t)=F(d,t)+\lambda\varphi(d,t)$, where λ is the Lagrange multiplier, is formed. Three unknowns d, t and λ are found from the joint solution of equation (1.10) and two equations:

$$\frac{\partial\Phi(d,t)}{\partial d}=0, \frac{\partial\Phi(d,t)}{\partial t}=0. (0.11)$$

In a radial active magnetic bearing, calculated by this method, the inner diameter d must be more than half of the outer diameter D. The pole width t, depending on the ratio between the selected values of the parameters B_{max}, j_{max} and δ can vary within certain limits [1].

After determining the geometric parameters of the bearing, it is necessary to calculate the winding. Each electrical circuit contains $p/4$pole coils. The number of circuits is equal to the number of electromagnets. The calculation is carried out under the condition that the maximum current i_{max} causes the maximum tractive effortF_{max} and the maximum magnetomotive force:

$$2\cdot n\cdot i_{\max}=j_{\max}\cdot k_{Cu}\cdot A, (0.12)$$

where$n$— is the number of windings per pole. At known i_{max}, we find this number of turns. Next comes the following sequence for determining the required values.

Conductor cross-section area

$$\alpha_0=\frac{i_{\max}}{j_{\max}}. (0.13)$$

The Ohmic resistance of the coil

$$R=\frac{\rho\cdot n\cdot l_m}{\alpha_0}. (0.14)$$

Heat loss power of one electromagnet

$$P_1=\frac{i^2\cdot R\cdot p}{4}. (0.15)$$

The task of thermal calculation is to determine the maximum temperature of the coil conductor and compare it with the allowable temperature for the used insulation class. The calculation is based on Ohm's law for a steady heat flux:

$$\Delta T=P\cdot R_T, (0.16)$$

where ΔT —isthe temperature difference on thermal conductor; P— is the dissipation power; R_T — is the thermal resistance.

The last step in the algorithm is to determine the inductance of the bearing. At this stage, we derive the expressions for self-inductances and mutual inductance as functions of the rotor displacement x and y. They (expressions) will help in describing electromagnetic processes in AMS.

The improved calculation method is essentially similar to the previously analyzed one, however, the stage of calculating the bearing geometry includes determining the geometric parameters of the bearing along the pack length (Fig. 5). Such an additional point of calculation helps us to avoid those cases in which the values of the pack length obtained by the general method may exceed the required ones.

In this technique, at the beginning of the calculation algorithm, the pack length l is set equal to zero and is increased step by step with a step equal to the error of the equipment on which the magnetic bearing will be manufactured. The pack length increases according to the law:

$$l=l+l_{pr}, (0.17)$$

where$l$— isthepacklength, l_{pr} — is the error of the device on which the magnetic bearing will be manufactured, as long as the condition is wrong:

$$F_{tuag}\geq F_{\max}, (0.18)$$

whereF_{tuag} — is the bearing tractive force. Once this condition is met, the calculation moves on to the next step — the "winding calculation" step.

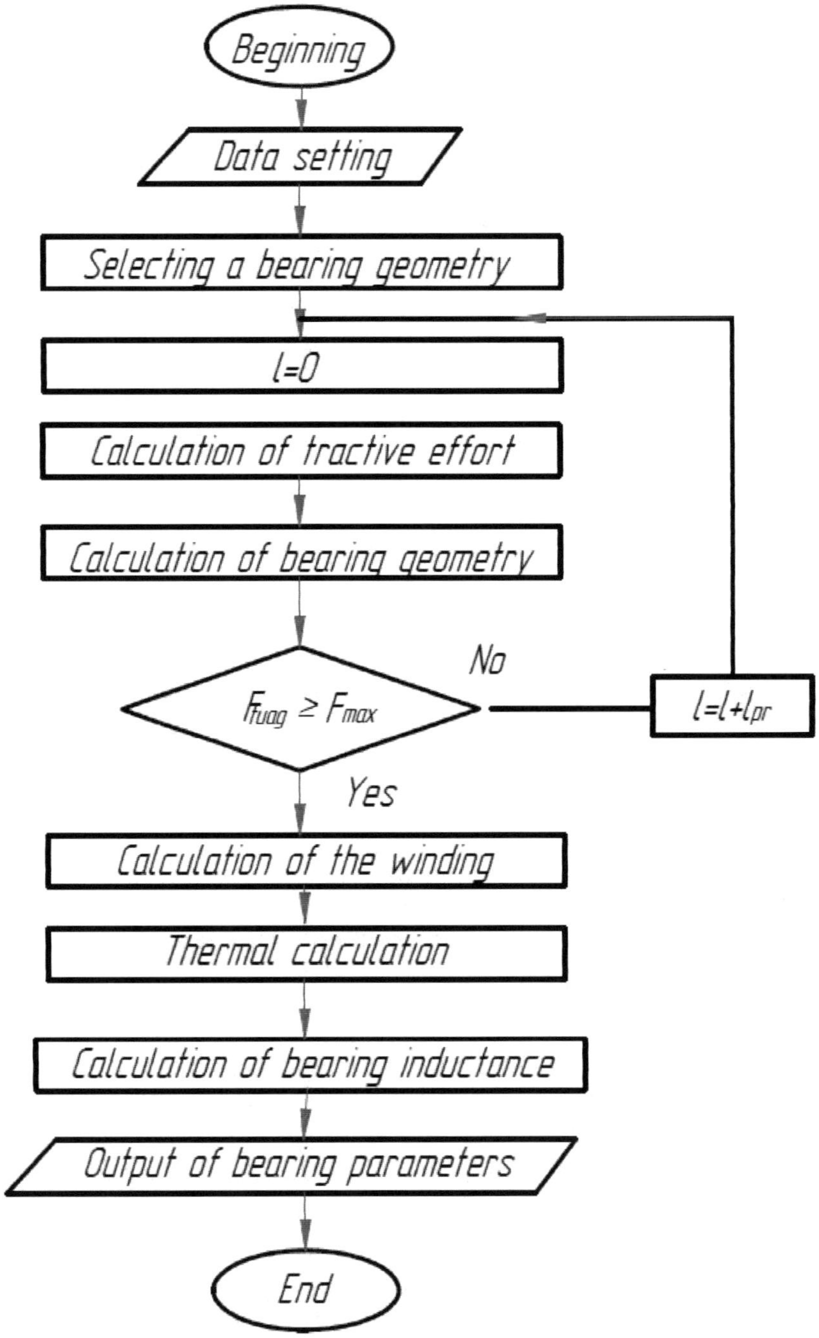

FIGURE.5. Block diagram of the improved general algorithm of AMS parameter determination.

In works [9–11], an algorithm for calculating forces in a radial permanent magnetic bearing is presented. The calculation is carried out using the Monte-Carlo method, which assumes a constant sample density at each iteration. This algorithm can be reduced to the following block diagram (Fig. 6):

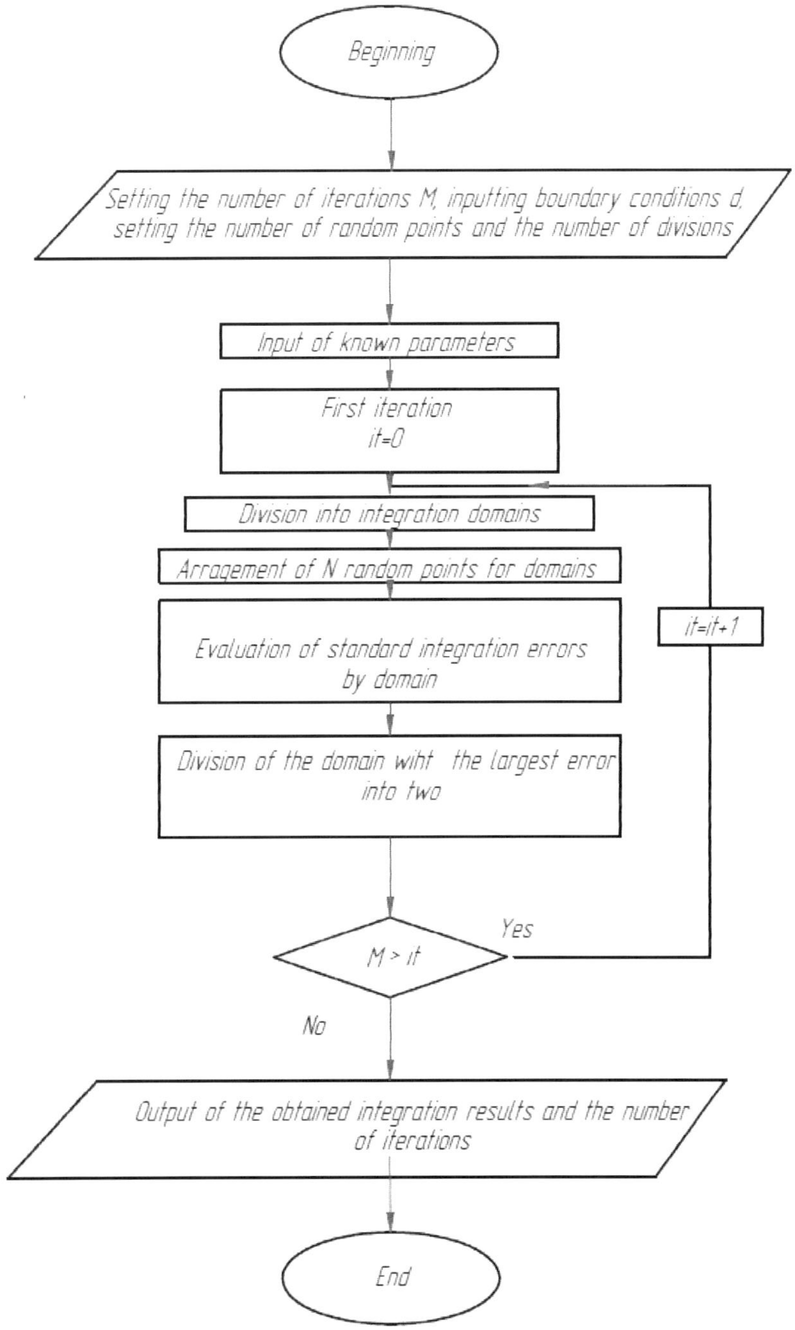

FIGURE.6. Block diagram of an algorithm for determining PMS parameters using the Monte-Carlo method.

Algorithm:

1. Setting the maximum number of iterations, introducing boundary conditions (d), setting the number of random points, and the number of divisions on the integration domain.
2. The input of known parameters.
3. Evaluation of standard integration errors by the domainof integration.

4. Find the domain in which the standard error of integration is maximum. Divide this domain into two.
5. Conductthenextiteration.
6. Repeat steps 2–5 until the number of iterations exceeds the specified one.

This technique allows one to determine the required magnetic forces in the supports to hold the rotor in the required position.

CONCLUSION

Magnetic bearings are widely used in technology due to many advantages over rolling and plain bearings:

- no mechanical contact between the rotor and the support;
- no probability of the lubricant penetration into external environments;
- nosealingrequired;
- almost unlimited resource of work;
- ability to reach rotor speeds, which are limited only by the strength properties of the material.

It should be noted that the use of AMS in vacuum technology has great prospects than PMS. This is justified by the following disadvantages of the PMS:

- there is no possibility to regulate the magnetic field;
- impossibility of implementation of a complete stable non-contact suspension using only permanent magnets;
- lowspecificcarryingcapacity;
- special dampers are required to dissipate energy from vibrations of the suspended body.

All these disadvantages make the use of PMS in mechanical vacuum pumps not only difficult to implement, but also undesirable.

In turn, the advantages of AMS make it highly demanded use in turbomolecular pumps. The main ones are:

- the ability to achieve extremely high rotor speeds;
- the almost complete absence of vibrations of the suspended body;
- eliminationofnoise;
- absence of hydrocarbons in TMP.

However, at the moment no algorithm allows calculating AMS for TMP in compliance with all the features of its operation, which makes this direction extremely promising.

References

[1] Gordeeva, U.S., Demikhov, K.E., Ochkov, A.A. Specifics of the application of mechanical vacuum pumps in vacuum distillation units used in chemical and oil and gas industries // AIP Conference Proceedings 2141, edited byA. V. Myshljavcev et al (OSTU, Omsk, 2019).

[2] E. V. Svichkar, N. K. Nikulin, V. S. Klyucharov and K. E. Demikhov, “Molecular-viscous vacuum pump (MVVP),” in Oil and Gas Engineering-2018, AIP Conference Proceedings 1976, edited by A. V. Myshljavcev et al., (OSTU, Omsk, 2017) pp. 1–5.

[3] Frolov, E.S., Demikhov, K.E. Theoretical analysis of the effect of the geometry of the groove on the performance curves for impellers in turbomolecular vacuum pumps // Chemical and Petroleum Engineering. 1970.

[4] E. V. Svichkar, N. K. Nikulin, K. E. Demikhov, Calculation method of pumping characteristics of high-vacuumsystem with turbomolecular vacuum pump, in Oil and Gas Engineering-2018, AIP Conference Proceedings 1976, edited byA. V. Myshljavcev et al (OSTU, Omsk, 2018) pp. 1–7.

[5] Gordeeva, U.S., Demikhov, K.E., Ochkov, A.A. Development of a calculation method of the main parameters of the multistage turbomolecular pump // IOP Conference Series: Materials Science and Engineering. — IOP Publishing, 2020. — V. 781 — S. 012010.

[6] ZhuravlevYu.N. Active magnetic bearings: Theory, calculation, application. —MPb.: Polytechnic, — 2003. — 206p.

[7] Zotov I.V., Lisienko V.G. Improvement of the calculation method for radial active magnetic bearings.

[8] Shchekleina I.L., Goryakova V.V., Bogatov E.A. Calculation of the winding of a radial magnetic bearing. In the book: Innovative technologies in the field of energy-saving: materials of the regional scientific-practical conference of students April 22, 2010, Yekaterinburg, 2010. P. 87–91.

[9] TapanSantra, Debabrata Roy, Amalendu B. Choudhury Calculation of passive magnetic force in a radial magnetic bearing using general division approach. — Progress In Electromagnetics Research M, Vol. 54, 91–102, 2017.

[10] Wang, F., J.Wang, Z. Kong, and F. Zhang, "Radial and axial force calculation of BLDC motor with passive magnetic bearing," *4th International Power Electronics and Motion Control Conference, IPEMC 2004*, Vol. 1, 290–293, Xi'an, China, Aug. 14–16, 2004.

[11] Pennanen, T. and M. Koivu, "An adaptive importance sampling technique," *Monte Carlo and Quasi-Monte Carlo Methods*, 443–455, Springer, 2004.

[12] Ivanov V.I. Vacuum technology: Teaching aid. — SPb.: ITMO University, 2016. — 129 p.

[13] Shchekleina I.L., Ugolnikov A.V., Stozhkov D.S. About active magnetic bearings.Active electromagnetic bearings for large power machines // Joint Company 'Research and Production Corporation'Space Monitoring Systems, Information & Control and ElectromechanicalComplexes' named after A.G. Iosifian'. Tech. Inform. — OFB.149.649. — M., 1988. — P. 10.

[14] Vereshchagin V.P., Klabukov V.A. Mathematical model of a magnetic bearing // Electromechanical issues. Proceedings of Joint Company 'Research and Production Corporation'Space Monitoring Systems, Information & Control and ElectromechanicalComplexes' named after A.G. Iosifian'. — 2009. — Vol. 112. — No. 5 — P. 17 — 22.

[15] Polyakhov N.D., Stotskaya A.D. On electromagnetic processes in active magnetic bearings // Soft calculations and measurements SCM'2012: abstracts of int. conf. (St. Petersburg, June 24–25, 2012), St. Petersburg, 2012 — St. Petersburg, 2012. — P. 143–145.

[16] SchweitzerG.,MaslenE. H. MagneticBearings. Theory, Design, and Applicationto Rotating Machinery. — Berlin: Springer-Verlag Berlin Heidelberg, 2009. — pp. 1–24.

[17] Kimman M. H., Langen H. H., Munnig Schmidt R. H. A miniature milling spindle with active magnetic bearings // Mechatronics. 2010. Vol. 20, No 2. P. 224–235.

[18] BogdanovaYu.V., Guskov A.M. Modeling the dynamics of the rotor of an electric spindle on magnetic bearings // Science and Education. Bauman Moscow State Technical University. Electronic journal. 2015.No 1.P. 201–220. DOI: 10.7463 / 0115.0753146

AUTHOR INDEX

Institute of Physics
Dirac House, Temple Back
Bristol BS1 6BE UK

ISSN: 1757-8981
ISBN 978-1-7138-3201-0
9 781713 832010

CURRAN ASSOCIATES INC.
proceedings.com